Student's Solutions Manual

Linear Algebra with Applications

Second Edition

Gareth Williams
Stetson University

Prepared by

Wanda J. Mourant

Wm. C. Brown Publishers

ISBN 0-697-11368-X

Printed in the United States of America by Wm. C. Brown Publishers, 2460 Kerper Blvd., Dubuque, Iowa, 52001

10 9 8 7 6 5 4 3 2

CONTENTS

PREFACE

This Student's Solutions Manual contains completely worked-out solutions to selected exercises at the ends of sections and all review exercises at the ends of chapters in the text Linear Algebra with Applications, 2e by Gareth Williams. The selected exercises at the ends of sections are the same exercises for which answers are provided in the back of the text, with the exception of the fractal images shown herein for exercises 45 and 46 in Section 7.2. All references in this manual refer to chapters, sections and exercises in the text.

It has been the intent of the solver to illustrate the techniques demonstrated in the text. All solutions have been checked. However the user should be aware that errors may exist.

INTRODUCTION

The purpose of this manual is to provide students with examples of completely solved problems of the various kinds given in the exercise sets in the text.

The solutions given here are intended to illustrate the techniques discussed and demonstrated in the text. However, in many cases choices have to be made in the course of solving a problem. Exchanging rows 1 and 2 of a matrix may be as reasonable as exchanging rows 1 and 3. These two different choices will produce different looking steps using the same basic technique. Writing the original equations or conditions in two different orders may produce different looking steps using the same technique. It is possible that more than one line of reasoning can be applied to a problem or that two very different techniques can be used in solving a problem. Thus a correct solution may not look at all like the solution in this manual, except for the answer.

The final answer is the same no matter what choices are made in the course of solving a problem. There are not different correct answers. It is often possible to check whether an answer is correct by checking whether it satisfies the given equations or conditions.

Students should make their best effort to understand the material in the text and to solve assigned homework problems themselves before looking at solutions in this manual.

Working through solutions in this manual or examples in the text can help students to understand a technique that at first seems confusing. Students who do this should then test their understanding by working similar problems independently.

Copying or mimicing solutions in this manual is not an adequate substitute for working the homework problems independently.

Chapter 1

Exercise Set 1.1, page 6

1. (a) $x = 3/2, y = -1/2$. (c) no solutions (d) General solution: $x = r,\ y = -r/2 + 2$.

2. (a) $x = 6/7, y = 1/7$. (c) $x = 4, y = 2$. (e) General solution: $x = 3r - 2, y = r$.

3. (a) $c = 3$. (c) $c = 2$. (e) $c = 7$.

4. (a) $k \neq 8$. $x = 5, y = 0$. (b) $k = 8$. General solution: $x = r, y = (5 - r)/4$. 5. $k = 6$.

7. The second equation minus 2 times the first gives $(d - 6)y = -2$.

   (a) $d \neq 6$. The solution is $x = 7 + 6/(d - 6), y = -2/(d - 6)$.

   (b) $d = 6$. This gives $0y = -2$ above, so there is no solution.

   There are no more values of d. For each value of d there is either a unique solution or no solution. Alternatively, the first line has x intercept 7 and the second line has x intercept 6, so they cannot be the same line, and there cannot be many solutions.

9. Multiply the first equation by c and the second by a to get

$$acx + bcy = ec \quad \text{and} \quad acx + ady = af.$$

   Subtract the first from the second and get

$$(ad - bc)y = af - ec.$$

   (a) $ad - bc \neq 0$. Then $y = (af - ec)/(ad - bc)$, and substituting in either equation,

   $x = (ed - bf)/(ad - bc)$. If $a = 1, b = 3, c = 3, d = -2, e = 10$, and $f = -3$, then

   $x = (10(-2) - 3(-3))/(1(-2) - 3(3)) = 1$ and $y = (1(-3) - 10(3))/(1(-2) - 3(3)) = 3$.

   (b) $ad - bc = 0$ and $af - ec \neq 0$ or $ad - bc = 0$ and $ed - bf \neq 0$. These conditions give $0y$ = a nonzero number or $0x$ = a nonzero number, and so there is no solution.

   (c) $a = kc, b = kd$, and $e = kf$ for some nonzero number k. In this case both equations represent the same relationship between x and y. One variable can be assigned the value r, and the equation can be solved for the other variable in terms of r.

Exercise Set 1.2, page 15

1. (a) 3 x 3 (c) 2 x 4 (e) 3 x 5

2. 1, 4, 9, −1, 3, 8

5. (a) $\begin{bmatrix} 1 & 3 \\ 2 & -5 \end{bmatrix}$ and $\begin{bmatrix} 1 & 3 & 7 \\ 2 & -5 & -3 \end{bmatrix}$ (c) $\begin{bmatrix} -1 & 3 & -5 \\ 2 & -2 & 4 \\ 1 & 3 & 0 \end{bmatrix}$ and $\begin{bmatrix} -1 & 3 & -5 & -3 \\ 2 & -2 & 4 & 8 \\ 1 & 3 & 0 & 6 \end{bmatrix}$

(e) $\begin{bmatrix} 5 & 2 & -4 \\ 0 & 4 & 3 \\ 1 & 0 & -1 \end{bmatrix}$ and $\begin{bmatrix} 5 & 2 & -4 & 8 \\ 0 & 4 & 3 & 0 \\ 1 & 0 & -1 & 7 \end{bmatrix}$

6. (b) $\begin{aligned} 7x_1 + 9x_2 &= 8 \\ 6x_1 + 4x_2 &= -3 \end{aligned}$

(d) $\begin{aligned} 8x_1 + 7x_2 + 5x_3 &= -1 \\ 4x_1 + 6x_2 + 2x_3 &= 4 \\ 9x_1 + 3x_2 + 7x_3 &= 6 \end{aligned}$

(f) $\begin{aligned} - 2x_2 &= 4 \\ 5x_1 + 7x_2 &= -3 \\ 6x_1 &= 8 \end{aligned}$

(h) $\begin{aligned} x_1 + 2x_2 - x_3 &= 6 \\ x_2 + 4x_3 &= 5 \\ x_3 &= -2 \end{aligned}$

7. (a) $\begin{bmatrix} 1 & 3 & -2 & 0 \\ 1 & 2 & -3 & 6 \\ 8 & 3 & 2 & 5 \end{bmatrix}$ (c) $\begin{bmatrix} 1 & 2 & 3 & -1 \\ 0 & 3 & 10 & 0 \\ 0 & -8 & -1 & -1 \end{bmatrix}$ (e) $\begin{bmatrix} 1 & 0 & 0 & -23 \\ 0 & 1 & 0 & 17 \\ 0 & 0 & 1 & 5 \end{bmatrix}$

8. (a) Elements in the first column , except for the leading 1, become zero. x_1 is eliminated from all equations except the first.

(c) The leading 1 in row 2 is moved to the left of the leading nonzero term in row 3. The second equation now contains x_2 with leading coefficient 1.

9. (a) Elements in the third column, except for the leading 1, become zero. x_3 is eliminated from all equations except the third.

(c) The leading nonzero element in row 3 becomes 1. Equation 3 is now solved for x_3.

10. (a) $$\begin{bmatrix} 1 & -2 & -8 \\ 2 & -3 & -11 \end{bmatrix} \underset{R2+(-2)R1}{\approx} \begin{bmatrix} 1 & -2 & -8 \\ 0 & 1 & 5 \end{bmatrix} \underset{R1+(2)R2}{\approx} \begin{bmatrix} 1 & 0 & 2 \\ 0 & 1 & 5 \end{bmatrix},$$

so the solution is $x_1 = 2$ and $x_2 = 5$.

(c) $$\begin{bmatrix} 1 & 0 & 1 & 3 \\ 0 & 2 & -2 & -4 \\ 0 & 1 & -2 & 5 \end{bmatrix} \underset{(1/2)R2}{\approx} \begin{bmatrix} 1 & 0 & 1 & 3 \\ 0 & 1 & -1 & -2 \\ 0 & 1 & -2 & 5 \end{bmatrix} \underset{R3+(-1)R2}{\approx} \begin{bmatrix} 1 & 0 & 1 & 3 \\ 0 & 1 & -1 & -2 \\ 0 & 0 & -1 & 7 \end{bmatrix}$$

$$\underset{(-1)R3}{\approx} \begin{bmatrix} 1 & 0 & 1 & 3 \\ 0 & 1 & -1 & -2 \\ 0 & 0 & 1 & -7 \end{bmatrix} \underset{\substack{R1+(-1)R3 \\ R2+R3}}{\approx} \begin{bmatrix} 1 & 0 & 0 & 10 \\ 0 & 1 & 0 & -9 \\ 0 & 0 & 1 & -7 \end{bmatrix},$$

so the solution is $x_1 = 10$, $x_2 = -9$, $x_3 = -7$.

(e) $$\begin{bmatrix} 1 & -1 & 3 & 3 \\ 2 & -1 & 2 & 2 \\ 3 & 1 & -2 & 2 \end{bmatrix} \underset{\substack{R2+(-2)R1 \\ R3+(-3)R1}}{\approx} \begin{bmatrix} 1 & -1 & 3 & 3 \\ 0 & 1 & -4 & -4 \\ 0 & 4 & -11 & -7 \end{bmatrix} \underset{\substack{R1+R2 \\ R3+(-4)R2}}{\approx} \begin{bmatrix} 1 & 0 & -1 & -1 \\ 0 & 1 & -4 & -4 \\ 0 & 0 & 5 & 9 \end{bmatrix}$$

$$\underset{(1/5)R3}{\approx} \begin{bmatrix} 1 & 0 & -1 & -1 \\ 0 & 1 & -4 & -4 \\ 0 & 0 & 1 & 9/5 \end{bmatrix} \underset{\substack{R1+R3 \\ R2+(4)R3}}{\approx} \begin{bmatrix} 1 & 0 & 0 & 4/5 \\ 0 & 1 & 0 & 16/5 \\ 0 & 0 & 1 & 9/5 \end{bmatrix},$$

so the solution is $x_1 = 4/5$, $x_2 = 16/5$, $x_3 = 9/5$.

11. (a) $$\begin{bmatrix} 0 & 2 & 4 & 8 \\ 2 & 2 & 0 & 6 \\ 1 & 1 & 1 & 5 \end{bmatrix} \underset{R1 \leftrightarrow R2}{\approx} \begin{bmatrix} 2 & 2 & 0 & 6 \\ 0 & 2 & 4 & 8 \\ 1 & 1 & 1 & 5 \end{bmatrix} \underset{(1/2)R1}{\approx} \begin{bmatrix} 1 & 1 & 0 & 3 \\ 0 & 2 & 4 & 8 \\ 1 & 1 & 1 & 5 \end{bmatrix}$$

$$\underset{R3+(-1)R1}{\approx}\begin{bmatrix}1&1&0&3\\0&2&4&8\\0&0&1&2\end{bmatrix}\underset{(1/2)R2}{\approx}\begin{bmatrix}1&1&0&3\\0&1&2&4\\0&0&1&2\end{bmatrix}\underset{R1\leftrightarrow(-1)R2}{\approx}\begin{bmatrix}1&0&-2&-1\\0&1&2&4\\0&0&1&2\end{bmatrix}$$

$$\underset{\substack{R1+(2)R3\\R2+(-2)R3}}{\approx}\begin{bmatrix}1&0&0&3\\0&1&0&0\\0&0&1&2\end{bmatrix},$$ so the solution is $x_1 = 3$, $x_2 = 0$, $x_3 = 2$.

(c) $$\begin{bmatrix}1&2&3&14\\2&5&8&36\\1&-1&0&-4\end{bmatrix}\underset{\substack{R2+(-2)R1\\R3+(-1)R1}}{\approx}\begin{bmatrix}1&2&3&14\\0&1&2&8\\0&-3&-3&-18\end{bmatrix}\underset{\substack{R1+(-2)R2\\R3+(3)R2}}{\approx}\begin{bmatrix}1&0&-1&-2\\0&1&2&8\\0&0&3&6\end{bmatrix}$$

$$\underset{(1/3)R3}{\approx}\begin{bmatrix}1&0&-1&-2\\0&1&2&8\\0&0&1&2\end{bmatrix}\underset{\substack{R1+R3\\R2+(-2)R3}}{\approx}\begin{bmatrix}1&0&0&0\\0&1&0&4\\0&0&1&2\end{bmatrix},$$

so the solution is $x_1 = 0$, $x_2 = 4$, $x_3 = 2$.

(e) $$\begin{bmatrix}2&2&-4&14\\3&1&1&8\\2&-1&2&-1\end{bmatrix}\underset{(1/2)R1}{\approx}\begin{bmatrix}1&1&-2&7\\3&1&1&8\\2&-1&2&-1\end{bmatrix}\underset{\substack{R2+(-3)R1\\R3+(-2)R1}}{\approx}\begin{bmatrix}1&1&-2&7\\0&-2&7&-13\\0&-3&6&-15\end{bmatrix}$$

$$\underset{R2\leftrightarrow R3}{\approx}\begin{bmatrix}1&1&-2&7\\0&-3&6&-15\\0&-2&7&-13\end{bmatrix}\underset{(-1/3)R2}{\approx}\begin{bmatrix}1&1&-2&7\\0&1&-2&5\\0&-2&7&-13\end{bmatrix}$$

$$\underset{\substack{R1+(-1)R2\\R3+(2)R2}}{\approx}\begin{bmatrix}1&0&0&2\\0&1&-2&5\\0&0&3&-3\end{bmatrix}\underset{(1/3)R3}{\approx}\begin{bmatrix}1&0&0&2\\0&1&-2&5\\0&0&1&-1\end{bmatrix}\underset{R2+(2)R3}{\approx}\begin{bmatrix}1&0&0&2\\0&1&0&3\\0&0&1&-1\end{bmatrix},$$

so the solution is $x_1 = 2$, $x_2 = 3$, $x_3 = -1$.

12. (b) $$\begin{bmatrix}-3&-6&-15&-3\\1&3/2&9/2&1/2\\-2&-7/2&-17/2&-2\end{bmatrix}\underset{\substack{(-1/3)R1\\(2)R2\\(2)R3}}{\approx}\begin{bmatrix}1&2&5&1\\2&3&9&1\\-4&-7&-17&-4\end{bmatrix}\underset{\substack{R2+(-2)R1\\R3+(4)R1}}{\approx}\begin{bmatrix}1&2&5&1\\0&-1&-1&-1\\0&1&3&0\end{bmatrix}$$

$$\underset{(-1)R2}{\approx}\begin{bmatrix} 1 & 2 & 5 & 1 \\ 0 & 1 & 1 & 1 \\ 0 & 1 & 3 & 0 \end{bmatrix} \underset{\substack{R1+(-2)R2 \\ R3+(-1)R2}}{\approx}\begin{bmatrix} 1 & 0 & 3 & -1 \\ 0 & 1 & 1 & 1 \\ 0 & 0 & 2 & -1 \end{bmatrix} \underset{(1/2)R3}{\approx}\begin{bmatrix} 1 & 0 & 3 & -1 \\ 0 & 1 & 1 & 1 \\ 0 & 0 & 1 & -1/2 \end{bmatrix}$$

$$\underset{\substack{R1+(-3)R3 \\ R2+(-1)R3}}{\approx}\begin{bmatrix} 1 & 0 & 0 & 1/2 \\ 0 & 1 & 0 & 3/2 \\ 0 & 0 & 1 & -1/2 \end{bmatrix},$$ so the solution is $x_1 = \frac{1}{2}$, $x_2 = \frac{3}{2}$, $x_3 = -\frac{1}{2}$.

(d) $$\begin{bmatrix} 1 & 2 & 2 & 5 & 11 \\ 2 & 4 & 2 & 8 & 14 \\ 1 & 3 & 4 & 8 & 19 \\ 1 & -1 & 1 & 0 & 2 \end{bmatrix} \underset{\substack{R2+(-2)R1 \\ R3+(-1)R1 \\ R4+(-1)R1}}{\approx}\begin{bmatrix} 1 & 2 & 2 & 5 & 11 \\ 0 & 0 & -2 & -2 & -8 \\ 0 & 1 & 2 & 3 & 8 \\ 0 & -3 & -1 & -5 & -9 \end{bmatrix}$$

$$\underset{R2 \leftrightarrow R3}{\approx}\begin{bmatrix} 1 & 2 & 2 & 5 & 11 \\ 0 & 1 & 2 & 3 & 8 \\ 0 & 0 & -2 & -2 & -8 \\ 0 & -3 & -1 & -5 & -9 \end{bmatrix} \underset{\substack{R1+(-2)R2 \\ R4+(3)R2}}{\approx}\begin{bmatrix} 1 & 0 & -2 & -1 & -5 \\ 0 & 1 & 2 & 3 & 8 \\ 0 & 0 & -2 & -2 & -8 \\ 0 & 0 & 5 & 4 & 15 \end{bmatrix}$$

$$\underset{(-1/2)R3}{\approx}\begin{bmatrix} 1 & 0 & -2 & -1 & -5 \\ 0 & 1 & 2 & 3 & 8 \\ 0 & 0 & 1 & 1 & 4 \\ 0 & 0 & 5 & 4 & 15 \end{bmatrix} \underset{\substack{R1+(2)R3 \\ R2+(-2)R3 \\ R4+(-5)R3}}{\approx}\begin{bmatrix} 1 & 0 & 0 & 1 & 3 \\ 0 & 1 & 0 & 1 & 0 \\ 0 & 0 & 1 & 1 & 4 \\ 0 & 0 & 0 & -1 & -5 \end{bmatrix}$$

$$\underset{(-1)R4}{\approx}\begin{bmatrix} 1 & 0 & 0 & 1 & 3 \\ 0 & 1 & 0 & 1 & 0 \\ 0 & 0 & 1 & 1 & 4 \\ 0 & 0 & 0 & 1 & 5 \end{bmatrix} \underset{\substack{R1+(-1)R4 \\ R2+(-1)R4 \\ R3+(-1)R4}}{\approx}\begin{bmatrix} 1 & 0 & 0 & 0 & -2 \\ 0 & 1 & 0 & 0 & -5 \\ 0 & 0 & 1 & 0 & -1 \\ 0 & 0 & 0 & 1 & 5 \end{bmatrix},$$

so the solution is $x_1 = -2$, $x_2 = -5$, $x_3 = -1$, $x_4 = 5$.

(e) $$\begin{bmatrix} 1 & 1 & 2 & 6 & 11 \\ 2 & 3 & 6 & 19 & 36 \\ 0 & 3 & 4 & 15 & 28 \\ 1 & -1 & -1 & -6 & -12 \end{bmatrix} \underset{\substack{R2+(-2)R1 \\ R4+(-1)R1}}{\approx}\begin{bmatrix} 1 & 1 & 2 & 6 & 11 \\ 0 & 1 & 2 & 7 & 14 \\ 0 & 3 & 4 & 15 & 28 \\ 0 & -2 & -3 & -12 & -23 \end{bmatrix}$$

$$\underset{\substack{R1+(-1)R2\\R3+(-3)R2\\R4+(2)R2}}{\approx}\begin{bmatrix}1&0&0&-1&-3\\0&1&2&7&14\\0&0&-2&-6&-14\\0&0&1&2&5\end{bmatrix}\underset{(-1/2)R3}{\approx}\begin{bmatrix}1&0&0&-1&-3\\0&1&2&7&14\\0&0&1&3&7\\0&0&1&2&5\end{bmatrix}$$

$$\underset{\substack{R2+(-2)R3\\R4+(-1)R3}}{\approx}\begin{bmatrix}1&0&0&-1&-3\\0&1&0&1&0\\0&0&1&3&7\\0&0&0&-1&-2\end{bmatrix}\underset{(-1)R4}{\approx}\begin{bmatrix}1&0&0&-1&-3\\0&1&0&1&0\\0&0&1&3&7\\0&0&0&1&2\end{bmatrix}$$

$$\underset{\substack{R1+R4\\R2+(-1)R4\\R3+(-3)R4}}{\approx}\begin{bmatrix}1&0&0&0&-1\\0&1&0&0&-2\\0&0&1&0&1\\0&0&0&1&2\end{bmatrix},$$ so the solution is $x_1 = -1$, $x_2 = -2$, $x_3 = 1$, $x_4 = 2$.

13. (a) $\begin{bmatrix}1&2&3&4&3\\3&5&8&9&7\end{bmatrix}\underset{R2+(-3)R1}{\approx}\begin{bmatrix}1&2&3&4&3\\0&-1&-1&-3&-2\end{bmatrix}\underset{(-1)R2}{\approx}\begin{bmatrix}1&2&3&4&3\\0&1&1&3&2\end{bmatrix}$

$\underset{R1+(-2)R2}{\approx}\begin{bmatrix}1&2&1&-2&-1\\0&1&1&3&2\end{bmatrix}$, so the solutions are in turn $x_1 = 1$, $x_2 = 1$;

$x_1 = -2$, $x_2 = 3$; and $x_1 = -1$, $x_2 = 2$.

(c) $\begin{bmatrix}1&-2&3&6&-5&4\\1&-1&2&5&-3&3\\2&-3&6&14&-8&9\end{bmatrix}\underset{\substack{R2+(-1)R1\\R3+(-2)R1}}{\approx}\begin{bmatrix}1&-2&3&6&-5&4\\0&1&-1&-1&2&-1\\0&1&0&12&2&1\end{bmatrix}$

$\underset{\substack{R1+(2)R2\\R3+(-1)R2}}{\approx}\begin{bmatrix}1&0&1&4&-1&2\\0&1&-1&-1&2&-1\\0&0&1&13&0&2\end{bmatrix}\underset{\substack{R1+(-1)R3\\R2+R3}}{\approx}\begin{bmatrix}1&0&0&-9&-1&0\\0&1&0&12&2&1\\0&0&1&13&0&2\end{bmatrix},$

so the solutions are in turn $x_1 = -9$, $x_2 = 12$, $x_3 = 13$; $x_1 = -1$, $x_2 = 2$, $x_3 = 0$; and $x_1 = 0$, $x_2 = 1$, $x_3 = 2$.

Exercise Set 1.3, page 25

1. (a) Yes.

(c) No. The second column contains a leading 1, so other elements in that column should be zero.

(e) Yes. (h) No. The second row does not have 1 as the first nonzero number.

2. (a) No. The leading 1 in row 2 is not to the right of the leading 1 in row 3.

(c) Yes. (e) No. The row containing all zeros should be at the bottom of the matrix.

(g) No. The leading 1 in row 2 is not to the right of the leading 1 in row 3. Also, since column 3 contains a leading 1, all other numbers in that column should be zero.

(i) Yes.

3. (a) $x_1 = 2,\ x_2 = 4,\ x_3 = -3.$ (c) $x_1 = -3r + 6, x_2 = r, x_3 = -2.$

(e) $x_1 = -5r + 3,\ x_2 = -6r - 2,\ x_3 = -2r - 4,\ x_4 = r.$

4. (a) $x_1 = -2r - 4s + 1,\ x_2 = 3r - 5s - 6,\ x_3 = r,\ x_4 = s.$

(c) $x_1 = 2r - 3s + 4,\ x_2 = r,\ x_3 = -2s + 9,\ x_4 = s,\ x_5 = 8.$

5. (a) $$\begin{bmatrix} 1 & 4 & 3 & 1 \\ 2 & 8 & 11 & 7 \\ 1 & 6 & 7 & 3 \end{bmatrix} \underset{\substack{R2+(-2)R1 \\ R3+(-1)R1}}{\approx} \begin{bmatrix} 1 & 4 & 3 & 1 \\ 0 & 0 & 5 & 5 \\ 0 & 2 & 4 & 2 \end{bmatrix} \underset{R2 \leftrightarrow R3}{\approx} \begin{bmatrix} 1 & 4 & 3 & 1 \\ 0 & 2 & 4 & 2 \\ 0 & 0 & 5 & 5 \end{bmatrix}$$

$$\underset{(1/2)R2}{\approx} \begin{bmatrix} 1 & 4 & 3 & 1 \\ 0 & 1 & 2 & 1 \\ 0 & 0 & 5 & 5 \end{bmatrix} \underset{R1+(-4)R2}{\approx} \begin{bmatrix} 1 & 0 & -5 & -3 \\ 0 & 1 & 2 & 1 \\ 0 & 0 & 5 & 5 \end{bmatrix} \underset{(1/5)R3}{\approx} \begin{bmatrix} 1 & 0 & -5 & -3 \\ 0 & 1 & 2 & 1 \\ 0 & 0 & 1 & 1 \end{bmatrix}$$

$$\underset{\substack{R1+(5)R3 \\ R2+(-2)R3}}{\approx} \begin{bmatrix} 1 & 0 & 0 & 2 \\ 0 & 1 & 0 & -1 \\ 0 & 0 & 1 & 1 \end{bmatrix},$$ so the solution is $x_1 = 2,\ x_2 = -1,\ x_3 = 1.$

(c) $\begin{bmatrix} 1 & 1 & 1 & 7 \\ 2 & 3 & 1 & 18 \\ -1 & 1 & -3 & 1 \end{bmatrix} \underset{\substack{R2+(-2)R1 \\ R3+R1}}{\approx} \begin{bmatrix} 1 & 1 & 1 & 7 \\ 0 & 1 & -1 & 4 \\ 0 & 2 & -2 & 8 \end{bmatrix} \underset{\substack{R1+(-1)R2 \\ R3+(-2)R2}}{\approx} \begin{bmatrix} 1 & 0 & 2 & 3 \\ 0 & 1 & -1 & 4 \\ 0 & 0 & 0 & 0 \end{bmatrix}$,

so $x_1 + 2x_3 = 3$ and $x_2 - x_3 = 4$.
Thus the general solution is $x_1 = 3 - 2r$, $x_2 = 4 + r$, $x_3 = r$.

(e) $\begin{bmatrix} 1 & -1 & 1 & 3 \\ 2 & -1 & 4 & 7 \\ 3 & -5 & -1 & 7 \end{bmatrix} \underset{\substack{R2+(-2)R1 \\ R3+(-3)R1}}{\approx} \begin{bmatrix} 1 & -1 & 1 & 3 \\ 0 & 1 & 2 & 1 \\ 0 & -2 & -4 & -2 \end{bmatrix} \underset{\substack{R1+R2 \\ R3+(2)R2}}{\approx} \begin{bmatrix} 1 & 0 & 3 & 4 \\ 0 & 1 & 2 & 1 \\ 0 & 0 & 0 & 0 \end{bmatrix}$,

so $x_1 + 3x_3 = 4$ and $x_2 + 2x_3 = 1$.
Thus the general solution is $x_1 = 4 - 3r$, $x_2 = 1 - 2r$, $x_3 = r$.

6. (a) $\begin{bmatrix} 3 & 6 & -3 & 6 \\ -2 & -4 & -3 & -1 \\ 3 & 6 & -2 & 10 \end{bmatrix} \underset{(1/3)R1}{\approx} \begin{bmatrix} 1 & 2 & -1 & 2 \\ -2 & -4 & -3 & -1 \\ 3 & 6 & -2 & 10 \end{bmatrix} \underset{\substack{R2+(2)R1 \\ R3+(-3)R1}}{\approx} \begin{bmatrix} 1 & 2 & -1 & 2 \\ 0 & 0 & -5 & 3 \\ 0 & 0 & 1 & 4 \end{bmatrix}$.

It is now clear that there is no solution. The last two rows give $-5x_3 = 3$ and $x_3 = 4$.

(c) $\begin{bmatrix} 1 & 2 & -1 & 3 \\ 2 & 4 & -2 & 6 \\ 3 & 6 & 2 & -1 \end{bmatrix} \underset{\substack{R2+(-2)R1 \\ R3+(-3)R1}}{\approx} \begin{bmatrix} 1 & 2 & -1 & 3 \\ 0 & 0 & 0 & 0 \\ 0 & 0 & 5 & -10 \end{bmatrix} \underset{R2 \leftrightarrow R3}{\approx} \begin{bmatrix} 1 & 2 & -1 & 3 \\ 0 & 0 & 5 & -10 \\ 0 & 0 & 0 & 0 \end{bmatrix}$

$\underset{(1/5)R2}{\approx} \begin{bmatrix} 1 & 2 & -1 & 3 \\ 0 & 0 & 1 & -2 \\ 0 & 0 & 0 & 0 \end{bmatrix} \underset{R1+R2}{\approx} \begin{bmatrix} 1 & 2 & 0 & 1 \\ 0 & 0 & 1 & -2 \\ 0 & 0 & 0 & 0 \end{bmatrix}$, so $x_1 + 2x_2 = 1$ and $x_3 = -2$.

Thus the general solution is $x_1 = 1 - 2r$, $x_2 = r$, $x_3 = -2$.

(e) $\begin{bmatrix} 0 & 1 & 2 & 5 \\ 1 & 2 & 5 & 13 \\ 1 & 0 & 2 & 4 \end{bmatrix} \underset{R1 \leftrightarrow R2}{\approx} \begin{bmatrix} 1 & 2 & 5 & 13 \\ 0 & 1 & 2 & 5 \\ 1 & 0 & 2 & 4 \end{bmatrix} \underset{R3+(-1)R1}{\approx} \begin{bmatrix} 1 & 2 & 5 & 13 \\ 0 & 1 & 2 & 5 \\ 0 & -2 & -3 & -9 \end{bmatrix}$

$$\underset{\substack{R1+(-2)R2\\R3+(2)R2}}{\approx}\begin{bmatrix}1&0&1&3\\0&1&2&5\\0&0&1&1\end{bmatrix}\underset{\substack{R1+(-1)R3\\R2+(-2)R3}}{\approx}\begin{bmatrix}1&0&0&2\\0&1&0&3\\0&0&1&1\end{bmatrix},$$

so the solution is $x_1 = 2,\ x_2 = 3,\ x_3 = 1$.

7. (a) $$\begin{bmatrix}1&1&-3&10\\-3&-2&4&-24\end{bmatrix}\underset{R2+(3)R1}{\approx}\begin{bmatrix}1&1&-3&10\\0&1&-5&6\end{bmatrix}\underset{R1+(-1)R2}{\approx}\begin{bmatrix}1&0&2&4\\0&1&-5&6\end{bmatrix},$$

so $x_1 + 2x_3 = 4$ and $x_2 - 5x_3 = 6$. Thus the general solution is
$x_1 = 4 - 2r,\ x_2 = 6 + 5r,\ x_3 = r$.

(c) $$\begin{bmatrix}1&2&-1&-1&0\\1&2&0&1&4\\-1&-2&2&4&5\end{bmatrix}\underset{\substack{R2+(-1)R1\\R3+R1}}{\approx}\begin{bmatrix}1&2&-1&-1&0\\0&0&1&2&4\\0&0&1&3&5\end{bmatrix}\underset{\substack{R1+R2\\R3+(-1)R2}}{\approx}\begin{bmatrix}1&2&0&1&4\\0&0&1&2&4\\0&0&0&1&1\end{bmatrix}$$

$$\underset{\substack{R1+(-1)R3\\R2+(-2)R3}}{\approx}\begin{bmatrix}1&2&0&0&3\\0&0&1&0&2\\0&0&0&1&1\end{bmatrix},$$ so $x_1 + 2x_2 = 3$, $x_3 = 2$, and $x_4 = 1$.

Thus the general solution is $x_1 = 3 - 2r,\ x_2 = r,\ x_3 = 2$, and $x_4 = 1$.

(e) $$\begin{bmatrix}0&1&-3&1&0\\1&1&-1&4&0\\-2&-2&2&-8&0\end{bmatrix}\underset{R1\leftrightarrow R2}{\approx}\begin{bmatrix}1&1&-1&4&0\\0&1&-3&1&0\\-2&-2&2&-8&0\end{bmatrix}$$

$$\underset{R3+(2)R1}{\approx}\begin{bmatrix}1&1&-1&4&0\\0&1&-3&1&0\\0&0&0&0&0\end{bmatrix}\underset{R1+(-1)R2}{\approx}\begin{bmatrix}1&0&2&3&0\\0&1&-3&1&0\\0&0&0&0&0\end{bmatrix},$$

so $x_1 + 2x_3 + 3x_4 = 0$ and $x_2 - 3x_3 + x_4 = 0$. Thus the general solution is
$x_1 = -2r - 3s,\ x_2 = 3r - s,\ x_3 = r,\ x_4 = s$.

8. (a) $$\begin{bmatrix} 1 & 1 & 1 & -1 & -3 \\ 2 & 3 & 1 & -5 & -9 \\ 1 & 3 & -1 & -6 & -7 \\ -1 & -1 & -1 & 0 & 1 \end{bmatrix} \underset{\substack{R2+(-2)R1 \\ R3+(-1)R1 \\ R4+R1}}{\approx} \begin{bmatrix} 1 & 1 & 1 & -1 & -3 \\ 0 & 1 & -1 & -3 & -3 \\ 0 & 2 & 0 & -5 & -4 \\ 0 & 0 & 0 & -1 & -2 \end{bmatrix}$$

$$\underset{\substack{R1+(-1)R2 \\ R3+(-2)R2}}{\approx} \begin{bmatrix} 1 & 0 & 2 & 2 & 0 \\ 0 & 1 & -1 & -3 & -3 \\ 0 & 0 & 2 & 1 & 2 \\ 0 & 0 & 0 & -1 & -2 \end{bmatrix} \underset{(1/2)R3}{\approx} \begin{bmatrix} 1 & 0 & 2 & 2 & 0 \\ 0 & 1 & -1 & -3 & -3 \\ 0 & 0 & 1 & 1/2 & 1 \\ 0 & 0 & 0 & -1 & -2 \end{bmatrix}$$

$$\underset{\substack{R1+(-2)R3 \\ R2+R3}}{\approx} \begin{bmatrix} 1 & 0 & 0 & 1 & -2 \\ 0 & 1 & 0 & -5/2 & -2 \\ 0 & 0 & 1 & 1/2 & 1 \\ 0 & 0 & 0 & -1 & -2 \end{bmatrix} \underset{(-1)R4}{\approx} \begin{bmatrix} 1 & 0 & 0 & 1 & -2 \\ 0 & 1 & 0 & -5/2 & -2 \\ 0 & 0 & 1 & 1/2 & 1 \\ 0 & 0 & 0 & 1 & 2 \end{bmatrix}$$

$$\underset{\substack{R1+(-1)R4 \\ R2+(5/2)R4 \\ R3+(-1/2)R4}}{\approx} \begin{bmatrix} 1 & 0 & 0 & 0 & -4 \\ 0 & 1 & 0 & 0 & 3 \\ 0 & 0 & 1 & 0 & 0 \\ 0 & 0 & 0 & 1 & 2 \end{bmatrix},$$ so the solution is

$x_1 = -4,\ x_2 = 3,\ x_3 = 0,\ x_4 = 2.$

(c) $$\begin{bmatrix} 2 & -4 & 16 & -14 & 10 \\ -1 & 5 & -17 & 19 & -2 \\ 1 & -3 & 11 & -11 & 4 \\ 3 & -4 & 18 & -13 & 17 \end{bmatrix} \underset{(1/2)R1}{\approx} \begin{bmatrix} 1 & -2 & 8 & -7 & 5 \\ -1 & 5 & -17 & 19 & -2 \\ 1 & -3 & 11 & -11 & 4 \\ 3 & -4 & 18 & -13 & 17 \end{bmatrix}$$

$$\underset{\substack{R2+R1 \\ R3+(-1)R1 \\ R4+(-3)R1}}{\approx} \begin{bmatrix} 1 & -2 & 8 & -7 & 5 \\ 0 & 3 & -9 & 12 & 3 \\ 0 & -1 & 3 & -4 & -1 \\ 0 & 2 & -6 & 8 & 2 \end{bmatrix} \underset{(1/3)R2}{\approx} \begin{bmatrix} 1 & -2 & 8 & -7 & 5 \\ 0 & 1 & -3 & 4 & 1 \\ 0 & -1 & 3 & -4 & -1 \\ 0 & 2 & -6 & 8 & 2 \end{bmatrix}$$

$$\underset{\substack{R1+(2)R2 \\ R3+R2 \\ R4+(-2)R2}}{\approx} \begin{bmatrix} 1 & 0 & 2 & 1 & 7 \\ 0 & 1 & -3 & 4 & 1 \\ 0 & 0 & 0 & 0 & 0 \\ 0 & 0 & 0 & 0 & 0 \end{bmatrix},$$ so $x_1 + 2x_3 + x_4 = 7$ and $x_2 - 3x_3 + 4x_4 = 1$.

Thus the general solution is $x_1 = 7 - 2r - s,\ x_2 = 1 + 3r - 4s,\ x_3 = r,\ x_4 = s.$

(d) $\begin{bmatrix} 1 & -1 & 2 & 0 & 7 \\ 2 & -2 & 2 & -4 & 12 \\ -1 & 1 & -1 & 2 & -4 \\ -3 & 1 & -8 & -10 & -29 \end{bmatrix} \underset{\substack{R2+(-2)R1 \\ R3+R1 \\ R4+(3)R1}}{\approx} \begin{bmatrix} 1 & -1 & 2 & 0 & 7 \\ 0 & 0 & -2 & -4 & -2 \\ 0 & 0 & 1 & 2 & 3 \\ 0 & -2 & -2 & -10 & -8 \end{bmatrix}$

$\underset{R2 \leftrightarrow R4}{\approx} \begin{bmatrix} 1 & -1 & 2 & 0 & 7 \\ 0 & -2 & -2 & -10 & -8 \\ 0 & 0 & 1 & 2 & 3 \\ 0 & 0 & -2 & -4 & -2 \end{bmatrix} \underset{(-1/2)R2}{\approx} \begin{bmatrix} 1 & -1 & 2 & 0 & 7 \\ 0 & 1 & 1 & 5 & 4 \\ 0 & 0 & 1 & 2 & 3 \\ 0 & 0 & -2 & -4 & -2 \end{bmatrix}$

$\underset{R4+(2)R3}{\approx} \begin{bmatrix} 1 & -1 & 2 & 0 & 7 \\ 0 & 1 & 1 & 5 & 4 \\ 0 & 0 & 1 & 2 & 3 \\ 0 & 0 & 0 & 0 & 4 \end{bmatrix}$. The last row gives 0 = 4, so there is no solution.

(g) $\begin{bmatrix} 1 & 1 & 2 \\ 2 & 3 & 3 \\ 1 & 3 & 0 \\ 1 & 2 & 1 \end{bmatrix} \underset{\substack{R2+(-2)R1 \\ R3+(-1)R1 \\ R4+(-1)R1}}{\approx} \begin{bmatrix} 1 & 1 & 2 \\ 0 & 1 & -1 \\ 0 & 2 & -2 \\ 0 & 1 & -1 \end{bmatrix} \underset{\substack{R1+(-1)R2 \\ R3+(-2)R2 \\ R4+(-1)R2}}{\approx} \begin{bmatrix} 1 & 0 & 3 \\ 0 & 1 & -1 \\ 0 & 0 & 0 \\ 0 & 0 & 0 \end{bmatrix}$,

so $x_1 = 3$, $x_2 = -1$.

9. (a) The system of equations

$$\begin{aligned} 3x_1 + 2x_2 - x_3 + x_4 &= 4 \\ 3x_1 + 2x_2 - x_3 + x_4 &= 1 \end{aligned}$$

clearly has no solution, since the equations are inconsistent. To make a system that is less obvious add another equation to the system and replace the second equation by the sum of the second equation and some multiple (2 in the example below) of the third equation:

$$\begin{aligned} 3x_1 + 2x_2 - x_3 + x_4 &= 4 \\ 5x_1 + 4x_2 - x_3 - x_4 &= 1 \\ x_1 + x_2 \quad\quad - x_4 &= 0 \end{aligned}$$

(b) Choose a solution, e.g., $x_1 = 1$, $x_2 = 2$. Now make up equations thinking of x_1 as 1 and x_2 as 2:

$$\begin{aligned} x_1 + x_2 &= 3 \\ x_1 + 2x_2 &= 5 \\ x_1 - 2x_2 &= -3 \end{aligned}$$

An easy way to ensure that there are no additional solutions is to include $x_1 = 1$ or $x_2 = 2$ as an equation in the system.

11. $\begin{bmatrix} a & b \\ c & d \end{bmatrix} \underset{(1/a)R1}{\approx} \begin{bmatrix} 1 & b/a \\ c & d \end{bmatrix} \underset{R2+(-c)R1}{\approx} \begin{bmatrix} 1 & b/a \\ 0 & (da-cb)/a \end{bmatrix} \underset{(a/(da-cb))R2}{\approx} \begin{bmatrix} 1 & b/a \\ 0 & 1 \end{bmatrix}$

$\underset{R1+(-b/a)R2}{\approx} \begin{bmatrix} 1 & 0 \\ 0 & 1 \end{bmatrix}$.

(a) $a \neq 0, da - cb \neq 0$. (b) $a \neq 0, b = 0, da - cb = 0$.

14. (a) and (b) If the first system of equations has a unique solution then the reduced echelon form of the matrix $[A:B_1]$ will be $[I_3:X]$. The reduced echelon form of $[A:B_2]$ must therefore be $[I_3:Y]$. So the second system must also have a unique solution.

15. (b) $\begin{bmatrix} 1 & 2 & 4 & 8 & 5 \\ 1 & 1 & 2 & 5 & 3 \\ 2 & 3 & 6 & 13 & 11 \end{bmatrix} \underset{\substack{R2+(-1)R1 \\ R3+(-2)R1}}{\approx} \begin{bmatrix} 1 & 2 & 4 & 8 & 5 \\ 0 & -1 & -2 & -3 & -2 \\ 0 & 1 & 2 & 3 & 5 \end{bmatrix} \underset{(-1)R2}{\approx} \begin{bmatrix} 1 & 2 & 4 & 8 & 5 \\ 0 & 1 & 2 & 3 & 2 \\ 0 & 1 & 2 & 3 & 5 \end{bmatrix}$

$\underset{\substack{R1+(-2)R2 \\ R3+(-1)R2}}{\approx} \begin{bmatrix} 1 & 0 & 0 & 2 & 1 \\ 0 & 1 & 2 & 3 & 2 \\ 0 & 0 & 0 & 0 & 3 \end{bmatrix}$, so the general solution to the first system is

$x_1 = 2, x_2 = 3 - 2r, x_3 = r$, and the second system has no solution.

17. A 3x4 matrix represents the equations of three planes. In order for there to be many solutions the three planes must have at least one line in common. For there to be no solutions either at least two of the three planes must be parallel or the line of intersection of two of the planes must lie in a plane which is parallel to the third plane. It is more likely that the three planes will meet in a single point, i.e., that there will be a unique solution. The reduced echelon form therefore will be $[I_3:X]$.

Exercise Set 1.4, page 33

1. (b) Yes. (d) Yes.

2. (a) Yes. (b) No. The leading 1 in row 3 is not to the right of the leading 1 in row 2.

3. $$\begin{bmatrix} 1 & 1 & 1 & 6 \\ 1 & -1 & 1 & 2 \\ 1 & 2 & 3 & 14 \end{bmatrix} \underset{\substack{R2+(-1)R1 \\ R3+(-1)R1}}{\approx} \begin{bmatrix} 1 & 1 & 1 & 6 \\ 0 & -2 & 0 & -4 \\ 0 & 1 & 2 & 8 \end{bmatrix} \underset{(-1/2)R2}{\approx} \begin{bmatrix} 1 & 1 & 1 & 6 \\ 0 & 1 & 0 & 2 \\ 0 & 1 & 2 & 8 \end{bmatrix}$$

$$\underset{R3+(-1)R2}{\approx} \begin{bmatrix} 1 & 1 & 1 & 6 \\ 0 & 1 & 0 & 2 \\ 0 & 0 & 2 & 6 \end{bmatrix} \underset{(1/2)R3}{\approx} \begin{bmatrix} 1 & 1 & 1 & 6 \\ 0 & 1 & 0 & 2 \\ 0 & 0 & 1 & 3 \end{bmatrix}.$$

(a) $x_1 + x_2 + x_3 = 6$, $x_2 = 2$, and $x_3 = 3$. Back substituting $x_2 = 2$ and $x_3 = 3$ into the first equation, $x_1 = 1$.

(b) $$\begin{bmatrix} 1 & 1 & 1 & 6 \\ 0 & 1 & 0 & 2 \\ 0 & 0 & 1 & 3 \end{bmatrix} \underset{R1+(-1)R3}{\approx} \begin{bmatrix} 1 & 1 & 0 & 3 \\ 0 & 1 & 0 & 2 \\ 0 & 0 & 1 & 3 \end{bmatrix} \underset{R1+(-1)R2}{\approx} \begin{bmatrix} 1 & 0 & 0 & 1 \\ 0 & 1 & 0 & 2 \\ 0 & 0 & 1 & 3 \end{bmatrix}$$, so again

$x_1 = 1, x_2 = 2, x_3 = 3$.

5. $$\begin{bmatrix} 1 & -1 & 2 & 3 \\ 2 & -2 & 5 & 4 \\ 1 & 2 & -1 & -3 \\ 0 & 2 & 2 & 1 \end{bmatrix} \underset{\substack{R2+(-2)R1 \\ R3+(-1)R1}}{\approx} \begin{bmatrix} 1 & -1 & 2 & 3 \\ 0 & 0 & 1 & -2 \\ 0 & 3 & -3 & -6 \\ 0 & 2 & 2 & 1 \end{bmatrix} \underset{R2 \leftrightarrow R3}{\approx} \begin{bmatrix} 1 & -1 & 2 & 3 \\ 0 & 3 & -3 & -6 \\ 0 & 0 & 1 & -2 \\ 0 & 2 & 2 & 1 \end{bmatrix}$$

$$\underset{(1/3)R2}{\approx} \begin{bmatrix} 1 & -1 & 2 & 3 \\ 0 & 1 & -1 & -2 \\ 0 & 0 & 1 & -2 \\ 0 & 2 & 2 & 1 \end{bmatrix} \underset{R4+(-2)R2}{\approx} \begin{bmatrix} 1 & -1 & 2 & 3 \\ 0 & 1 & -1 & -2 \\ 0 & 0 & 1 & -2 \\ 0 & 0 & 4 & 5 \end{bmatrix}$$, and there is no reason to

continue. The last two rows give inconsistent equations, so there is no solution.

6. $$\begin{bmatrix} 1 & -1 & 1 & 2 & -2 & 1 \\ 2 & -1 & -1 & 3 & -1 & 3 \\ -1 & -1 & 5 & 0 & -4 & -3 \end{bmatrix} \underset{\substack{R2+(-2)R1 \\ R3+R1}}{\approx} \begin{bmatrix} 1 & -1 & 1 & 2 & -2 & 1 \\ 0 & 1 & -3 & -1 & 3 & 1 \\ 0 & -2 & 6 & 2 & -6 & -2 \end{bmatrix}$$

$$\underset{R3+2R2}{\approx} \begin{bmatrix} 1 & -1 & 1 & 2 & -2 & 1 \\ 0 & 1 & -3 & -1 & 3 & 1 \\ 0 & 0 & 0 & 0 & 0 & 0 \end{bmatrix}.$$

(a) $x_1 - x_2 + x_3 + 2x_4 - 2x_5 = 1$ and $x_2 - 3x_3 - x_4 + 3x_5 = 1$.
Back substituting $x_3 = r$, $x_4 = s$, $x_5 = t$ into the second equation, $x_2 = 1 + 3r + s - 3t$, and substituting for x_2, x_3, x_4, and x_5 in the first equation, $x_1 = 2 + 2r - s - t$.

(b) $$\begin{bmatrix} 1 & -1 & 1 & 2 & -2 & 1 \\ 0 & 1 & -3 & -1 & 3 & 1 \\ 0 & 0 & 0 & 0 & 0 & 0 \end{bmatrix} \underset{R1+R2}{\approx} \begin{bmatrix} 1 & 0 & -2 & 1 & 1 & 2 \\ 0 & 1 & -3 & -1 & 3 & 1 \\ 0 & 0 & 0 & 0 & 0 & 0 \end{bmatrix},$$ and again the general solution is $x_1 = 2 + 2r - s - t$, $x_2 = 1 + 3r + s - 3t$, $x_3 = r$, $x_4 = s$, $x_5 = t$.

7. The augmented matrix is 4x6.

Gauss-Jordan elimination:

$$\begin{bmatrix} * & * & * & * & * & * \\ * & * & * & * & * & * \\ * & * & * & * & * & * \\ * & * & * & * & * & * \end{bmatrix} \underset{5 \text{ mults}}{\approx} \begin{bmatrix} 1 & * & * & * & * & * \\ * & * & * & * & * & * \\ * & * & * & * & * & * \\ * & * & * & * & * & * \end{bmatrix} \underset{\substack{15 \text{ mults} \\ 15 \text{ adds}}}{\approx} \begin{bmatrix} 1 & * & * & * & * & * \\ 0 & * & * & * & * & * \\ 0 & * & * & * & * & * \\ 0 & * & * & * & * & * \end{bmatrix}$$

$$\underset{4 \text{ mults}}{\approx} \begin{bmatrix} 1 & * & * & * & * & * \\ 0 & 1 & * & * & * & * \\ 0 & * & * & * & * & * \\ 0 & * & * & * & * & * \end{bmatrix} \underset{\substack{12 \text{ mults} \\ 12 \text{ adds}}}{\approx} \begin{bmatrix} 1 & 0 & * & * & * & * \\ 0 & 1 & * & * & * & * \\ 0 & 0 & * & * & * & * \\ 0 & 0 & * & * & * & * \end{bmatrix}$$

$$\underset{3 \text{ mults}}{\approx} \begin{bmatrix} 1 & 0 & * & * & * & * \\ 0 & 1 & * & * & * & * \\ 0 & 0 & 1 & * & * & * \\ 0 & 0 & * & * & * & * \end{bmatrix} \underset{\substack{9 \text{ mults} \\ 9 \text{ adds}}}{\approx} \begin{bmatrix} 1 & 0 & 0 & * & * & * \\ 0 & 1 & 0 & * & * & * \\ 0 & 0 & 1 & * & * & * \\ 0 & 0 & 0 & * & * & * \end{bmatrix}$$

$$\underset{\text{2 mults}}{\approx}\begin{bmatrix} 1 & 0 & 0 & \ast & \ast & \ast \\ 0 & 1 & 0 & \ast & \ast & \ast \\ 0 & 0 & 1 & \ast & \ast & \ast \\ 0 & 0 & 0 & 1 & \ast & \ast \end{bmatrix} \underset{\substack{\text{6 mults} \\ \text{6 adds}}}{\approx} \begin{bmatrix} 1 & 0 & 0 & 0 & \ast & \ast \\ 0 & 1 & 0 & 0 & \ast & \ast \\ 0 & 0 & 1 & 0 & \ast & \ast \\ 0 & 0 & 0 & 1 & \ast & \ast \end{bmatrix}.$$

Number of mults: 5 + 4 + 3 + 2 + 3(5 + 4 + 3 + 2). Number of adds: 3(5 + 4 + 3 + 2).
Total number of operations: 7(5 + 4 + 3 + 2) = 98.

Gaussian elimination:

$$\begin{bmatrix} \ast & \ast & \ast & \ast & \ast & \ast \\ \ast & \ast & \ast & \ast & \ast & \ast \\ \ast & \ast & \ast & \ast & \ast & \ast \\ \ast & \ast & \ast & \ast & \ast & \ast \end{bmatrix} \underset{\text{5 mults}}{\approx} \begin{bmatrix} 1 & \ast & \ast & \ast & \ast & \ast \\ \ast & \ast & \ast & \ast & \ast & \ast \\ \ast & \ast & \ast & \ast & \ast & \ast \\ \ast & \ast & \ast & \ast & \ast & \ast \end{bmatrix} \underset{\substack{\text{15 mults} \\ \text{15 adds}}}{\approx} \begin{bmatrix} 1 & \ast & \ast & \ast & \ast & \ast \\ 0 & \ast & \ast & \ast & \ast & \ast \\ 0 & \ast & \ast & \ast & \ast & \ast \\ 0 & \ast & \ast & \ast & \ast & \ast \end{bmatrix}$$

$$\underset{\text{4 mults}}{\approx}\begin{bmatrix} 1 & \ast & \ast & \ast & \ast & \ast \\ 0 & 1 & \ast & \ast & \ast & \ast \\ 0 & \ast & \ast & \ast & \ast & \ast \\ 0 & \ast & \ast & \ast & \ast & \ast \end{bmatrix} \underset{\substack{\text{8 mults} \\ \text{8 adds}}}{\approx} \begin{bmatrix} 1 & \ast & \ast & \ast & \ast & \ast \\ 0 & 1 & \ast & \ast & \ast & \ast \\ 0 & 0 & \ast & \ast & \ast & \ast \\ 0 & 0 & \ast & \ast & \ast & \ast \end{bmatrix}$$

$$\underset{\text{3 mults}}{\approx}\begin{bmatrix} 1 & \ast & \ast & \ast & \ast & \ast \\ 0 & 1 & \ast & \ast & \ast & \ast \\ 0 & 0 & 1 & \ast & \ast & \ast \\ 0 & 0 & \ast & \ast & \ast & \ast \end{bmatrix} \underset{\substack{\text{3 mults} \\ \text{3 adds}}}{\approx} \begin{bmatrix} 1 & \ast & \ast & \ast & \ast & \ast \\ 0 & 1 & \ast & \ast & \ast & \ast \\ 0 & 0 & 1 & \ast & \ast & \ast \\ 0 & 0 & 0 & \ast & \ast & \ast \end{bmatrix}$$

$$\underset{\text{2 mults}}{\approx}\begin{bmatrix} 1 & \ast & \ast & \ast & \ast & \ast \\ 0 & 1 & \ast & \ast & \ast & \ast \\ 0 & 0 & 1 & \ast & \ast & \ast \\ 0 & 0 & 0 & 1 & \ast & \ast \end{bmatrix}.$$

Number of mults: 5 + 4 + 3 + 2 + 15 + 8 + 3. Number of adds: 15 + 8 + 3.
Total number of operations: 66.

Back substitution:

$$\begin{bmatrix} 1 & \ast & \ast & \ast & \ast & \ast \\ 0 & 1 & \ast & \ast & \ast & \ast \\ 0 & 0 & 1 & \ast & \ast & \ast \\ 0 & 0 & 0 & 1 & \ast & \ast \end{bmatrix} \underset{\substack{\text{6 mults} \\ \text{6 adds}}}{\approx} \begin{bmatrix} 1 & \ast & \ast & 0 & \ast & \ast \\ 0 & 1 & \ast & 0 & \ast & \ast \\ 0 & 0 & 1 & 0 & \ast & \ast \\ 0 & 0 & 0 & 1 & \ast & \ast \end{bmatrix} \underset{\substack{\text{4 mults} \\ \text{4 adds}}}{\approx} \begin{bmatrix} 1 & \ast & 0 & 0 & \ast & \ast \\ 0 & 1 & 0 & 0 & \ast & \ast \\ 0 & 0 & 1 & 0 & \ast & \ast \\ 0 & 0 & 0 & 1 & \ast & \ast \end{bmatrix}$$

$$\underset{\substack{2\text{ mults}\\ 2\text{ adds}}}{\approx}\begin{bmatrix} 1 & 0 & 0 & 0 & * & * \\ 0 & 1 & 0 & 0 & * & * \\ 0 & 0 & 1 & 0 & * & * \\ 0 & 0 & 0 & 1 & * & * \end{bmatrix}.$$

Number of mults: 6 + 4 + 2. Number of adds: 6 + 4 + 2.
Total number of operations: 24.

Eight operations are saved using Gaussian elimination. One addition and one multiplication are saved in getting zeros in each of the positions above the diagonal in column 3 and two additions and two multiplications are saved in getting the zero above the diagonal in column 2. The basic reason for the saving is that in back substituting one is working from right to left. This provides the same savings in operations above the diagonal that one has below the diagonal working from left to right.

For Gauss-Jordan elimination the number of operations is 98.
For Gaussian elimination the total number of operations is 66 + 24 = 90.

9. Yes. Consider a 2x3 matrix with no zeros in the first column. Find the echelon form leaving the rows in the original order, then do it switching the order of the rows first.

Exercise Set 1.5, page 42

Exercises 1 and 3 can be solved simultaneously since the coefficient matrices are the same for both.

$$\begin{aligned} a_0 + a_1 + a_2 &= b_1 \\ a_0 + 2a_1 + 4a_2 &= b_2 \\ a_0 + 3a_1 + 9a_2 &= b_3 \end{aligned}$$, where b_1, b_2, b_3 are the y values 2, 2, 4 in exercise 1

and 5, 7, 9 in exercise 3.

$$\begin{bmatrix} 1 & 1 & 1 & 2 & 5 \\ 1 & 2 & 4 & 2 & 7 \\ 1 & 3 & 9 & 4 & 9 \end{bmatrix} \underset{\substack{R2+(-1)R1\\ R3+(-1)R1}}{\approx} \begin{bmatrix} 1 & 1 & 1 & 2 & 5 \\ 0 & 1 & 3 & 0 & 2 \\ 0 & 2 & 8 & 2 & 4 \end{bmatrix} \underset{\substack{R1+(-1)R2\\ R3+(-2)R2}}{\approx} \begin{bmatrix} 1 & 0 & -2 & 2 & 3 \\ 0 & 1 & 3 & 0 & 2 \\ 0 & 0 & 2 & 2 & 0 \end{bmatrix}$$

$$\underset{(1/2)R3}{\approx}\begin{bmatrix} 1 & 0 & -2 & 2 & 3 \\ 0 & 1 & 3 & 0 & 2 \\ 0 & 0 & 1 & 1 & 0 \end{bmatrix} \underset{\substack{R1+(2)R3 \\ R2+(-3)R3}}{\approx} \begin{bmatrix} 1 & 0 & 0 & 4 & 3 \\ 0 & 1 & 0 & -3 & 2 \\ 0 & 0 & 1 & 1 & 0 \end{bmatrix}$$

, so the values of a_0, a_1, a_2 are 4, −3, 1 for exercise 1 and 3, 2, 0 for exercise 3. Thus the equations of the polynomials are: 1. $4 - 3x + x^2 = y$ 3. $3 + 2x = y$

4. $$\begin{aligned} a_0 + a_1 + a_2 &= 8 \\ a_0 + 3a_1 + 9a_2 &= 26 \\ a_0 + 5a_1 + 25a_2 &= 60 \end{aligned}$$

$$\begin{bmatrix} 1 & 1 & 1 & 8 \\ 1 & 3 & 9 & 26 \\ 1 & 5 & 25 & 60 \end{bmatrix} \underset{\substack{R2+(-1)R1 \\ R3+(-1)R1}}{\approx} \begin{bmatrix} 1 & 1 & 1 & 8 \\ 0 & 2 & 8 & 18 \\ 0 & 4 & 24 & 52 \end{bmatrix}$$

$$\underset{(1/2)R2}{\approx}\begin{bmatrix} 1 & 1 & 1 & 8 \\ 0 & 1 & 4 & 9 \\ 0 & 4 & 24 & 52 \end{bmatrix} \underset{\substack{R1+(-1)R2 \\ R3+(-4)R2}}{\approx} \begin{bmatrix} 1 & 0 & -3 & -1 \\ 0 & 1 & 4 & 9 \\ 0 & 0 & 8 & 16 \end{bmatrix} \underset{(1/8)R3}{\approx} \begin{bmatrix} 1 & 0 & -3 & -1 \\ 0 & 1 & 4 & 9 \\ 0 & 0 & 1 & 2 \end{bmatrix}$$

$$\underset{\substack{R1+(3)R3 \\ R2+(-4)R3}}{\approx} \begin{bmatrix} 1 & 0 & 0 & 5 \\ 0 & 1 & 0 & 1 \\ 0 & 0 & 1 & 2 \end{bmatrix}$$

, so $a_0 = 5$, $a_1 = 1$, $a_2 = 2$, and the equation is

$5 + x + 2x^2 = y$. When $x = 2$, $y = 5 + 2 + 8 = 15$.

6. $$\begin{aligned} a_0 + a_1 + a_2 + a_3 &= -3 \\ a_0 + 2a_1 + 4a_2 + 8a_3 &= -1 \\ a_0 + 3a_1 + 9a_2 + 27a_3 &= 9 \\ a_0 + 4a_1 + 16a_2 + 64a_3 &= 33 \end{aligned}$$

$$\begin{bmatrix} 1 & 1 & 1 & 1 & -3 \\ 1 & 2 & 4 & 8 & -1 \\ 1 & 3 & 9 & 27 & 9 \\ 1 & 4 & 16 & 64 & 33 \end{bmatrix}$$

$$\underset{\substack{R2+(-1)R1 \\ R3+(-1)R1 \\ R4+(-1)R1}}{\approx} \begin{bmatrix} 1 & 1 & 1 & 1 & -3 \\ 0 & 1 & 3 & 7 & 2 \\ 0 & 2 & 8 & 26 & 12 \\ 0 & 3 & 15 & 63 & 36 \end{bmatrix} \underset{\substack{R1+(-1)R2 \\ R3+(-2)R2 \\ R4+(-3)R2}}{\approx} \begin{bmatrix} 1 & 0 & -2 & -6 & -5 \\ 0 & 1 & 3 & 7 & 2 \\ 0 & 0 & 2 & 12 & 8 \\ 0 & 0 & 6 & 42 & 30 \end{bmatrix}$$

$$\underset{\substack{(1/2)R3 \\ (1/6)R4}}{\approx} \begin{bmatrix} 1 & 0 & -2 & -6 & -5 \\ 0 & 1 & 3 & 7 & 2 \\ 0 & 0 & 1 & 6 & 4 \\ 0 & 0 & 1 & 7 & 5 \end{bmatrix} \underset{\substack{R1+(2)R3 \\ R2+(-3)R3 \\ R4+(-1)R3}}{\approx} \begin{bmatrix} 1 & 0 & 0 & 6 & 3 \\ 0 & 1 & 0 & -11 & -10 \\ 0 & 0 & 1 & 6 & 4 \\ 0 & 0 & 0 & 1 & 1 \end{bmatrix}$$

$$\underset{\substack{R1+(-6)R4\\R2+(11)R4\\R3+(-6)R4}}{\approx}\begin{bmatrix}1&0&0&0&-3\\0&1&0&0&1\\0&0&1&0&-2\\0&0&0&1&1\end{bmatrix}$$, so $a_0 = -3, a_1 = 1, a_2 = -2, a_3 = 1$ and the equation is

$-3 + x - 2x^2 + x^3 = y.$

7.
$$\begin{aligned} I_1 + I_2 - I_3 &= 0 \\ 2I_1 \qquad + 4I_3 &= 34 \\ 4I_2 + 4I_3 &= 28 \end{aligned}$$

so that $I_1 = 5,\ I_2 = 1, I_3 = 6.$

9.
$$\begin{aligned} I_1 + I_2 - I_3 &= 0 \\ 3I_3 &= 9 \\ 4I_2 + 3I_3 &= 13 \end{aligned}$$

so that $I_1 = 2,\ I_2 = 1, I_3 = 3.$

11.
$$\begin{aligned} I_1 - I_2 - I_3 &= 0 \\ I_1 + 3I_2 \qquad &= 31 \\ I_1 \qquad + 7I_3 &= 31 \end{aligned}$$

so that $I_1 = 10,\ I_2 = 7, I_3 = 3.$

13.
$$\begin{aligned} I_1 - I_2 - I_3 \qquad\qquad &= 0 \\ I_3 - I_4 + I_5 &= 0 \\ I_1 + I_2 \qquad\qquad &= 4 \\ I_1 \qquad + 2I_4 \qquad &= 4 \\ 2I_4 + 2I_5 &= 2 \end{aligned}$$

so that $I_1 = 7/3,\ I_2 = 5/3, I_3 = 2/3, I_4 = 5/6, I_5 = 1/6.$

15. (a) Let I_1 be the current in the direction from the16-volt battery to A; let I_2 be the current from A to B; and let I_3 be the current in the direction from C to B.

$$I_1 - I_2 - I_3 = 0$$
$$5I_1 + I_2 = 16$$
$$-I_2 + 5I_3 = 9$$

$$\begin{bmatrix} 1 & -1 & -1 & 0 \\ 5 & 1 & 0 & 16 \\ 0 & -1 & 5 & 9 \end{bmatrix} \underset{R2+(-5)R1}{\approx} \begin{bmatrix} 1 & -1 & -1 & 0 \\ 0 & 6 & 5 & 16 \\ 0 & -1 & 5 & 9 \end{bmatrix} \underset{R2 \leftrightarrow R3}{\approx} \begin{bmatrix} 1 & -1 & -1 & 0 \\ 0 & -1 & 5 & 9 \\ 0 & 6 & 5 & 16 \end{bmatrix}$$

$$\underset{(-1)R2}{\approx} \begin{bmatrix} 1 & -1 & -1 & 0 \\ 0 & 1 & -5 & -9 \\ 0 & 6 & 5 & 16 \end{bmatrix} \underset{\substack{R1+R2 \\ R3+(-6)R2}}{\approx} \begin{bmatrix} 1 & 0 & -6 & -9 \\ 0 & 1 & -5 & -9 \\ 0 & 0 & 35 & 70 \end{bmatrix} \underset{(1/35)R3}{\approx} \begin{bmatrix} 1 & 0 & -6 & -9 \\ 0 & 1 & -5 & -9 \\ 0 & 0 & 1 & 2 \end{bmatrix}$$

$$\underset{\substack{R1+(6)R3 \\ R2+(5)R3}}{\approx} \begin{bmatrix} 1 & 0 & 0 & 3 \\ 0 & 1 & 0 & 1 \\ 0 & 0 & 1 & 2 \end{bmatrix}, \text{ so } I_1 = 3,\ I_2 = 1,\ I_3 = 2.$$

17. A: $x_1 - x_4 = 100$ B: $x_1 - x_2 = 200$

C: $-x_2 + x_3 = 150$ D: $x_3 - x_4 = 50$

Solving these equations simultaneously gives

$$x_1 = x_4 + 100,\ x_2 = x_4 - 100,\ x_3 = x_4 + 50.$$

$x_2 = 0$ is theoretically possible. In that case $x_4 = 100$, $x_1 = 200$, $x_3 = 150$.

This flow is not likely to be realized in practice unless branch BC is completely closed.

Chapter 1 Review Exercises, page 44

1. (a) Both equations are equivalent to $x + 2y = 1$, so the general solution is $x = 1 - 2r$, $y = r$.

   (b) $x = -1$, $y = 3$ (c) There is no solution. The equations are inconsistent.

2. (a) $c = -3$ (b) $c = 4$

3. The second equation minus 2 times the first gives $(d - 2)y = 2d - 4$.

(a) $d \neq 2$. The solution is $x = 0, y = 2$.

(b) $d = 2$. The general solution is $x = 2 - r, y = r$.

There are no more values of d. For each value of d either there is a unique solution or there are many solutions.

4. (a)
$$\begin{bmatrix} 1 & -2 & -6 & -17 \\ 2 & -6 & -16 & -46 \\ 1 & 2 & -1 & -5 \end{bmatrix} \underset{\substack{R2+(-2)R1 \\ R3+(-1)R1}}{\approx} \begin{bmatrix} 1 & -2 & -6 & -17 \\ 0 & -2 & -4 & -12 \\ 0 & 4 & 5 & 12 \end{bmatrix} \underset{(-1/2)R2}{\approx} \begin{bmatrix} 1 & -2 & -6 & -17 \\ 0 & 1 & 2 & 6 \\ 0 & 4 & 5 & 12 \end{bmatrix}$$

$$\underset{\substack{R1+(2)R2 \\ R3+(-4)R2}}{\approx} \begin{bmatrix} 1 & 0 & -2 & -5 \\ 0 & 1 & 2 & 6 \\ 0 & 0 & -3 & -12 \end{bmatrix} \underset{(-1/3)R3}{\approx} \begin{bmatrix} 1 & 0 & -2 & -5 \\ 0 & 1 & 2 & 6 \\ 0 & 0 & 1 & 4 \end{bmatrix}$$

$$\underset{\substack{R1+(2)R3 \\ R2+(-2)R3}}{\approx} \begin{bmatrix} 1 & 0 & 0 & 3 \\ 0 & 1 & 0 & -2 \\ 0 & 0 & 1 & 4 \end{bmatrix}$$, so that $x_1 = 3, x_2 = -2, x_3 = 4$.

(b)
$$\begin{bmatrix} 0 & 1 & 2 & 6 & 21 \\ 1 & -1 & 1 & 5 & 12 \\ 1 & -1 & -1 & -4 & -9 \\ 3 & -2 & 0 & -6 & -4 \end{bmatrix} \underset{R1 \leftrightarrow R2}{\approx} \begin{bmatrix} 1 & -1 & 1 & 5 & 12 \\ 0 & 1 & 2 & 6 & 21 \\ 1 & -1 & -1 & -4 & -9 \\ 3 & -2 & 0 & -6 & -4 \end{bmatrix}$$

$$\underset{\substack{R3+(-1)R1 \\ R4+(-3)R1}}{\approx} \begin{bmatrix} 1 & -1 & 1 & 5 & 12 \\ 0 & 1 & 2 & 6 & 21 \\ 0 & 0 & -2 & -9 & -21 \\ 0 & 1 & -3 & -21 & -40 \end{bmatrix} \underset{\substack{R1+R2 \\ R4+(-1)R2}}{\approx} \begin{bmatrix} 1 & 0 & 3 & 11 & 33 \\ 0 & 1 & 2 & 6 & 21 \\ 0 & 0 & -2 & -9 & -21 \\ 0 & 0 & -5 & -27 & -61 \end{bmatrix}$$

$$\underset{(-1/2)R3}{\approx} \begin{bmatrix} 1 & 0 & 3 & 11 & 33 \\ 0 & 1 & 2 & 6 & 21 \\ 0 & 0 & 1 & 9/2 & 21/2 \\ 0 & 0 & -5 & -27 & -61 \end{bmatrix} \underset{\substack{R1+(-3)R3 \\ R2+(-2)R3 \\ R4+(5)R3}}{\approx} \begin{bmatrix} 1 & 0 & 0 & -5/2 & 3/2 \\ 0 & 1 & 0 & -3 & 0 \\ 0 & 0 & 1 & 9/2 & 21/2 \\ 0 & 0 & 0 & -9/2 & -17/2 \end{bmatrix}$$

$$\underset{(-2/9)R4}{\approx}\begin{bmatrix} 1 & 0 & 0 & -5/2 & 3/2 \\ 0 & 1 & 0 & -3 & 0 \\ 0 & 0 & 1 & 9/2 & 21/2 \\ 0 & 0 & 0 & 1 & 17/9 \end{bmatrix} \underset{\substack{R1+(5/2)R4 \\ R2+(3)R4 \\ R3+(-9/2)R4}}{\approx} \begin{bmatrix} 1 & 0 & 0 & 0 & 56/9 \\ 0 & 1 & 0 & 0 & 17/3 \\ 0 & 0 & 1 & 0 & 2 \\ 0 & 0 & 0 & 1 & 17/9 \end{bmatrix},$$

so that $x_1 = 56/9$, $x_2 = 17/3$, $x_3 = 2$, $x_4 = 17/9$.

5. (a) Yes. (b) Yes.

   (c) No. The second column contains a leading 1, so all other elements in that column should be zero.

6. (a) $$\begin{bmatrix} 1 & -1 & 1 & 3 \\ -2 & 3 & 1 & -8 \\ 4 & -2 & 10 & 10 \end{bmatrix} \underset{\substack{R2+(2)R1 \\ R3+(-4)R1}}{\approx} \begin{bmatrix} 1 & -1 & 1 & 3 \\ 0 & 1 & 3 & -2 \\ 0 & 2 & 6 & -2 \end{bmatrix} \underset{\substack{R1+R2 \\ R3+(-2)R2}}{\approx} \begin{bmatrix} 1 & 0 & 4 & 1 \\ 0 & 1 & 3 & -2 \\ 0 & 0 & 0 & 2 \end{bmatrix}.$$

   There is no need to continue. The last row gives 0 = 2, so there is no solution.

   (b) $$\begin{bmatrix} 1 & 3 & 6 & -2 & -7 \\ -2 & -5 & -10 & 3 & 10 \\ 1 & 2 & 4 & 0 & 0 \\ 0 & 1 & 2 & -3 & -10 \end{bmatrix} \underset{\substack{R2+(2)R1 \\ R3+(-1)R1}}{\approx} \begin{bmatrix} 1 & 3 & 6 & -2 & -7 \\ 0 & 1 & 2 & -1 & -4 \\ 0 & -1 & -2 & 2 & 7 \\ 0 & 1 & 2 & -3 & -10 \end{bmatrix}$$

$$\underset{\substack{R1+(-3)R2 \\ R3+R2 \\ R4+(-1)R2}}{\approx} \begin{bmatrix} 1 & 0 & 0 & 1 & 5 \\ 0 & 1 & 2 & -1 & -4 \\ 0 & 0 & 0 & 1 & 3 \\ 0 & 0 & 0 & -2 & -6 \end{bmatrix} \underset{\substack{R1+(-1)R3 \\ R2+R3 \\ R4+(2)R3}}{\approx} \begin{bmatrix} 1 & 0 & 0 & 0 & 2 \\ 0 & 1 & 2 & 0 & -1 \\ 0 & 0 & 0 & 1 & 3 \\ 0 & 0 & 0 & 0 & 0 \end{bmatrix},$$

   so there are many solutions, and the general solution is

   $x_1 = 2$, $x_2 = -1 - 2r$, $x_3 = r$, $x_4 = 3$.

7. $$\begin{bmatrix} 1 & -2 & 3 & 1 \\ 3 & -4 & 5 & 3 \\ 2 & -3 & 4 & 2 \end{bmatrix} \underset{\substack{R2+(-3)R1 \\ R3+(-2)R1}}{\approx} \begin{bmatrix} 1 & -2 & 3 & 1 \\ 0 & 2 & -4 & 0 \\ 0 & 1 & -2 & 0 \end{bmatrix} \underset{(1/2)R2}{\approx} \begin{bmatrix} 1 & -2 & 3 & 1 \\ 0 & 1 & -2 & 0 \\ 0 & 1 & -2 & 0 \end{bmatrix}$$

back substituting

$$\underset{R3+(-1)R2}{\approx} \begin{bmatrix} 1 & -2 & 3 & 1 \\ 0 & 1 & -2 & 0 \\ 0 & 0 & 0 & 0 \end{bmatrix} \underset{R1+(2)R2}{\approx} \begin{bmatrix} 1 & 0 & -1 & 1 \\ 0 & 1 & -2 & 0 \\ 0 & 0 & 0 & 0 \end{bmatrix},$$

so $x_1 - x_3 = 1$ and $x_2 - 2x_3 = 0$. Thus there are many solutions, and the general

solution is $x_1 = 1 + r$, $x_2 = 2r$, $x_3 = r$.

8. If a matrix A is in reduced echelon form, it is clear from the definition that the leading 1 in any row cannot be to the left of the diagonal element in that row. Therefore if $A \neq I_n$, there must be some row which has its leading 1 to the right of the diagonal element in that row. Suppose row j is such a row and the leading 1 is in position (j, k) where $j < k \leq n$. Then if rows $j + 1, j + 2, \ldots, j + (n - k) < n$ all contain nonzero terms, the leading 1 in these rows must be at least as far to the right as columns $k + 1, k + 2, \ldots, k + (n - k) = n$, respectively. The leading 1 in row $j + (n - k) + 1$ must then be to the right of column n. But there is no column to the right of column n, so row $j + (n - k) + 1$ must consist of all zeros.

9. Let E be the reduced echelon form of A. Since B is row equivalent to A, B is also row equivalent to E. But since E is in reduced echelon form, it must be the reduced echelon form of B.

10. The equation is of the form $a_0 + a_1x + a_2x^2 = y$, so the system of equations to be solved is

$$\begin{aligned} a_0 + a_1 + a_2 &= 3 \\ a_0 + 2a_1 + 4a_2 &= 6 \\ a_0 + 3a_1 + 9a_2 &= 13 \end{aligned}$$

$$\begin{bmatrix} 1 & 1 & 1 & 3 \\ 1 & 2 & 4 & 6 \\ 1 & 3 & 9 & 13 \end{bmatrix} \underset{\substack{R2+(-1)R1 \\ R3+(-1)R1}}{\approx} \begin{bmatrix} 1 & 1 & 1 & 3 \\ 0 & 1 & 3 & 3 \\ 0 & 2 & 8 & 10 \end{bmatrix} \underset{\substack{R1+(-1)R2 \\ R3+(-2)R2}}{\approx} \begin{bmatrix} 1 & 0 & -2 & 0 \\ 0 & 1 & 3 & 3 \\ 0 & 0 & 2 & 4 \end{bmatrix}$$

$$\underset{(1/2)R3}{\approx}\begin{bmatrix} 1 & 0 & -2 & 0 \\ 0 & 1 & 3 & 3 \\ 0 & 0 & 1 & 2 \end{bmatrix}\underset{\substack{R1+(2)R3 \\ R2+(-3)R3}}{\approx}\begin{bmatrix} 1 & 0 & 0 & 4 \\ 0 & 1 & 0 & -3 \\ 0 & 0 & 1 & 2 \end{bmatrix}$$, so $a_0 = 4$, $a_1 = -3$, $a_2 = 2$, and the

equation is $4 - 3x + 2x^2 = y$.

11. $$\begin{aligned} I_1 - I_2 - I_3 &= 0 \\ 2I_1 \qquad + I_3 &= 7 \\ 3I_2 - I_3 &= 5 \end{aligned}$$

gives $I_1 = 3$, $I_2 = 2$, $I_3 = 1$.

Chapter 2

Exercise Set 2.1, page 56

1. (a) $A+B = \begin{bmatrix} 5-3 & 4+0 \\ -1+4 & 7+2 \\ 9+5 & -3-7 \end{bmatrix} = \begin{bmatrix} 2 & 4 \\ 3 & 9 \\ 14 & -10 \end{bmatrix}$.

   (c) $-D = \begin{bmatrix} -9 & 5 \\ -3 & 0 \end{bmatrix}$.

   (e) A + D does not exist.

   (g) $A - B = \begin{bmatrix} 8 & 4 \\ -5 & 5 \\ 4 & 4 \end{bmatrix}$.

2. (a) A + B does not exist.

   (d) $B - 3C = \begin{bmatrix} 0-3 & -1-6 & 4+15 \\ 6+21 & -8-27 & 2-9 \\ -4-15 & 5+12 & 9-0 \end{bmatrix} = \begin{bmatrix} -3 & -7 & 19 \\ 27 & -35 & -7 \\ -19 & 17 & 9 \end{bmatrix}$.

   (f) $3A + 2D = \begin{bmatrix} 27-6 \\ 6+0 \\ -3+4 \end{bmatrix} = \begin{bmatrix} 21 \\ 6 \\ 1 \end{bmatrix}$.

3. (a) $AB = \begin{bmatrix} (1x0)+(0x-2) & (1x1)+(0x5) \\ (0x0)+(1x-2) & (0x1)+(1x5) \end{bmatrix} = \begin{bmatrix} 0 & 1 \\ -2 & 5 \end{bmatrix} = B$.

   (d) CA does not exist.

   (e) AD = D.

   (g) $BD = \begin{bmatrix} (0x-1)+(1x5) & (0x0)+(1x7) & (0x3)+(1x2) \\ (-2x-1)+(5x5) & (-2x0)+(5x7) & (-2x3)+(5x2) \end{bmatrix} = \begin{bmatrix} 5 & 7 & 2 \\ 27 & 35 & 4 \end{bmatrix}$.

4. (a) $BA = \begin{bmatrix} 27 \\ 23 \\ 9 \end{bmatrix}$. (d) $CA = [27]$. (f) DB does not exist.

(h) $B^2 = \begin{bmatrix} (0x0)+(1x3)+(5x2) & (0x1)+(1x-7)+(5x3) & (0x5)+(1x8)+(5x1) \\ (3x0)+(-7x3)+(8x2) & (3x1)+(-7x-7)+(8x3) & (3x5)+(-7x8)+(8x1) \\ (2x0)+(3x3)+(1x2) & (2x1)+(3x-7)+(1x3) & (2x5)+(3x8)+(1x1) \end{bmatrix}$

$= \begin{bmatrix} 13 & 8 & 13 \\ -5 & 76 & -33 \\ 11 & -16 & 35 \end{bmatrix}$.

5. (a) $2A - 3(BC) = \begin{bmatrix} -12 & 2 \\ -9 & -24 \\ -20 & -24 \end{bmatrix}$. (c) $AC - BD = \begin{bmatrix} 8 & 2 \\ 4 & 1 \\ 0 & 6 \end{bmatrix}$.

(e) BA does not exist. (g) $C^3 + 2(D^2) = \begin{bmatrix} -14 & 0 \\ 12 & 10 \end{bmatrix}$.

6. (a) 3x2 (c) does not exist (e) 4x2 (g) does not exist

7. (b) 2x3 (d) does not exist (e) 3x2 (g) 2x2

9. (a) $c_{31} = (1x-1)+(0x5)+(-2x0) = -1$. (c) $d_{12} = (-1x3)+(2x6)+(-3x0) = 9$.

10. (b) $r_{33} = (-1x3)+(3x4) = 9$. (d) s_{23} does not exist. S is a 2x2 matrix.

11. (a) $d_{12} = (1x2)+(-3x0) + 2x- 4 = - 6$.

12. (a) $d_{11} = 2[(1x1) + (-3x3) + (0x-1)] + (2x2) + (0x4) + (-2x1) = -14.$

13. (a) $\begin{bmatrix} 2 & 3 \\ 3 & -8 \end{bmatrix}\begin{bmatrix} x_1 \\ x_2 \end{bmatrix} = \begin{bmatrix} 4 \\ -1 \end{bmatrix}.$

14. (b) $\begin{bmatrix} 5 & 2 \\ 4 & 3 \\ 3 & 1 \end{bmatrix}\begin{bmatrix} x_1 \\ x_2 \end{bmatrix} = \begin{bmatrix} 6 \\ -2 \\ 9 \end{bmatrix}.$

(d) $\begin{bmatrix} 2 & 5 & -3 & 4 \\ 1 & 0 & 9 & 5 \\ 3 & -3 & -8 & 5 \end{bmatrix}\begin{bmatrix} x_1 \\ x_2 \\ x_3 \\ x_4 \end{bmatrix} = \begin{bmatrix} 4 \\ 12 \\ -2 \end{bmatrix}.$

15. The third row of AB is the third row of A times each of the columns of B in turn. Since the third row of A is all zeros, each of the products is zero.

19. $A_2B = \begin{bmatrix} 4 & 0 & 3 \end{bmatrix}\begin{bmatrix} 8 & 1 & 3 \\ 2 & 1 & 0 \\ 4 & 6 & 3 \end{bmatrix} = \begin{bmatrix} 44 & 22 & 21 \end{bmatrix}.$

21. (a) $m_1 = (a_{12} - a_{22})(b_{21} + b_{22}) = (-3)(4) = -12.$ $m_5 = a_{11}(b_{12} - b_{22}) = (1)(-2) = -2.$

$m_2 = (a_{11} + a_{22})(b_{11} + b_{22}) = (4)(5) = 20.$ $m_6 = a_{22}(b_{21} - b_{11}) = (3)(-1) = -3.$

$m_3 = (a_{11} - a_{21})(b_{11} + b_{12}) = (-1)(3) = -3.$ $m_7 = (a_{21} + a_{22})b_{11} = (5)(1) = 5.$

$m_4 = (a_{11} + a_{12})b_{22} = (1)(4) = 4.$

$$AB = \begin{bmatrix} -12+20-4+(-3) & 4+(-2) \\ -3+5 & 20-(-3)+(-2)-5 \end{bmatrix} = \begin{bmatrix} 1 & 2 \\ 2 & 16 \end{bmatrix}.$$

(c) $m_1 = (-1)(3) = -3$, $m_2 = (2)(6) = 12$, $m_3 = (-1)(6) = -6$, $m_4 = (1)(1) = 1$, $m_5 = (2)(0) = 0$, $m_6 = (0)(-3) = 0$, $m_7 = (3)(5) = 15$.

$$AB = \begin{bmatrix} -3+12-1+0 & 1+0 \\ 0+15 & 12-(-6)+0-15 \end{bmatrix} = \begin{bmatrix} 8 & 1 \\ 15 & 3 \end{bmatrix}.$$

to make 4x4 matrices.

$$\begin{matrix} 1 & -1 \\ 2 & 1 \end{matrix}\Bigg]\begin{bmatrix} 0 & 0 \\ 3 & 0 \end{bmatrix}\Bigg]\Bigg[\begin{bmatrix} 1 & 1 \\ 3 & 2 \end{bmatrix}\begin{bmatrix} 2 & 0 \\ -1 & 0 \end{bmatrix}\Bigg]$$

$$\begin{matrix} 0 & 4 \\ 0 & 0 \end{matrix}\Bigg]\begin{bmatrix} 1 & 0 \\ 0 & 0 \end{bmatrix}\Bigg]\Bigg[\begin{bmatrix} 0 & 1 \\ 0 & 0 \end{bmatrix}\begin{bmatrix} 1 & 0 \\ 0 & 0 \end{bmatrix}\Bigg]$$

$$\Bigg] \quad \Bigg[\begin{bmatrix} 1 & -1 \\ 2 & 1 \end{bmatrix}\begin{bmatrix} 2 & 0 \\ -1 & 0 \end{bmatrix} + \begin{bmatrix} 0 & 0 \\ 3 & 0 \end{bmatrix}\begin{bmatrix} 1 & 0 \\ 0 & 0 \end{bmatrix}\Bigg]$$

$$\Bigg] \quad \Bigg[\begin{bmatrix} 0 & 4 \\ 0 & 0 \end{bmatrix}\begin{bmatrix} 2 & 0 \\ -1 & 0 \end{bmatrix} + \begin{bmatrix} 1 & 0 \\ 0 & 0 \end{bmatrix}\begin{bmatrix} 1 & 0 \\ 0 & 0 \end{bmatrix}\Bigg]$$

$$\begin{bmatrix} 0 & 0 \\ 3 & 0 \end{bmatrix}\Bigg] \quad \begin{bmatrix} 1 & 0 \\ 0 & 0 \end{bmatrix}\Bigg] = \Bigg[\begin{bmatrix} -2 & -1 \\ 5 & 7 \end{bmatrix}\begin{bmatrix} 3 & 0 \\ 6 & 0 \end{bmatrix} \quad \begin{bmatrix} 12 & 9 \\ 0 & 0 \end{bmatrix}\begin{bmatrix} -3 & 0 \\ 0 & 0 \end{bmatrix}\Bigg]$$

$$\begin{matrix} -1 & 3 \\ 7 & 6 \\ 9 & -3 \end{matrix}\Bigg].$$ Of course each of the multiplications

ing the method demonstrated in the solution to

1. (a) $AB = \begin{bmatrix} 4 & 7 & 10 \\ 0 & -5 & -4 \end{bmatrix}$. BA does not exist.

(c) $AD = \begin{bmatrix} 4 & 4 \\ -2 & 2 \end{bmatrix}$. $DA = \begin{bmatrix} 4 & 4 \\ -2 & 2 \end{bmatrix}$.

2. $A(BC) = \begin{bmatrix} 1 & 2 \\ -1 & 0 \\ 1 & 1 \end{bmatrix}\begin{bmatrix} 10 \\ 4 \end{bmatrix} = \begin{bmatrix} 18 \\ -10 \\ 14 \end{bmatrix}$. $(AB)C = \begin{bmatrix} -2 & 10 \\ -2 & -4 \\ 0 & 7 \end{bmatrix}\begin{bmatrix} 1 \\ 2 \end{bmatrix} = \begin{bmatrix} 18 \\ -10 \\ 14 \end{bmatrix}$.

4. (a) $AC^2 = \begin{bmatrix} 1 & 2 \\ 3 & 4 \end{bmatrix}\begin{bmatrix} 4 & 0 \\ 6 & 16 \end{bmatrix} = \begin{bmatrix} 16 & 32 \\ 36 & 64 \end{bmatrix}$.

(c) $BC^2 + B^3 = \begin{bmatrix} 2 & -3 \\ 0 & 1 \end{bmatrix}\begin{bmatrix} 4 & 0 \\ 6 & 16 \end{bmatrix} + \begin{bmatrix} 8 & -21 \\ 0 & 1 \end{bmatrix}$

$$= \begin{bmatrix} -10 & -48 \\ 6 & 16 \end{bmatrix} + \begin{bmatrix} 8 & -21 \\ 0 & 1 \end{bmatrix} = \begin{bmatrix} -2 & -69 \\ 6 & 17 \end{bmatrix}.$$

5. (a) $(AB)^2 = \begin{bmatrix} -2 & 2 \\ 11 & 19 \end{bmatrix}^2 = \begin{bmatrix} 26 & 34 \\ 187 & 383 \end{bmatrix}$.

(c) $A^2B + 2C^3 = \begin{bmatrix} 4 & 0 \\ -7 & 25 \end{bmatrix}\begin{bmatrix} -1 & 1 \\ 2 & 4 \end{bmatrix} + 2\begin{bmatrix} 27 & 76 \\ 0 & 8 \end{bmatrix}$

$$= \begin{bmatrix} -4 & 4 \\ 57 & 93 \end{bmatrix} + \begin{bmatrix} 54 & 152 \\ 0 & 16 \end{bmatrix} = \begin{bmatrix} 50 & 156 \\ 57 & 109 \end{bmatrix}.$$

6. (a) does not exist (c) 3x6

7. (a) 3x3 (c) does not exist, since PR does not exist (e) 2x3

8. The number of multiplications required to compute AB is mrn and the number required to compute BC is rns. (AB)C is the product of an mxn matrix and an nxs matrix, so the number of multiplications required is mns + mrn.

9. (a) 2x3x7 = 42. (c) 1x9x27 = 243.

11. (a) $2x4x3 + 2x3x1 = 30$ for (AB)C. $4x3x1 + 2x4x1 = 20$ for A(BC).

(c) $6x2x5 + 6x5x3 = 150$ for (AB)C. $2x5x3 + 6x2x3 = 66$ for A(BC).

(e) $7x97x2 + 7x2x3 = 1400$ for (AB)C. $97x2x3 + 7x97x3 = 2619$ for A(BC).

13. Each of the mn elements of AB is computed by multiplying a row of A times a column of B. This requires adding r terms, i.e., doing r−1 additions. Thus the total number of additions required to compute AB is mn(r−1).

To calculate (AB)C requires mn(r−1) additions for AB and ms(n−1) additions for the second product. Thus the total is mn(r−1) + ms(n−1). To calculate A(BC) requires rs(n−1) additions for BC and ms(r−1) for the second product. Thus the total is rs(n−1) + ms(r−1).

If A, B, and C are 2x2, 2x3, and 3x1, respectively, the number of additions is $2x3x1 + 2x1x2 = 10$ for (AB)C and $2x1x2 + 2x1x1 = 6$ for A(BC).

15. (a) The (i,j)th element of A + (B + C) is $a_{ij} + (b_{ij} + c_{ij})$. The (i,j)th element of (A + B) + C is $(a_{ij} + b_{ij}) + c_{ij} = a_{ij} + (b_{ij} + c_{ij})$. Since their elements are the same, A + (B + C) = (A + B) + C.

17. The (i,j)th element of cA is ca_{ij}. If $cA = O_{mn}$, then $ca_{ij} = 0$ for all i and j. So either c = 0 or all $a_{ij} = 0$, in which case $A = O_{mn}$.

18. (a) $A(A - 4B) + 2B(A + B) - A^2 + 7B^2 + 3AB = A^2 - 4AB + 2BA + 2B^2 - A^2 + 7B^2 + 3AB$
$= 9B^2 - AB + 2BA.$

(c) $(A - B)(A + B) - (A + B)^2 = A^2 - BA + AB - B^2 - (A^2 + AB + BA + B^2)$
$= -2BA - 2B^2.$

19. (b) $A(A - B)B + B2AB - 3A^2 = A^2B - AB^2 + 2BAB - 3A^2$.

(c) $(A + B)^3 - 2A^3 - 3ABA - A3B^2 - B^3$
$= A^3 + A^2B + BA^2 + BAB + ABA + AB^2 + B^2A + B^3 - 2A^3 - 3ABA - 3AB^2 - B^3$
$= - A^3 + A^2B + BA^2 + BAB - 2ABA - 2AB^2 + B^2A.$

20. (a) $\begin{bmatrix} 1 & 0 \\ -1 & 0 \end{bmatrix}\begin{bmatrix} a & b \\ c & d \end{bmatrix} = \begin{bmatrix} a & b \\ -a & -b \end{bmatrix}$ and $\begin{bmatrix} a & b \\ c & d \end{bmatrix}\begin{bmatrix} 1 & 0 \\ -1 & 0 \end{bmatrix} = \begin{bmatrix} a-b & 0 \\ c-d & 0 \end{bmatrix}$.

For equality it is necessary to have $a = a-b$, $b = 0$, and $-a = c-d$. Thus those matrices that commute with the given matrix are all matrices of the form

$$\begin{bmatrix} a & 0 \\ c & c+a \end{bmatrix}$$

where a and c can take any real values.

21. $AX_1 = AX_2$ does not imply that $X_1 = X_2$.

23. $(A + B)^2 = (A + B)(A + B) = A(A + B) + B(A + B)$ (distributive law)
$= A^2 + AB + BA + B^2$.

If $AB = BA$ then $(A + B)^2 = A^2 + 2AB + B^2$.

27. (a) The (i,j)th element of A + B is $a_{ij} + b_{ij} = 0 + 0$ if $i \neq j$, so A + B is a diagonal matrix.

30. (a) Yes, $\begin{bmatrix} 1 & 0 \\ 0 & 1 \end{bmatrix}^2 = \begin{bmatrix} 1 & 0 \\ 0 & 1 \end{bmatrix}$. (c) No, $\begin{bmatrix} 0 & 1 \\ 1 & 0 \end{bmatrix}^2 = \begin{bmatrix} 1 & 0 \\ 0 & 1 \end{bmatrix}$.

(e) Yes, $\begin{bmatrix} 1 & 2 & 2 \\ 0 & 0 & -1 \\ 0 & 0 & 1 \end{bmatrix}^2 = \begin{bmatrix} 1 & 2 & 2 \\ 0 & 0 & -1 \\ 0 & 0 & 1 \end{bmatrix}$.

31. $\begin{bmatrix} 1 & b \\ c & d \end{bmatrix}^2 = \begin{bmatrix} 1+bc & b+bd \\ c+cd & bc+d^2 \end{bmatrix}$, so the matrix will be idempotent if

$$1 = 1 + bc, \quad b = b + bd, \quad c = c + cd, \quad \text{and} \quad d = bc + d^2.$$

The first equation implies that $bc = 0$ and substituting $bc = 0$ in the last equation gives $d = d^2$. Thus d must be either zero or 1. If one of b and c is nonzero, the second and third equations give d = 0. If both b and c are zero then d can be either zero or 1. Thus the idempotent matrices of the given form are

$$\begin{bmatrix} 1 & 0 \\ 0 & 0 \end{bmatrix}, \begin{bmatrix} 1 & 0 \\ 0 & 1 \end{bmatrix}, \begin{bmatrix} 1 & b \\ 0 & 0 \end{bmatrix}, \text{ and } \begin{bmatrix} 1 & 0 \\ c & 0 \end{bmatrix}.$$

33. $(AB)^2 = (AB)(AB) = A(BA)B$. If $AB = BA$, then $(AB)^2 = A(AB)B = A^2B^2 = AB$. Thus AB is idempotent.

Exercise Set 2.3, page 74

1. (a) $A^t = \begin{bmatrix} -1 & 2 \\ 2 & -3 \end{bmatrix}$. symmetric (c) $C^t = \begin{bmatrix} 3 & 2 \\ -1 & 4 \end{bmatrix}$. not symmetric

(d) $D^t = \begin{bmatrix} 4 & -2 & 7 \\ 5 & 3 & 0 \end{bmatrix}$. not symmetric (f) $F^t = \begin{bmatrix} 1 & -1 & 3 \\ -1 & 2 & 0 \\ 3 & 0 & 4 \end{bmatrix}$. symmetric

(h) $H^t = \begin{bmatrix} 1 & 4 & -2 \\ -2 & 5 & 6 \\ 3 & 6 & 7 \end{bmatrix}$. not symmetric

2. (a) $\begin{bmatrix} 1 & 2 & 4 \\ 2 & 6 & 5 \\ 4 & 5 & 2 \end{bmatrix}$ 3. (a) 4x2 (c) does not exist (e) 4x3

5. (a) $(A + B + C)^t = (A + (B + C))^t = A^t + (B + C)^t = A^t + B^t + C^t$.

9. If A is symmetric then $A = A^t$. From Theorem 2.4 and exercise 4(c) this means $A^t = A = (A^t)^t$. So A^t is symmetric.

11. (a) $\begin{bmatrix} 0 & -1 \\ 1 & 0 \end{bmatrix}$

(c) If A and B are antisymmetric, then $A + B = (-A^t) + (-B^t) = -(A^t + B^t) = -(A + B)^t$, so $A + B$ is antisymmetric.

13. $B = (1/2)(A + A^t)$ is symmetric and $C = (1/2)(A - A^t)$ is antisymmetric. $A = B + C$.

15. (a) $2 + (-4) = -2$. (c) $0 + 5 - 7 + 1 = -1$.

17. $\text{tr}(A + B + C) = \text{tr}(A + (B + C)) = \text{tr}(A) + \text{tr}(B + C) = \text{tr}(A) + \text{tr}(B) + \text{tr}(C)$.

21. $AB = \begin{bmatrix} 5(-2+i)+(3-i)(3-i) & 5(5+2i)+(3-i)(4+3i) \\ (2+3i)(-2+i)+(-5i)(3-i) & (2+3i)(5+2i)+(-5i)(4+3i) \end{bmatrix} = \begin{bmatrix} -2-i & 40+15i \\ -12-19i & 19-i \end{bmatrix}.$

23. $\overline{A} = \begin{bmatrix} 2+3i & -5i \\ 2 & 5+4i \end{bmatrix}$ $\qquad A^* = \overline{A}^t = \begin{bmatrix} 2+3i & 2 \\ -5i & 5+4i \end{bmatrix}$

$\overline{D} = \begin{bmatrix} -2 & 3+5i \\ 3-5i & 9 \end{bmatrix}$ $\qquad D^* = \overline{D}^t = \begin{bmatrix} -2 & 3-5i \\ 3+5i & 9 \end{bmatrix}$ hermitian

25. (a) The (i,j)th element of $A^* + B^*$ is $\overline{a_{ji}} + \overline{b_{ji}} = \overline{a_{ji} + b_{ji}}$, the (i,j)th element of $(A+B)^*$.

27. (a) $G^t = (AA^t)^t = (A^t)^tA^t = AA^t = G$, and $P^t = (A^tA)^t = A^t(A^t)^t = A^tA = P$.

28. (a) $G = \begin{bmatrix} 1 & 0 & 1 \\ 0 & 1 & 1 \\ 1 & 1 & 2 \end{bmatrix}$. $P = \begin{bmatrix} 2 & 1 \\ 1 & 2 \end{bmatrix}$.

$g_{12} = 0$, $g_{13} = 1$, $g_{23} = 1$, so $1 \to 3 \to 2$ or $2 \to 3 \to 1$.

$p_{12} = 1$, so $1 \to 2$ or $2 \to 1$, gives no information.

(c) $G = \begin{bmatrix} 2 & 1 & 2 \\ 1 & 3 & 3 \\ 2 & 3 & 4 \end{bmatrix}$. $P = \begin{bmatrix} 2 & 1 & 2 & 1 \\ 1 & 2 & 2 & 2 \\ 2 & 2 & 3 & 2 \\ 1 & 2 & 2 & 2 \end{bmatrix}$

$g_{12} = 1$, $g_{13} = 2$, $g_{23} = 3$, so $1 \to 3 \to 2$ or $2 \to 3 \to 1$.

$p_{12} = 1$, $p_{13} = 2$, $p_{14} = 1$, $p_{23} = 2$, $p_{24} = 2$, $p_{34} = 2$, so $1 \to 3 \to \{2 \leftrightarrow 4\}$ or $\{2 \leftrightarrow 4\} \to 3 \to 1$.

(e) $G = \begin{bmatrix} 2 & 1 & 0 & 1 \\ 1 & 1 & 0 & 0 \\ 0 & 0 & 2 & 1 \\ 1 & 0 & 1 & 2 \end{bmatrix}$. $P = \begin{bmatrix} 2 & 0 & 1 & 0 \\ 0 & 2 & 1 & 1 \\ 1 & 1 & 2 & 0 \\ 0 & 1 & 0 & 1 \end{bmatrix}$.

$g_{12} = 1$, $g_{13} = 0$, $g_{14} = 1$, $g_{23} = 0$, $g_{24} = 0$, $g_{34} = 1$, so $2 \to 1 \to 4 \to 3$ or $3 \to 4 \to 1 \to 2$.

$p_{12} = 0$, $p_{13} = 1$, $p_{14} = 0$, $p_{23} = 1$, $p_{24} = 1$, $p_{34} = 0$, so $1 \to 3 \to 2 \to 4$ or $4 \to 2 \to 3 \to 1$.

Exercise Set 2.4, page 84

1. (a) $AB = BA = I_2$, so B is the inverse of A.

(c) $AB = \begin{bmatrix} 2 & -4 \\ -5 & 3 \end{bmatrix}\begin{bmatrix} 3 & 1 \\ 5 & 2 \end{bmatrix} = \begin{bmatrix} -14 & -6 \\ 0 & 1 \end{bmatrix}$, so B is not the inverse of A.

2. (a) $AB = BA = I_3$, so B is the inverse of A.

(c) $AB = \begin{bmatrix} 0 & 1 & -1 \\ 2 & -2 & -1 \\ -1 & 1 & 1 \end{bmatrix}\begin{bmatrix} 1 & 2 & 3 \\ 1 & 1 & 2 \\ 0 & 1 & 1 \end{bmatrix} = \begin{bmatrix} 1 & 0 & 1 \\ 0 & 1 & 1 \\ 0 & 0 & 0 \end{bmatrix}$, so B is not the inverse of A.

3. (a) $\begin{bmatrix} 1 & 0 & 1 & 0 \\ 2 & 1 & 0 & 1 \end{bmatrix} \underset{R2+(-2)R1}{\approx} \begin{bmatrix} 1 & 0 & 1 & 0 \\ 0 & 1 & -2 & 1 \end{bmatrix}$, and the inverse matrix is $\begin{bmatrix} 1 & 0 \\ -2 & 1 \end{bmatrix}$.

(c) $\begin{bmatrix} 2 & 1 & 1 & 0 \\ 4 & 3 & 0 & 1 \end{bmatrix} \underset{(1/2)R1}{\approx} \begin{bmatrix} 1 & 1/2 & 1/2 & 0 \\ 4 & 3 & 0 & 1 \end{bmatrix} \underset{R2+(-4)R1}{\approx} \begin{bmatrix} 1 & 1/2 & 1/2 & 0 \\ 0 & 1 & -2 & 1 \end{bmatrix}$

$\underset{R1+(-1/2)R2}{\approx} \begin{bmatrix} 1 & 0 & 3/2 & -1/2 \\ 0 & 1 & -2 & 1 \end{bmatrix}$, and the inverse is $\begin{bmatrix} 3/2 & -1/2 \\ -2 & 1 \end{bmatrix}$.

(e) $\begin{bmatrix} 1 & 2 & 1 & 0 \\ 3 & 6 & 0 & 1 \end{bmatrix} \underset{R2+(-3)R1}{\approx} \begin{bmatrix} 1 & 2 & 1 & 0 \\ 0 & 0 & -3 & 1 \end{bmatrix}$, and the inverse does not exist.

4. (a) $\begin{bmatrix} 1 & 2 & 3 & 1 & 0 & 0 \\ 0 & 1 & 2 & 0 & 1 & 0 \\ 4 & 5 & 3 & 0 & 0 & 1 \end{bmatrix} \underset{R2+(-4)R1}{\approx} \begin{bmatrix} 1 & 2 & 3 & 1 & 0 & 0 \\ 0 & 1 & 2 & 0 & 1 & 0 \\ 0 & -3 & -9 & -4 & 0 & 1 \end{bmatrix}$

$\underset{\substack{R1+(-2)R2 \\ R3+(3)R2}}{\approx} \begin{bmatrix} 1 & 0 & -1 & 1 & -2 & 0 \\ 0 & 1 & 2 & 0 & 1 & 0 \\ 0 & 0 & -3 & -4 & 3 & 1 \end{bmatrix} \underset{(-1/3)R3}{\approx} \begin{bmatrix} 1 & 0 & -1 & 1 & -2 & 0 \\ 0 & 1 & 2 & 0 & 1 & 0 \\ 0 & 0 & 1 & 4/3 & -1 & -1/3 \end{bmatrix}$

$\underset{\substack{R1+R3 \\ R2+(-2)R3}}{\approx} \begin{bmatrix} 1 & 0 & 0 & 7/3 & -3 & -1/3 \\ 0 & 1 & 0 & -8/3 & 3 & 2/3 \\ 0 & 0 & 1 & 4/3 & -1 & -1/3 \end{bmatrix}$, and the inverse is $\begin{bmatrix} 7/3 & -3 & -1/3 \\ -8/3 & 3 & 2/3 \\ 4/3 & -1 & -1/3 \end{bmatrix}$.

(c) $\begin{bmatrix} 1 & 2 & -3 & 1 & 0 & 0 \\ 1 & -2 & 1 & 0 & 1 & 0 \\ 5 & -2 & -3 & 0 & 0 & 1 \end{bmatrix} \underset{\substack{R2+(-1)R1 \\ R3+(-5)R1}}{\approx} \begin{bmatrix} 1 & 2 & -3 & 1 & 0 & 0 \\ 0 & -4 & 4 & -1 & 1 & 0 \\ 0 & -12 & 12 & -5 & 0 & 1 \end{bmatrix}$

$$\underset{(-1/4)R2}{\approx}\begin{bmatrix} 1 & 2 & -3 & 1 & 0 & 0 \\ 0 & 1 & -1 & 1/4 & -1/4 & 0 \\ 0 & -12 & 12 & -5 & 0 & 1 \end{bmatrix} \underset{\substack{R1+(-2)R2 \\ R3+(12)R2}}{\approx}\begin{bmatrix} 1 & 0 & -1 & 1/2 & 1/2 & 0 \\ 0 & 1 & -1 & 1/4 & -1/4 & 0 \\ 0 & 0 & 0 & -2 & -3 & 1 \end{bmatrix},$$

so the inverse does not exist.

5. (a) $$\begin{bmatrix} 1 & 2 & 3 & 1 & 0 & 0 \\ 2 & -1 & 4 & 0 & 1 & 0 \\ 0 & -1 & 1 & 0 & 0 & 1 \end{bmatrix} \underset{R2+(-2)R1}{\approx}\begin{bmatrix} 1 & 2 & 3 & 1 & 0 & 0 \\ 0 & -5 & -2 & -2 & 1 & 0 \\ 0 & -1 & 1 & 0 & 0 & 1 \end{bmatrix}$$

$$\underset{R2 \leftrightarrow R3}{\approx}\begin{bmatrix} 1 & 2 & 3 & 1 & 0 & 0 \\ 0 & -1 & 1 & 0 & 0 & 1 \\ 0 & -5 & -2 & -2 & 1 & 0 \end{bmatrix} \underset{(-1)R2}{\approx}\begin{bmatrix} 1 & 2 & 3 & 1 & 0 & 0 \\ 0 & 1 & -1 & 0 & 0 & -1 \\ 0 & -5 & -2 & -2 & 1 & 0 \end{bmatrix}$$

$$\underset{\substack{R1+(-2)R2 \\ R3+(5)R2}}{\approx}\begin{bmatrix} 1 & 0 & 5 & 1 & 0 & 2 \\ 0 & 1 & -1 & 0 & 0 & -1 \\ 0 & 0 & -7 & -2 & 1 & -5 \end{bmatrix} \underset{(-1/7)R3}{\approx}\begin{bmatrix} 1 & 0 & 5 & 1 & 0 & 2 \\ 0 & 1 & -1 & 0 & 0 & -1 \\ 0 & 0 & 1 & 2/7 & -1/7 & 5/7 \end{bmatrix}$$

$$\underset{\substack{R1+(-5)R3 \\ R2+R3}}{\approx}\begin{bmatrix} 1 & 0 & 0 & -3/7 & 5/7 & -11/7 \\ 0 & 1 & 0 & 2/7 & -1/7 & -2/7 \\ 0 & 0 & 1 & 2/7 & -1/7 & 5/7 \end{bmatrix},$$

and the inverse is $\begin{bmatrix} -3/7 & 5/7 & -11/7 \\ 2/7 & -1/7 & -2/7 \\ 2/7 & -1/7 & 5/7 \end{bmatrix}$.

(c) $$\begin{bmatrix} 1 & -2 & -1 & 1 & 0 & 0 \\ -2 & 4 & 6 & 0 & 1 & 0 \\ 0 & 0 & 5 & 0 & 0 & 1 \end{bmatrix} \underset{R2+(2)R1}{\approx}\begin{bmatrix} 1 & -2 & -1 & 1 & 0 & 0 \\ 0 & 0 & 4 & 2 & 1 & 0 \\ 0 & 0 & 5 & 0 & 0 & 1 \end{bmatrix},$$

and the inverse does not exist.

6. (a) $$\begin{bmatrix} -3 & -1 & 1 & -2 & 1 & 0 & 0 & 0 \\ -1 & 3 & 2 & 1 & 0 & 1 & 0 & 0 \\ 1 & 2 & 3 & -1 & 0 & 0 & 1 & 0 \\ -2 & 1 & -1 & -3 & 0 & 0 & 0 & 1 \end{bmatrix} \underset{R1 \leftrightarrow R3}{\approx}\begin{bmatrix} 1 & 2 & 3 & -1 & 0 & 0 & 1 & 0 \\ -1 & 3 & 2 & 1 & 0 & 1 & 0 & 0 \\ -3 & -1 & 1 & -2 & 1 & 0 & 0 & 0 \\ -2 & 1 & -1 & -3 & 0 & 0 & 0 & 1 \end{bmatrix}$$

$$\underset{\substack{R2+R1\\R3+(3)R1\\R4+(2)R1}}{\approx}\begin{bmatrix}1&2&3&-1&0&0&1&0\\0&5&5&0&0&1&1&0\\0&5&10&-5&1&0&3&0\\0&5&5&-5&0&0&2&1\end{bmatrix}\underset{(1/5)R2}{\approx}\begin{bmatrix}1&2&3&-1&0&0&1&0\\0&1&1&0&0&1/5&1/5&0\\0&5&10&-5&1&0&3&0\\0&5&5&-5&0&0&2&1\end{bmatrix}$$

$$\underset{\substack{R1+(-2)R2\\R3+(-5)R2\\R4+(-5)R2}}{\approx}\begin{bmatrix}1&0&1&-1&0&-2/5&3/5&0\\0&1&1&0&0&1/5&1/5&0\\0&0&5&-5&1&-1&2&0\\0&0&0&-5&0&-1&1&1\end{bmatrix}$$

$$\underset{\substack{(1/5)R3\\(-1/5)R4}}{\approx}\begin{bmatrix}1&0&1&-1&0&-2/5&3/5&0\\0&1&1&0&0&1/5&1/5&0\\0&0&1&-1&1/5&-1/5&2/5&0\\0&0&0&1&0&1/5&-1/5&-1/5\end{bmatrix}$$

$$\underset{\substack{R1+(-1)R3\\R2+(-1)R3}}{\approx}\begin{bmatrix}1&0&0&0&-1/5&-1/5&1/5&0\\0&1&0&1&-1/5&2/5&-1/5&0\\0&0&1&-1&1/5&-1/5&2/5&0\\0&0&0&1&0&1/5&-1/5&-1/5\end{bmatrix}$$

$$\underset{\substack{R2+(-1)R4\\R3+R4}}{\approx}\begin{bmatrix}1&0&0&0&-1/5&-1/5&1/5&0\\0&1&0&0&-1/5&1/5&0&1/5\\0&0&1&0&1/5&0&1/5&-1/5\\0&0&0&1&0&1/5&-1/5&-1/5\end{bmatrix},$$

so the inverse is $\begin{bmatrix}-1/5&-1/5&1/5&0\\-1/5&1/5&0&1/5\\1/5&0&1/5&-1/5\\0&1/5&-1/5&-1/5\end{bmatrix}$.

(c) $$\begin{bmatrix}-1&0&-1&-1&1&0&0&0\\-3&-1&0&-1&0&1&0&0\\5&0&4&3&0&0&1&0\\3&0&3&2&0&0&0&1\end{bmatrix}\underset{(-1)R1}{\approx}\begin{bmatrix}1&0&1&1&-1&0&0&0\\-3&-1&0&-1&0&1&0&0\\5&0&4&3&0&0&1&0\\3&0&3&2&0&0&0&1\end{bmatrix}$$

$$\underset{\substack{R2+(3)R1\\R3+(-5)R1\\R4+(-3)R1}}{\approx}\begin{bmatrix}1&0&1&1&-1&0&0&0\\0&-1&3&2&-3&1&0&0\\0&0&-1&-2&5&0&1&0\\0&0&0&-1&3&0&0&1\end{bmatrix}\underset{\substack{(-1)R2\\(-1)R3\\(-1)R4}}{\approx}\begin{bmatrix}1&0&1&1&-1&0&0&0\\0&1&-3&-2&3&-1&0&0\\0&0&1&2&-5&0&-1&0\\0&0&0&1&-3&0&0&-1\end{bmatrix}$$

$$\underset{\substack{R1+(-1)R3 \\ R2+(3)R3}}{\approx}\begin{bmatrix} 1 & 0 & 0 & -1 & 4 & 0 & 1 & 0 \\ 0 & 1 & 0 & 4 & -12 & -1 & -3 & 0 \\ 0 & 0 & 1 & 2 & -5 & 0 & -1 & 0 \\ 0 & 0 & 0 & 1 & -3 & 0 & 0 & -1 \end{bmatrix}$$

$$\underset{\substack{R1+R4 \\ R2+(-4)R4 \\ R3+(-2)R4}}{\approx}\begin{bmatrix} 1 & 0 & 0 & 0 & 1 & 0 & 1 & -1 \\ 0 & 1 & 0 & 0 & 0 & -1 & -3 & 4 \\ 0 & 0 & 1 & 0 & 1 & 0 & -1 & 2 \\ 0 & 0 & 0 & 1 & -3 & 0 & 0 & -1 \end{bmatrix}, \text{ so the inverse is } \begin{bmatrix} 1 & 0 & 1 & -1 \\ 0 & -1 & -3 & 4 \\ 1 & 0 & -1 & 2 \\ -3 & 0 & 0 & -1 \end{bmatrix}.$$

7. (a) The inverse of the coefficient matrix is $\begin{bmatrix} -5 & 2 \\ 3 & -1 \end{bmatrix}$.

$$\begin{bmatrix} x_1 \\ x_2 \end{bmatrix} = \begin{bmatrix} -5 & 2 \\ 3 & -1 \end{bmatrix}\begin{bmatrix} 2 \\ 4 \end{bmatrix} = \begin{bmatrix} -2 \\ 2 \end{bmatrix}.$$

(c) $$\begin{bmatrix} x_1 \\ x_2 \end{bmatrix} = \begin{bmatrix} -1/5 & 3/5 \\ 2/5 & -1/5 \end{bmatrix}\begin{bmatrix} 5 \\ 10 \end{bmatrix} = \begin{bmatrix} 5 \\ 0 \end{bmatrix}.$$

(e) $$\begin{bmatrix} x_1 \\ x_2 \end{bmatrix} = \begin{bmatrix} 2 & -1 \\ -3/4 & 1/2 \end{bmatrix}\begin{bmatrix} 6 \\ 1 \end{bmatrix} = \begin{bmatrix} 11 \\ -4 \end{bmatrix}.$$

8. (a) $$\begin{bmatrix} x_1 \\ x_2 \\ x_3 \end{bmatrix} = \begin{bmatrix} 1/9 & 3/9 & 5/9 \\ 3/9 & 0 & -3/9 \\ -2/9 & 3/9 & -1/9 \end{bmatrix}\begin{bmatrix} 2 \\ 0 \\ 1 \end{bmatrix} = \begin{bmatrix} 7/9 \\ 3/9 \\ -5/9 \end{bmatrix}.$$

(c) $$\begin{bmatrix} x_1 \\ x_2 \\ x_3 \end{bmatrix} = \begin{bmatrix} -40 & 16 & 9 \\ 13 & -5 & -3 \\ 5 & -2 & -1 \end{bmatrix}\begin{bmatrix} 1 \\ 3 \\ 15 \end{bmatrix} = \begin{bmatrix} 143 \\ -47 \\ -16 \end{bmatrix}.$$

(e) $$\begin{bmatrix} x_1 \\ x_2 \\ x_3 \end{bmatrix} = \begin{bmatrix} 2 & 3 & 1 \\ 3 & 3 & 1 \\ 2 & 4 & 1 \end{bmatrix}\begin{bmatrix} 5 \\ -2 \\ 1 \end{bmatrix} = \begin{bmatrix} 5 \\ 10 \\ 3 \end{bmatrix}.$$

9. $\begin{bmatrix} x_1 \\ x_2 \\ x_3 \\ x_4 \end{bmatrix} = \begin{bmatrix} -14/17 & 8/17 & 4/17 & 5/17 \\ 3/17 & -9/17 & 4/17 & 5/17 \\ 5/17 & 2/17 & 1/17 & -3/17 \\ 18/17 & -3/17 & -10/17 & -4/17 \end{bmatrix} \begin{bmatrix} 5 \\ 6 \\ 1 \\ 7 \end{bmatrix} = \begin{bmatrix} 1 \\ 0 \\ 1 \\ 2 \end{bmatrix}.$

11. $\begin{bmatrix} a & b \\ c & d \end{bmatrix} \frac{1}{ad-bc} \begin{bmatrix} d & -b \\ -c & a \end{bmatrix} = \frac{1}{ad-bc} \begin{bmatrix} a & b \\ c & d \end{bmatrix} \begin{bmatrix} d & -b \\ -c & a \end{bmatrix}$

$= \frac{1}{ad-bc} \begin{bmatrix} ad-bc & 0 \\ 0 & ad-bc \end{bmatrix} = \begin{bmatrix} 1 & 0 \\ 0 & 1 \end{bmatrix}.$

Likewise $\frac{1}{ad-bc} \begin{bmatrix} d & -b \\ -c & a \end{bmatrix} \begin{bmatrix} a & b \\ c & d \end{bmatrix} = \begin{bmatrix} 1 & 0 \\ 0 & 1 \end{bmatrix}$. Thus $\frac{1}{ad-bc} \begin{bmatrix} d & -b \\ -c & a \end{bmatrix} = A^{-1}$.

The inverses of the given matrices are:

(a) $\begin{bmatrix} 3 & -8 \\ -1 & 3 \end{bmatrix}$, (c) $\frac{1}{2} \begin{bmatrix} 4 & -6 \\ -3 & 5 \end{bmatrix} = \begin{bmatrix} 2 & -3 \\ -3/2 & 5/2 \end{bmatrix}$.

12. (a) $I = (cA)(cA)^{-1}$. Multiply both sides by $\frac{1}{c}A^{-1}$.

$\frac{1}{c}A^{-1} = \frac{1}{c}A^{-1}(cA)(cA)^{-1} = \frac{1}{c}cA^{-1}A(cA)^{-1} = I\,(cA)^{-1} = (cA)^{-1}.$

14. $A = 2 \begin{bmatrix} -3 & 2 \\ -10 & 6 \end{bmatrix}^{-1} = 2 \begin{bmatrix} 3 & -1 \\ 5 & -3/2 \end{bmatrix} = \begin{bmatrix} 6 & -2 \\ 10 & -3 \end{bmatrix}.$

15. (a) $(3A)^{-1} = \frac{1}{3} \begin{bmatrix} 2 & -1 \\ -5 & 3 \end{bmatrix} = \begin{bmatrix} 2/3 & -1/3 \\ -5/3 & 3/3 \end{bmatrix}$. (c) $A^{-2} = (A^{-1})^2 = \begin{bmatrix} 9 & -5 \\ -25 & 14 \end{bmatrix}$.

16. (a) $(2A^t)^{-1} = \frac{1}{2}(A^t)^{-1} = \frac{1}{2}(A^{-1})^t = \frac{1}{2} \begin{bmatrix} 2 & -9 \\ -1 & 5 \end{bmatrix}.$

(c) $(AA^t)^{-1} = (A^t)^{-1}A^{-1} = (A^{-1})^tA^{-1} = \begin{bmatrix} 2 & -9 \\ -1 & 5 \end{bmatrix}\begin{bmatrix} 2 & -1 \\ -9 & 5 \end{bmatrix} = \begin{bmatrix} 85 & -47 \\ -47 & 26 \end{bmatrix}$.

17. $\begin{bmatrix} 2 & -7 \\ -1 & 4 \end{bmatrix}^{-1} = \begin{bmatrix} 4 & 7 \\ 1 & 2 \end{bmatrix}$, so $2x = 4$ and $x = 2$.

19. $4A^t = \begin{bmatrix} 2 & 3 \\ -4 & -4 \end{bmatrix}^{-1} = \frac{1}{4}\begin{bmatrix} -4 & -3 \\ 4 & 2 \end{bmatrix}$, so $A^t = \frac{1}{16}\begin{bmatrix} -4 & -3 \\ 4 & 2 \end{bmatrix} = \begin{bmatrix} -1/4 & -3/16 \\ 1/4 & 1/8 \end{bmatrix}$ and

$A = \begin{bmatrix} -1/4 & 1/4 \\ -3/16 & 1/8 \end{bmatrix}$.

21. $(A^tB^t)^{-1} = (B^t)^{-1}(A^t)^{-1} = (B^{-1})^t(A^{-1})^t = (A^{-1}B^{-1})^t$.

24. (a) $AB = AC$. Multiply by A^{-1} on the left. $A^{-1}AB = A^{-1}AC$; i.e., $B = C$.

27. (a) multiplications: $\frac{5^3}{3} + 5^2 - \frac{5}{3} = \frac{125+75-5}{3} = 65$;

additions: $\frac{5^3}{3} + \frac{25}{2} - \frac{25}{6} = \frac{250+75-25}{6} = 50$

(b) multiplications: $5^3 + 5^2 = 150$; additions: $5^3 - 5^2 = 100$

29.

R	E	T	R	E	A	T
18	5	20	18	5	1	20

The vectors are $\begin{bmatrix} 18 \\ 5 \end{bmatrix}, \begin{bmatrix} 20 \\ 18 \end{bmatrix}, \begin{bmatrix} 5 \\ 1 \end{bmatrix}$, and $\begin{bmatrix} 20 \\ 27 \end{bmatrix}$. The vectors obtained on multiplication

by $\begin{bmatrix} 4 & -3 \\ 3 & -2 \end{bmatrix}$ are $\begin{bmatrix} 57 \\ 44 \end{bmatrix}, \begin{bmatrix} 26 \\ 24 \end{bmatrix}, \begin{bmatrix} 17 \\ 13 \end{bmatrix}$, and $\begin{bmatrix} -1 \\ 6 \end{bmatrix}$. The coded message is, therefore,

57, 44, 26, 24, 17, 13, −1, 6.

31. The decoding matrix is $\begin{bmatrix} 4 & -3 \\ 3 & -2 \end{bmatrix}^{-1} = \begin{bmatrix} -2 & 3 \\ -3 & 4 \end{bmatrix}$.

$\begin{bmatrix} -2 & 3 \\ -3 & 4 \end{bmatrix}\begin{bmatrix} 49 & -5 & -61 \\ 38 & -3 & -39 \end{bmatrix} = \begin{bmatrix} 16 & 1 & 5 \\ 5 & 3 & 17 \end{bmatrix}$, and the message is PEACE.

Exercise Set 2.5, page 91

1. (a) $a_{32} = .25$ (c) electrical industry ($a_{43} = .30$)

   (e) steel industry ($a_{21} = .40$)

2. $(I - A)^{-1} = \begin{bmatrix} 15/8 & 5/4 \\ 5/6 & 5/3 \end{bmatrix}$. We do all the multiplications at once.

   $\begin{bmatrix} 15/8 & 5/4 \\ 5/6 & 5/3 \end{bmatrix}\begin{bmatrix} 24 & 8 & 0 \\ 12 & 6 & 12 \end{bmatrix} = \begin{bmatrix} 60 & 45/2 & 15 \\ 40 & 50/3 & 20 \end{bmatrix}$, so the values of X, in turn, are

   $\begin{bmatrix} 60 \\ 40 \end{bmatrix}$, $\begin{bmatrix} 45/2 \\ 50/3 \end{bmatrix}$, and $\begin{bmatrix} 15 \\ 20 \end{bmatrix}$.

4. $(I - A)^{-1} = \begin{bmatrix} 15/7 & 10/7 \\ 5/6 & 5/3 \end{bmatrix}$.

   $\begin{bmatrix} 15/7 & 10/7 \\ 5/6 & 5/3 \end{bmatrix}\begin{bmatrix} 42 & 0 & 14 & 42 \\ 84 & 10 & 7 & 42 \end{bmatrix} = \begin{bmatrix} 210 & 100/7 & 40 & 150 \\ 175 & 50/3 & 70/3 & 105 \end{bmatrix}$, so the values of X,

   in turn, are $\begin{bmatrix} 210 \\ 175 \end{bmatrix}$, $\begin{bmatrix} 100/7 \\ 50/3 \end{bmatrix}$, $\begin{bmatrix} 40 \\ 70/3 \end{bmatrix}$, and $\begin{bmatrix} 150 \\ 105 \end{bmatrix}$.

5. $(I - A)^{-1} = \begin{bmatrix} 5/4 & 5/8 & 5/8 \\ 0 & 2 & 1 \\ 0 & 1 & 3 \end{bmatrix}$.

$\begin{bmatrix} 5/4 & 5/8 & 5/8 \\ 0 & 2 & 1 \\ 0 & 1 & 3 \end{bmatrix}\begin{bmatrix} 4 & 0 & 8 \\ 8 & 8 & 24 \\ 8 & 16 & 8 \end{bmatrix} = \begin{bmatrix} 15 & 15 & 30 \\ 24 & 32 & 56 \\ 32 & 56 & 48 \end{bmatrix}$, so the values of X, in turn, are

$\begin{bmatrix} 15 \\ 24 \\ 32 \end{bmatrix}, \begin{bmatrix} 15 \\ 32 \\ 56 \end{bmatrix}$, and $\begin{bmatrix} 30 \\ 56 \\ 48 \end{bmatrix}$.

7. $(I - A)X = \begin{bmatrix} .8 & -.4 \\ -.5 & .9 \end{bmatrix}\begin{bmatrix} 8 \\ 10 \end{bmatrix} = \begin{bmatrix} 2.4 \\ 5 \end{bmatrix} = D.$

9. $(I - A)X = \begin{bmatrix} .9 & -.1 & -.2 \\ -.2 & .9 & -.3 \\ -.4 & -.3 & .85 \end{bmatrix}\begin{bmatrix} 6 \\ 4 \\ 5 \end{bmatrix} = \begin{bmatrix} 4 \\ .9 \\ .65 \end{bmatrix} = D.$

Exercise Set 2.6, page 98

1. (a) stochastic

   (c) The elements in column 2 do not add up to 1, so the matrix is not stochastic.

   (e) stochastic

3. $\begin{bmatrix} 1 & 0 \\ 0 & 1 \end{bmatrix}$ and $\begin{bmatrix} 1/2 & 1/4 & 1/4 \\ 1/4 & 1/2 & 1/4 \\ 1/4 & 1/4 & 1/2 \end{bmatrix}$ are doubly stochastic matrices.

   If A and B are doubly stochastic matrices, then AB is stochastic, and A^t and B^t are stochastic, so $B^tA^t = (AB)^t$ is stochastic. Thus AB is doubly stochastic.

4. (a) $p_{38} = .02$ (c) $p_{1\,10} = .25$; low-density residential

5. (a) $P^2 = \begin{bmatrix} .96 & .01 \\ .04 & .99 \end{bmatrix}^2 = \begin{bmatrix} .922 & .0195 \\ .0780 & .9805 \end{bmatrix}$, so the probability of moving from city to suburb in two years is .0780.

7. 1986: $\begin{bmatrix} .96 & .01 & .015 \\ .03 & .98 & .005 \\ .01 & .01 & .98 \end{bmatrix}\begin{bmatrix} 60 \\ 125 \\ 55 \end{bmatrix} = \begin{bmatrix} 59.675 \\ 124.575 \\ 55.75 \end{bmatrix}$.

City 59.675 million, Suburb 124.575 million, Nonmetro 55.75 million

1987: $\begin{bmatrix} .96 & .01 & .015 \\ .03 & .98 & .005 \\ .01 & .01 & .98 \end{bmatrix}\begin{bmatrix} 59.675 \\ 124.575 \\ 55.75 \end{bmatrix} = \begin{bmatrix} 59.37 \\ 124.1525 \\ 56.4775 \end{bmatrix}$.

City 59.37 million, Suburb 124.1525 million, Nonmetro 56.4775 million

$\begin{bmatrix} .96 & .01 & .015 \\ .03 & .98 & .005 \\ .01 & .01 & .98 \end{bmatrix}^2 = \begin{bmatrix} .92205 & .01955 & .02915 \\ .05825 & .96075 & .01025 \\ .0197 & .0197 & .9606 \end{bmatrix}$, so the probability that a person living in the city in 1985 will be living in a nonmetropolitan area in 1987 is .0197.

8. $X_1 = 1.002\begin{bmatrix} .96 & .01 \\ .04 & .99 \end{bmatrix}\begin{bmatrix} 60 \\ 125 \end{bmatrix} = \begin{bmatrix} .96192 & .01002 \\ .04008 & .99198 \end{bmatrix}\begin{bmatrix} 60 \\ 125 \end{bmatrix}$. Likewise

$X_2 = \begin{bmatrix} .96192 & .01002 \\ .04008 & .99198 \end{bmatrix} X_1 = \begin{bmatrix} .96192 & .01002 \\ .04008 & .99198 \end{bmatrix}^2 \begin{bmatrix} 60 \\ 125 \end{bmatrix}$

and in general $X_n = \begin{bmatrix} .96192 & .01002 \\ .04008 & .99198 \end{bmatrix}^n \begin{bmatrix} 60 \\ 125 \end{bmatrix}$.

$X_5 = \begin{bmatrix} .96192 & .01002 \\ .04008 & .99198 \end{bmatrix}^5 \begin{bmatrix} 60 \\ 125 \end{bmatrix} = \begin{bmatrix} 55.347127 \\ 131.510287 \end{bmatrix}$, so the predicted 1990 city population is 55,347,127 and the predicted 1990 suburban population is 131,510,287.

10. (a) .2 (b) $\begin{bmatrix} 1 & .2 \\ 0 & .8 \end{bmatrix}\begin{bmatrix} 10000 \\ 20000 \end{bmatrix} = \begin{bmatrix} 14000 \\ 16000 \end{bmatrix}\begin{matrix} \text{white collar} \\ \text{manual} \end{matrix}$

12. $\begin{bmatrix} .8 & .4 \\ .2 & .6 \end{bmatrix}^4\begin{bmatrix} 40000 \\ 50000 \end{bmatrix} = \begin{bmatrix} 59488 \\ 30512 \end{bmatrix}\begin{matrix} \text{small} \\ \text{large} \end{matrix}$

14. $\begin{matrix} \text{AA} & \text{Aa} & \text{aa} \end{matrix}$

$$\begin{bmatrix} 1/2 & 1/4 & 0 \\ 1/2 & 1/2 & 1/2 \\ 0 & 1/4 & 1/2 \end{bmatrix}\begin{matrix} \text{AA} \\ \text{Aa} \\ \text{aa} \end{matrix}$$

after one generation

$$\begin{bmatrix} 1/2 & 1/4 & 0 \\ 1/2 & 1/2 & 1/2 \\ 0 & 1/4 & 1/2 \end{bmatrix}\begin{bmatrix} 1/3 \\ 1/3 \\ 1/3 \end{bmatrix} = \begin{bmatrix} 1/4 \\ 1/2 \\ 1/4 \end{bmatrix}\begin{matrix} \text{AA} \\ \text{Aa} \\ \text{aa} \end{matrix}$$

after two generations

$$\begin{bmatrix} 1/2 & 1/4 & 0 \\ 1/2 & 1/2 & 1/2 \\ 0 & 1/4 & 1/2 \end{bmatrix}\begin{bmatrix} 1/4 \\ 1/2 \\ 1/4 \end{bmatrix} = \begin{bmatrix} 1/4 \\ 1/2 \\ 1/4 \end{bmatrix}\begin{matrix} \text{AA} \\ \text{Aa} \\ \text{aa} \end{matrix}$$

after three generations

$$\begin{bmatrix} 1/2 & 1/4 & 0 \\ 1/2 & 1/2 & 1/2 \\ 0 & 1/4 & 1/2 \end{bmatrix}\begin{bmatrix} 1/4 \\ 1/2 \\ 1/4 \end{bmatrix} = \begin{bmatrix} 1/4 \\ 1/2 \\ 1/4 \end{bmatrix}\begin{matrix} \text{AA} \\ \text{Aa} \\ \text{aa} \end{matrix}$$

<u>Exercise Set 2.7</u>, page 105

1. adjacency matrix / distance matrix

(a) $\begin{bmatrix} 0 & 1 & 0 \\ 1 & 0 & 1 \\ 1 & 0 & 0 \end{bmatrix}$ $\begin{bmatrix} 0 & 1 & 2 \\ 1 & 0 & 1 \\ 1 & 2 & 0 \end{bmatrix}$

(c) $\begin{bmatrix} 0 & 1 & 1 & 1 & 1 \\ 0 & 0 & 0 & 0 & 0 \\ 0 & 0 & 0 & 0 & 0 \\ 0 & 0 & 0 & 0 & 0 \\ 0 & 0 & 0 & 0 & 0 \end{bmatrix}$ $\begin{bmatrix} 0 & 1 & 1 & 1 & 1 \\ x & 0 & x & x & x \\ x & x & 0 & x & x \\ x & x & x & 0 & x \\ x & x & x & x & 0 \end{bmatrix}$

2. (a) 2 (c) undefined

3. (a)

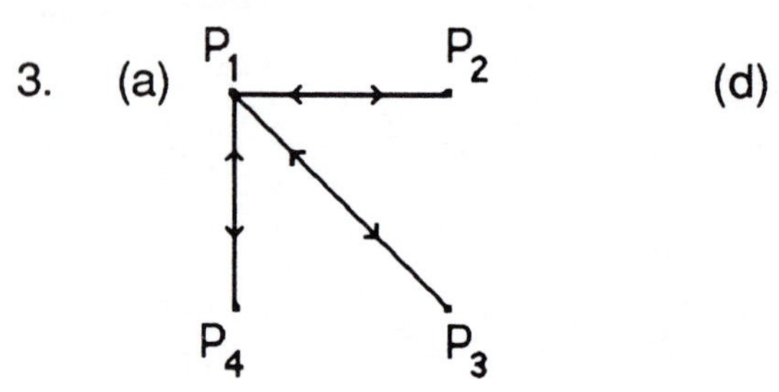

(d)

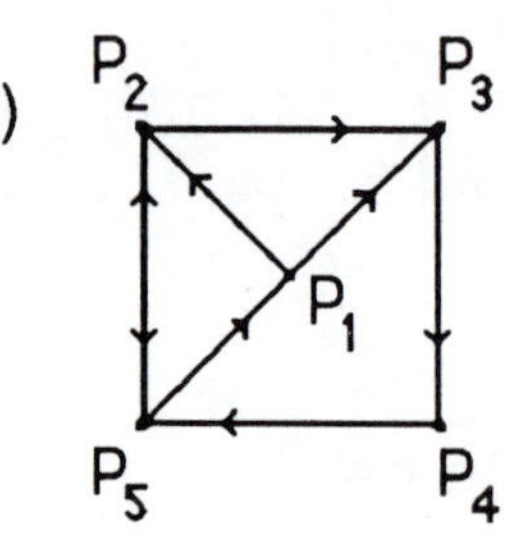

5.

	R	B	D	G	I
Raccoon	0	0	0	0	1
Bird	0	0	0	1	1
Deer	0	0	0	1	0
Grass	0	0	0	0	0
Insect	0	0	0	1	0

7.

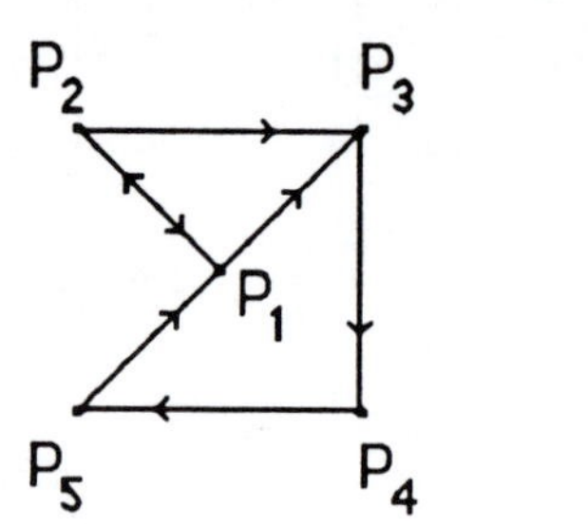

(a) $P_2 \to P_3 \to P_4 \to P_5$

length = 3

(b) $P_3 \to P_4 \to P_5 \to P_1 \to P_2$

length = 4

distance matrix

$$\begin{bmatrix} 0 & 1 & 1 & 2 & 3 \\ 1 & 0 & 1 & 2 & 3 \\ 3 & 4 & 0 & 1 & 2 \\ 2 & 3 & 3 & 0 & 1 \\ 1 & 2 & 2 & 3 & 0 \end{bmatrix}$$

8. (a) $(a_{22})^2 = 1$. There is one 2-path from P_2 to P_2.
$(a_{24})^2 = 0$. There is no 2-path from P_2 to P_4.
$(a_{31})^2 = 1$. There is one 2-path from P_3 to P_1.

$(a_{42})^2 = 1$. There is one 2-path from P_4 to P_2.

$(a_{12})^3 = 1$. There is one 3-path from P_1 to P_2.
$(a_{24})^3 = 0$. There is no 3-path from P_2 to P_4.
$(a_{32})^3 = 1$. There is one 3-path from P_3 to P_2.
$(a_{41})^3 = 1$. There is one 3-path from P_4 to P_1.

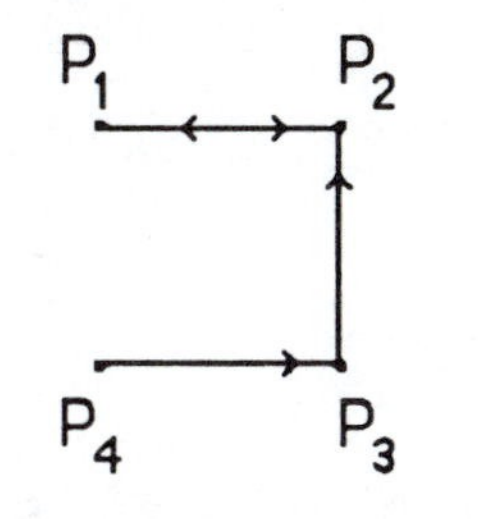

9. (a) There are no arcs from P_3 to any other vertex.
   (c) There are three arcs from P_5.
   (f) No 4-paths lead to P_3.

10. (a) There are no arcs from P_2 to any other vertex.
    (c) There is an arc from P_4 to every other vertex.
    (e) There are five arcs from P_3.
    (g) There are seven arcs in the digraph.
    (i) Four 5-paths lead to P_3.

11. From A^2 it is known that there is one 2-path from P_1 to P_2, one 2-path from P_2 to P_3, one 2-path from P_3 to P_4, and one 2-path from P_4 to P_2. The possible 2-paths are

$P_1 \to P_3 \to P_2$ or $P_1 \to P_4 \to P_2$

$P_2 \to P_1 \to P_3$ or $P_2 \to P_4 \to P_3$

$P_3 \to P_1 \to P_4$ or $P_3 \to P_2 \to P_4$

$P_4 \to P_1 \to P_2$ or $P_4 \to P_3 \to P_2$

The only combination of these that does not produce additional 2-paths is

$P_1 \to P_3 \to P_2, P_2 \to P_4 \to P_3, P_3 \to P_2 \to P_4, P_4 \to P_3 \to P_2.$

digraph

P_1 P_2

P_4 P_3

$$A^3 = \begin{bmatrix} 0 & 0 & 0 & 1 \\ 0 & 1 & 0 & 0 \\ 0 & 0 & 1 & 0 \\ 0 & 0 & 0 & 1 \end{bmatrix}$$

13. (a)

M_1 M_2

M_4 M_3

$$D = \begin{bmatrix} 0 & 1 & 2 & 2 \\ 2 & 0 & 1 & 1 \\ x & x & 0 & x \\ 1 & 2 & 3 & 0 \end{bmatrix} \begin{matrix} 5 \\ 4 \\ 3x \\ 6 \end{matrix}$$

most to least influential: M_2, M_1, M_4, M_3

15.

M_1 M_2 M_3

M_5 M_4

In this digraph there is one clique:

M_1 M_2

M_5

17. If the digraph of A contains an arc from P_i to P_j, then the digraph of A^t contains an arc from P_j to P_i. If the digraph of A does not contain an arc from P_i to P_j, then the digraph of A^t does not contain an arc from P_j to P_i.

19. If a path from P_i to P_j contains P_k twice,

$$P_i \to \ldots \to P_k \to \ldots \to P_k \to \ldots \to P_j,$$

then there is a shorter path from P_i to P_j obtained by omitting the path from P_k to P_k.

21. $c_{ij} = a_{i1}a_{j1} + a_{i2}a_{j2} + \ldots + a_{in}a_{jn}$. $a_{ik}a_{jk} = 1$ if station k can receive messages directly from both station i and station j and $a_{ik}a_{jk} = 0$ otherwise. Thus c_{ij} is the number of stations that can receive messages directly from both station i and station j.

22. (a) $\begin{bmatrix} 1 & 1 & 0 & 0 \\ 1 & 1 & 0 & 0 \\ 1 & 1 & 1 & 0 \\ 1 & 1 & 1 & 1 \end{bmatrix}$ (c) $\begin{bmatrix} 1 & 1 & 1 & 1 \\ 1 & 1 & 1 & 1 \\ 1 & 1 & 1 & 1 \\ 1 & 1 & 1 & 1 \end{bmatrix}$

23. For each vertex, the adjacency matrix gives the vertices reachable by arcs or 1-paths, its square gives all vertices reachable by 2-paths, etc. Since there are n vertices, any vertex can be reached by a path of length at most n − 1 or else it cannot be reached at all. Thus all the information needed will be contained in the first n − 1 powers of the adjacency matrix.

24. (b) No. The digraph which has adjacency matrix $\begin{bmatrix} 0 & 1 & 0 \\ 0 & 0 & 1 \\ 1 & 1 & 0 \end{bmatrix}$ has reachability matrix $\begin{bmatrix} 1 & 1 & 1 \\ 1 & 1 & 1 \\ 1 & 1 & 1 \end{bmatrix}$.

28. (c) Yes. If $A = \begin{bmatrix} 0 & 1 & 0 \\ 0 & 0 & 1 \\ 0 & 0 & 0 \end{bmatrix}$, then $R = \begin{bmatrix} 1 & 1 & 1 \\ 0 & 1 & 1 \\ 0 & 0 & 1 \end{bmatrix}$. In fact R will always contain more 1s than A.

Chapter 2 Review Exercises, page 110

1. (a) $2AB = \begin{bmatrix} 28 & 0 \\ 100 & -6 \end{bmatrix}.$ (b) $AB + C$ does not exist.

(c) $BA + AB = \begin{bmatrix} 28 & 0 \\ 69 & -6 \end{bmatrix}.$ (d) $AD - 3D = \begin{bmatrix} -6 \\ 26 \end{bmatrix}.$

(e) $AC + BC = \begin{bmatrix} 54 & -9 & 27 \\ 46 & -6 & 14 \end{bmatrix}.$ (f) $2DA + B$ does not exist.

2. (a) 2x2 (b) 2x3 (c) 2x2 (d) does not exist (e) 3x2

(f) does not exist (g) 2x2

3. (a) $d_{12} = 2(1x2 + (-3)x0) - 3(-4) = 16.$

(b) $d_{23} = 2(0x(-3) + 4x(-1)) - 3x0 = -8.$

4. $m_1 = (a_{12} - a_{22})(b_{21} + b_{22}) = (8)(-2) = -16,$ $m_5 = a_{11}(b_{12} - b_{22}) = (2)(7) = 14,$

$m_2 = (a_{11} + a_{22})(b_{11} + b_{22}) = (-3)(-3) = 9,$ $m_6 = a_{22}(b_{21} - b_{11}) = (-5)(1) = -5,$

$m_3 = (a_{11} - a_{21})(b_{11} + b_{12}) = (1)(4) = 4,$ $m_7 = (a_{21} + a_{22})b_{11} = (-4)(1) = -4,$

$m_4 = (a_{11} + a_{12})b_{22} = (5)(-4) = -20,$

so $AB = \begin{bmatrix} -16+9-(-20)+(-5) & -20+(14) \\ -5+(-4) & 9-(4)+(14)-(-4) \end{bmatrix} = \begin{bmatrix} 8 & -6 \\ -9 & 23 \end{bmatrix}.$

5. (a) $(A^t)^2 = (A^2)^t = \begin{bmatrix} 9 & 5 \\ 0 & 4 \end{bmatrix}^t = \begin{bmatrix} 9 & 0 \\ 5 & 4 \end{bmatrix}.$

(b) $A^t - B^2 = \begin{bmatrix} 3 & 0 \\ 1 & 2 \end{bmatrix} - \begin{bmatrix} 7 & -1 \\ -3 & 4 \end{bmatrix} = \begin{bmatrix} -4 & 1 \\ 4 & -2 \end{bmatrix}.$

(c) $AB^3 - 2C^2 = \begin{bmatrix} 3 & 1 \\ 0 & 2 \end{bmatrix}\begin{bmatrix} -17 & 6 \\ 18 & 1 \end{bmatrix} - 2\begin{bmatrix} 3 & 2 \\ 6 & 7 \end{bmatrix} = \begin{bmatrix} -39 & 15 \\ 24 & -12 \end{bmatrix}.$

(d) $A^2 - 3A + 4I_2 = \begin{bmatrix} 9 & 5 \\ 0 & 4 \end{bmatrix} - \begin{bmatrix} 9 & 3 \\ 0 & 6 \end{bmatrix} + \begin{bmatrix} 4 & 0 \\ 0 & 4 \end{bmatrix} = \begin{bmatrix} 4 & 2 \\ 0 & 2 \end{bmatrix}.$

6. (a) 2x2x6 = 24. (b) 4x2x3 = 24. (c) 1x7x25 = 175

(d) 9x5x11 = 495.

7. (a) $\begin{bmatrix} 1 & 4 & 1 & 0 \\ 2 & -1 & 0 & 1 \end{bmatrix} \underset{R2+(-2)R1}{\approx} \begin{bmatrix} 1 & 4 & 1 & 0 \\ 0 & -9 & -2 & 1 \end{bmatrix} \underset{(-1/9)R2}{\approx} \begin{bmatrix} 1 & 4 & 1 & 0 \\ 0 & 1 & 2/9 & -1/9 \end{bmatrix}$

$\underset{R1+(-4)R2}{\approx} \begin{bmatrix} 1 & 0 & 1/9 & 4/9 \\ 0 & 1 & 2/9 & -1/9 \end{bmatrix}$, so the inverse is $\begin{bmatrix} 1/9 & 4/9 \\ 2/9 & -1/9 \end{bmatrix}.$

(b) $\begin{bmatrix} 0 & 3 & 3 & 1 & 0 & 0 \\ 1 & 2 & 3 & 0 & 1 & 0 \\ 1 & 4 & 6 & 0 & 0 & 1 \end{bmatrix} \underset{R1 \leftrightarrow R2}{\approx} \begin{bmatrix} 1 & 2 & 3 & 0 & 1 & 0 \\ 0 & 3 & 3 & 1 & 0 & 0 \\ 1 & 4 & 6 & 0 & 0 & 1 \end{bmatrix}$

$\underset{R3+(-1)R1}{\approx} \begin{bmatrix} 1 & 2 & 3 & 0 & 1 & 1 \\ 0 & 3 & 3 & 1 & 0 & 0 \\ 0 & 2 & 3 & 0 & -1 & 1 \end{bmatrix} \underset{(1/3)R2}{\approx} \begin{bmatrix} 1 & 2 & 3 & 0 & 1 & 0 \\ 0 & 1 & 1 & 1/3 & 0 & 0 \\ 0 & 2 & 3 & 0 & -1 & 1 \end{bmatrix}$

$\underset{\substack{R1+(-2)R2 \\ R3+(-2)R2}}{\approx} \begin{bmatrix} 1 & 0 & 1 & -2/3 & 1 & 0 \\ 0 & 1 & 1 & 1/3 & 0 & 0 \\ 0 & 0 & 1 & -2/3 & -1 & 1 \end{bmatrix} \underset{\substack{R1+(-1)R3 \\ R2+(-1)R3}}{\approx} \begin{bmatrix} 1 & 0 & 0 & 0 & 2 & -1 \\ 0 & 1 & 0 & 1 & 1 & -1 \\ 0 & 0 & 1 & -2/3 & -1 & 1 \end{bmatrix},$

so the inverse is $\begin{bmatrix} 0 & 2 & -1 \\ 1 & 1 & -1 \\ -2/3 & -1 & 1 \end{bmatrix}.$

(c) $\begin{bmatrix} 1 & 2 & 3 & 1 & 0 & 0 \\ 2 & 5 & 3 & 0 & 1 & 0 \\ 1 & 0 & 8 & 0 & 0 & 1 \end{bmatrix} \underset{\substack{R2+(-2)R1 \\ R3+(-1)R1}}{\approx} \begin{bmatrix} 1 & 2 & 3 & 1 & 0 & 0 \\ 0 & 1 & -3 & -2 & 1 & 0 \\ 0 & -2 & 5 & -1 & 0 & 1 \end{bmatrix}$

$$\underset{\substack{R1+(-2)R2\\R3+(2)R2}}{\approx}\begin{bmatrix} 1 & 0 & 9 & 5 & -2 & 0\\ 0 & 1 & -3 & -2 & 1 & 0\\ 0 & 0 & -1 & -5 & 2 & 1\end{bmatrix}\underset{(-1)R3}{\approx}\begin{bmatrix} 1 & 0 & 9 & 5 & -2 & 0\\ 0 & 1 & -3 & -2 & 1 & 0\\ 0 & 0 & 1 & 5 & -2 & -1\end{bmatrix}$$

$$\underset{\substack{R1+(-9)R3\\R2+(3)R3}}{\approx}\begin{bmatrix} 1 & 0 & 0 & -40 & 16 & 9\\ 0 & 1 & 0 & 13 & -5 & -3\\ 0 & 0 & 1 & 5 & -2 & -1\end{bmatrix}, \text{ so the inverse is } \begin{bmatrix} -40 & 16 & 9\\ 13 & -5 & -3\\ 5 & -2 & -1\end{bmatrix}.$$

8. The inverse of the coefficient matrix is $\begin{bmatrix} 14 & -8 & -1\\ -17 & 10 & 1\\ -19 & 11 & 1\end{bmatrix}$.

$$\begin{bmatrix} x_1\\ x_2\\ x_3\end{bmatrix} = \begin{bmatrix} 14 & -8 & -1\\ -17 & 10 & 1\\ -19 & 11 & 1\end{bmatrix}\begin{bmatrix} 1\\ 5\\ 7\end{bmatrix} = \begin{bmatrix} -33\\ 40\\ 43\end{bmatrix}.$$

9. $A = 3\begin{bmatrix} 5 & -6\\ -2 & 3\end{bmatrix}^{-1} = 3\begin{bmatrix} 1 & 2\\ 2/3 & 5/3\end{bmatrix} = \begin{bmatrix} 3 & 6\\ 2 & 5\end{bmatrix}$.

10. The (i,j)th element of A(BC) is $A_i(BC_j)$, where A_i is the ith row of A and C_j is the jth column of C. Likewise, the (i,j)th element of (AB)C is $(A_iB)C_j$. We show that these elements are the same.

$$A_i(BC_j) = [\, a_{i1}\ a_{i2} \ldots a_{in}] \begin{bmatrix} b_{11}c_{1j}+b_{12}c_{2j}+ \ldots + b_{1r}c_{rj}\\ b_{21}c_{1j}+ b_{22}c_{2j}+ \ldots + b_{2r}c_{rj}\\ \vdots \\ b_{n1}c_{1j}+ b_{n2}c_{2j}+ \ldots + b_{nr}c_{rj}\end{bmatrix}.$$

$$= a_{i1}(b_{11}c_{1j}+ b_{12}c_{2j}+ \ldots + b_{1r}c_{rj}) + a_{i2}(b_{21}c_{1j}+ b_{22}c_{2j}+ \ldots + b_{2r}c_{rj}) + \quad . \quad . \quad . \quad + a_{in}(b_{n1}c_{1j}+ b_{n2}c_{2j}+ \ldots + b_{nr}c_{rj})$$

$$= (a_{i1}\, b_{11}+ a_{i2}\, b_{21}+ \ldots + a_{in}b_{n1})c_{1j} + (a_{i1}\, b_{12}+ a_{i2}\, b_{22}+ \ldots + a_{in}b_{n2})c_{2j} + \quad . \quad . \quad . \quad + (a_{i1}\, b_{1r}+ a_{i2}\, b_{2r}+ \ldots + a_{in}b_{nr})c_{rj}$$

$$= (A_iB)C_j.$$

11. $A(AB + A^2) - B(A^2 + AB) + 3ABA - 4AB^2 = A^2B + A^3 - BA^2 - BAB + 3ABA - 4AB^2$.

12. $(cA)^n = (cA)(cA)...(cA) = c^nA^n$.

13. The ith diagonal element of AA^t is

$$a_{i1}a_{i1} + a_{i2}a_{i2} + \ldots + a_{in}a_{in} = (a_{i1})^2 + (a_{i2})^2 + \ldots + (a_{in})^2.$$

Since each term is a square, the sum can equal zero only if each term is zero, i.e. if each $a_{ij} = 0$. Thus if $AA^t = O$ then $A = O$.

14. If A is a symmetric matrix then $A = A^t$, so $AA^t = A^2 = A^tA$ and A is normal.

15. If $A = A^2$ then $A^t = (A^2)^t = (A^t)^2$, so A^t is idempotent.

16. From exercise 8 in section 2.3, $(A^t)^n = (A^n)^t$. If $n < p$, $A^n \neq O$, so $(A^n)^t \neq O$, so $(A^t)^n \neq O$. $A^p = O$ so $(A^p)^t = (A^t)^p = O$. Thus A^t is nilpotent with degree of nilpotency = p.

17. $A = A^t$, so $A^{-1} = (A^t)^{-1} = (A^{-1})^t$, so A^{-1} is symmetric.

18. If row i of A is all zeros then row i of AB is all zeros for any matrix B. The ith diagonal term of I_n is 1, so there is no matrix B for which $AB = I_n$. Likewise, if column j of A is all zeros then column j of BA is all zeros for any B, so there is no matrix B with $BA = I_n$.

19. $A + B = \begin{bmatrix} 5+i & 5-5i \\ 6+10i & -3+i \end{bmatrix}$. $\quad AB = \begin{bmatrix} 35+24i & -3+6i \\ 7+6i & 12-6i \end{bmatrix}$.

$\overline{A} = \begin{bmatrix} 2 & 4+3i \\ 4-3i & -1 \end{bmatrix}$, so $A^* = \begin{bmatrix} 2 & 4-3i \\ 4+3i & -1 \end{bmatrix} = A$; i.e., A is hermitian.

20. If A is a real symmetric matrix, then $\overline{A} = A = A^t$ so $A^* = A^t = A$. Thus A is hermitian.

21. $G = AA^t = \begin{bmatrix} 1 & 1 & 0 & 0 & 0 \\ 1 & 2 & 0 & 0 & 1 \\ 0 & 0 & 2 & 1 & 1 \\ 0 & 0 & 1 & 1 & 0 \\ 0 & 1 & 1 & 0 & 2 \end{bmatrix}$ $P = A^tA = \begin{bmatrix} 2 & 1 & 0 & 0 \\ 1 & 2 & 1 & 0 \\ 0 & 1 & 2 & 1 \\ 0 & 0 & 1 & 2 \end{bmatrix}$

$g_{12} = 1$, $g_{13} = 0$, $g_{14} = 0$, $g_{15} = 0$,
$g_{23} = 0$, $g_{24} = 0$, $g_{25} = 1$,
$g_{34} = 1$, $g_{35} = 1$, $g_{45} = 0$,
so $1 \to 2 \to 5 \to 3 \to 4$
or $4 \to 3 \to 5 \to 2 \to 1$.

$p_{12} = 1$, $p_{13} = 0$, $p_{14} = 0$,
$p_{23} = 1$, $p_{24} = 0$, $p_{34} = 1$,
so $1 \to 2 \to 3 \to 4$ or $4 \to 3 \to 2 \to 1$.

22. $P^2 = \begin{bmatrix} .9 & .25 \\ .1 & .75 \end{bmatrix}\begin{bmatrix} .9 & .25 \\ .1 & .75 \end{bmatrix} = \begin{bmatrix} .835 & .4125 \\ .165 & .5875 \end{bmatrix}$.

$\begin{bmatrix} .835 & .4125 \\ .165 & .5875 \end{bmatrix}\begin{bmatrix} 300000 \\ 750000 \end{bmatrix} = \begin{bmatrix} 559875 \\ 490125 \end{bmatrix}$, so in two generations the distribution will

be 559,875 college-educated and 490,125 noncollege-educated.

The probability is .4125 that a couple with no college education will have at least one grandchild with a college education.

23. (a) No arcs lead to vertex 4.
(b) There are two arcs from vertex 3.
(c) There are four 3-paths from vertex 2.
(d) No 2-paths lead to vertex 3.
(e) There are two 3-paths from vertex 4 to vertex 4.
(f) There are three pairs of vertices joined by 4-paths.

Chapter 3

Exercise Set 3.1, page 120

1. (a) $\begin{vmatrix} 2 & 1 \\ 3 & 5 \end{vmatrix} = (2x5) - (1x3) = 7.$ (c) $\begin{vmatrix} 4 & 1 \\ -2 & 3 \end{vmatrix} = (4x3) - (1x{-2}) = 14.$

2. (a) $\begin{vmatrix} 1 & -5 \\ 0 & 3 \end{vmatrix} = (1x3) - (-5x0) = 3.$ (c) $\begin{vmatrix} -3 & 1 \\ 2 & -5 \end{vmatrix} = (-3x{-5}) - (1x2) = 13.$

3. (a) $M_{11} = \begin{vmatrix} 0 & 6 \\ 1 & -4 \end{vmatrix} = (0x{-4}) - (6x1) = -6.$ $C_{11} = (-1)^{1+1}M_{11} = (-1)^2(-6) = -6.$

   (c) $M_{23} = \begin{vmatrix} 1 & 2 \\ 7 & 1 \end{vmatrix} = (1x1) - (2x7) = -13.$ $C_{23} = (-1)^{2+3}M_{23} = (-1)^5(-13) = 13.$

4. (a) $M_{13} = \begin{vmatrix} -2 & 3 \\ 0 & -6 \end{vmatrix} = (-2x{-6}) - (3x0) = 12.$ $C_{13} = (-1)^{1+3}M_{13} = (-1)^4(12) = 12.$

   (c) $M_{31} = \begin{vmatrix} 0 & 1 \\ 3 & 7 \end{vmatrix} = (0x7) - (1x3) = -3.$ $C_{31} = (-1)^{3+1}M_{31} = (-1)^4(-3) = -3.$

5. (a) $M_{12} = \begin{vmatrix} 8 & 2 & 1 \\ 4 & -5 & 0 \\ 1 & 8 & 2 \end{vmatrix}$

   $= (8x{-5}x2) + (2x0x1) + (1x4x8) - (1x{-5}x1) - (8x0x8) - (2x4x2)$

   $= -80 + 0 + 32 - (-5) - 0 - 16 = -59.$

   $C_{12} = (-1)^{1+2}M_{12} = (-1)^3(-59) = 59.$

   (c) $M_{33} = \begin{vmatrix} 2 & 0 & -5 \\ 8 & -1 & 1 \\ 1 & 4 & 2 \end{vmatrix}$

$= (2 \times -1 \times 2) + (0 \times 1 \times 1) + (-5 \times 8 \times 4) - (-5 \times -1 \times 1) - (2 \times 1 \times 4) - (0 \times 8 \times 2)$

$= -4 + 0 + (-160) - 5 - 8 - 0 = -177.$

$C_{33} = (-1)^{3+3} M_{33} = (-1)^6(-177) = -177.$

6. (a) $\begin{vmatrix} 1 & 2 & 4 \\ 4 & -1 & 5 \\ -2 & 2 & 1 \end{vmatrix} = \begin{vmatrix} -1 & 5 \\ 2 & 1 \end{vmatrix} - 2 \begin{vmatrix} 4 & 5 \\ -2 & 1 \end{vmatrix} + 4 \begin{vmatrix} 4 & -1 \\ -2 & 2 \end{vmatrix}$

$= [(-1 \times 1) - (5 \times 2)] - 2[(4 \times 1) - (5 \times -2)] + 4[(4 \times 2) - (-1 \times -2)]$

$= -11 - 2(14) + 4(6) = -15.$

"diagonals" method:

$(1 \times -1 \times 1) + (2 \times 5 \times -2) + (4 \times 4 \times 2) - (4 \times -1 \times -2) - (1 \times 5 \times 2) - (2 \times 4 \times 1)$

$= -1 + (-20) + 32 - 8 - 10 - 8 = -15.$

(c) $\begin{vmatrix} 4 & 1 & -2 \\ 5 & 3 & -1 \\ 2 & 4 & 1 \end{vmatrix} = 4 \begin{vmatrix} 3 & -1 \\ 4 & 1 \end{vmatrix} - \begin{vmatrix} 5 & -1 \\ 2 & 1 \end{vmatrix} + (-2) \begin{vmatrix} 5 & 3 \\ 2 & 4 \end{vmatrix}$

$= 4[(3 \times 1) - (-1 \times 4)] - [(5 \times 1) - (-1 \times 2)] + (-2)[(5 \times 4) - (3 \times 2)]$

$= 4(7) - 7 + (-2)(14) = -7.$

"diagonals" method:

$(4 \times 3 \times 1) + (1 \times -1 \times 2) + (-2 \times 5 \times 4) - (-2 \times 3 \times 2) - (4 \times -1 \times 4) - (1 \times 5 \times 1)$

$= 12 + (-2) + (-40) - (-12) - (-16) - 5 = -7.$

7. (a) $\begin{vmatrix} 2 & 0 & 7 \\ 8 & -1 & -2 \\ 5 & 6 & 1 \end{vmatrix} = 2 \begin{vmatrix} -1 & -2 \\ 6 & 1 \end{vmatrix} - 0 \begin{vmatrix} 8 & -2 \\ 5 & 1 \end{vmatrix} + 7 \begin{vmatrix} 8 & -1 \\ 5 & 6 \end{vmatrix}$

$= 2[(-1 \times 1) - (-2 \times 6)] - 0 + 7[(8 \times 6) - (-1 \times 5)] = 2(11) - 0 + 7(53) = 393.$

"diagonals" method:

$(2x{-}1x1) + (0x{-}2x5) + (7x8x6) - (7x{-}1x5) - (2x{-}2x6) - (0x8x1)$

$= -2 + 0 + (336) - (-35) - (-24) - 0 = 393.$

(c) $\begin{vmatrix} 0 & 0 & 5 \\ 1 & 1 & 1 \\ 2 & 2 & 2 \end{vmatrix} = 0\begin{vmatrix} 1 & 1 \\ 2 & 2 \end{vmatrix} - 0\begin{vmatrix} 1 & 1 \\ 2 & 2 \end{vmatrix} + 5\begin{vmatrix} 1 & 1 \\ 2 & 2 \end{vmatrix}$

$= 0 - 0 + 5[(1x2) - (1x2)] = 0$

"diagonals" method:

$(0x1x2) + (0x1x2) + (5x1x2) - (5x1x2) - (0x1x2) - (0x1x2) = 0.$

8. (a) $\begin{vmatrix} 0 & 3 & 2 \\ 1 & 5 & 7 \\ -2 & -6 & -1 \end{vmatrix}$

using row 2:

$= -\begin{vmatrix} 3 & 2 \\ -6 & -1 \end{vmatrix} + 5\begin{vmatrix} 0 & 2 \\ -2 & -1 \end{vmatrix} - 7\begin{vmatrix} 0 & 3 \\ -2 & -6 \end{vmatrix} = -9 + 5x4 - 7x6 = -31.$

using column 1:

$= 0\begin{vmatrix} 5 & 7 \\ -6 & -1 \end{vmatrix} - \begin{vmatrix} 3 & 2 \\ -6 & -1 \end{vmatrix} + (-2)\begin{vmatrix} 3 & 2 \\ 5 & 7 \end{vmatrix} = 0 - 9 + (-2)(11) = -31.$

(c) $\begin{vmatrix} 5 & -1 & 2 \\ 3 & 0 & 6 \\ -4 & 3 & 1 \end{vmatrix}$

using row 1:

$= 5\begin{vmatrix} 0 & 6 \\ 3 & 1 \end{vmatrix} - (-1)\begin{vmatrix} 3 & 6 \\ -4 & 1 \end{vmatrix} + 2\begin{vmatrix} 3 & 0 \\ -4 & 3 \end{vmatrix} = -90 - (-27) + 18 = -45.$

using row 3:

$$= (-4)\begin{vmatrix} -1 & 2 \\ 0 & 6 \end{vmatrix} - 3\begin{vmatrix} 5 & 2 \\ 3 & 6 \end{vmatrix} + \begin{vmatrix} 5 & -1 \\ 3 & 0 \end{vmatrix} = 24 - (72) + (3) = -45.$$

9. (a) $\begin{vmatrix} 1 & 3 & -1 \\ 2 & 0 & 5 \\ 1 & 4 & 3 \end{vmatrix}$

using row 2:

$$= -2\begin{vmatrix} 3 & -1 \\ 4 & 3 \end{vmatrix} + 0\begin{vmatrix} 1 & -1 \\ 1 & 3 \end{vmatrix} - 5\begin{vmatrix} 1 & 3 \\ 1 & 4 \end{vmatrix} = -26 + 0 - 5 = -31.$$

using column 1:

$$= \begin{vmatrix} 0 & 5 \\ 4 & 3 \end{vmatrix} - 2\begin{vmatrix} 3 & -1 \\ 4 & 3 \end{vmatrix} + \begin{vmatrix} 3 & -1 \\ 0 & 5 \end{vmatrix} = -20 - 26 + 15 = -31.$$

(c) $\begin{vmatrix} 1 & 0 & 2 \\ 3 & -2 & 1 \\ 4 & 0 & 2 \end{vmatrix}$

using column 1:

$$= \begin{vmatrix} -2 & 1 \\ 0 & 2 \end{vmatrix} - 3\begin{vmatrix} 0 & 2 \\ 0 & 2 \end{vmatrix} + 4\begin{vmatrix} 0 & 2 \\ -2 & 1 \end{vmatrix} = -4 - 0 + 16 = 12.$$

using column 2:

$$= -0\begin{vmatrix} 3 & 1 \\ 4 & 2 \end{vmatrix} + (-2)\begin{vmatrix} 1 & 2 \\ 4 & 2 \end{vmatrix} - 0\begin{vmatrix} 1 & 2 \\ 3 & 1 \end{vmatrix} = 0 + 12 - 0 = 12.$$

10. (a) Using column 3, $\begin{vmatrix} 1 & -2 & 3 \\ 1 & 4 & 0 \\ 2 & -1 & 0 \end{vmatrix} = 3\begin{vmatrix} 1 & 4 \\ 2 & -1 \end{vmatrix} = -27.$

(c) Using row 3, $\begin{vmatrix} 9 & 2 & 1 \\ -3 & 2 & 6 \\ 0 & 0 & -3 \end{vmatrix} = (-3)\begin{vmatrix} 9 & 2 \\ -3 & 2 \end{vmatrix} = -72.$

11. (a) Using column 4, $\begin{vmatrix} 1 & -2 & 3 & 0 \\ 4 & 0 & 5 & 0 \\ 7 & -3 & 8 & 4 \\ -3 & 0 & 4 & 0 \end{vmatrix} = -4 \begin{vmatrix} 1 & -2 & 3 \\ 4 & 0 & 5 \\ -3 & 0 & 4 \end{vmatrix}$

(using column 2 of the 3x3 matrix) $= -4(-(-2))\begin{vmatrix} 4 & 5 \\ -3 & 4 \end{vmatrix} = -8 \times 31 = -248.$

(c) Using column 2, $\begin{vmatrix} 9 & 3 & 7 & -8 \\ 1 & 0 & 4 & 2 \\ 1 & 0 & 0 & -1 \\ -2 & 0 & -1 & 3 \end{vmatrix} = -3 \begin{vmatrix} 1 & 4 & 2 \\ 1 & 0 & -1 \\ -2 & -1 & 3 \end{vmatrix}$

(using row 2 of the 3x3 matrix) $= -3\left(-\begin{vmatrix} 4 & 2 \\ -1 & 3 \end{vmatrix} - (-1)\begin{vmatrix} 1 & 4 \\ -2 & -1 \end{vmatrix}\right)$

$= -3(-14 - (-7)) = 21.$

12. $\begin{vmatrix} x+1 & x \\ 3 & x-2 \end{vmatrix} = (x+1)(x-2) - 3x = x^2 - x - 2 - 3x = 3$, so $x^2 - 4x - 5 = 0$.

$(x-5)(x+1) = 0$, so there are two solutions, $x = 5$ and $x = -1$.

14. $\begin{vmatrix} x-1 & -2 \\ x-2 & x-1 \end{vmatrix} = (x-1)(x-1) - (-2)(x-2) = x^2 - 2x + 1 - (-2x) - 4 = 0,$

so $x^2 - 3 = 0$, and there are two solutions, $\sqrt{3}$ and $-\sqrt{3}$.

16. The cofactor expansion of each determinant using the third column gives $-3\begin{vmatrix} 4 & -1 \\ 2 & 1 \end{vmatrix}$.

18. (a) $4213 \rightarrow 4123 \rightarrow 1423 \rightarrow 1243 \rightarrow 1234$; even

(c) $3214 \rightarrow 3124 \rightarrow 1324 \rightarrow 1234$; odd

(e) $4321 \rightarrow 4312 \rightarrow 3412 \rightarrow 3142 \rightarrow 1342 \rightarrow 1324 \rightarrow 1234$; even

19. (a) $35241 \to 32541 \to 32451 \to 32415 \to 32145 \to 31245 \to 13245 \to 12345$; odd

(c) $54312 \to 45312 \to 43512 \to 43152 \to 43125 \to 34125 \to 31425 \to 31245 \to 13245 \to 12345$; odd

(e) $32514 \to 23514 \to 23154 \to 21354 \to 12354 \to 12345$; odd

Exercise Set 3.2, page 126

1. (a) $$\begin{vmatrix} 1 & 0 & 3 \\ 2 & 0 & 1 \\ 1 & -1 & 1 \end{vmatrix} \underset{C2+(-2)C1}{=} \begin{vmatrix} 1 & 0 & 3 \\ 2 & 0 & 1 \\ 1 & -1 & 1 \end{vmatrix} = -(-1)\begin{vmatrix} 1 & 3 \\ 2 & 1 \end{vmatrix} = -5.$$

(c) $$\begin{vmatrix} 2 & 1 & -1 \\ 3 & -1 & 1 \\ 1 & 4 & -4 \end{vmatrix} \underset{C2+C3}{=} \begin{vmatrix} 2 & 0 & -1 \\ 3 & 0 & 1 \\ 1 & 0 & -4 \end{vmatrix} = 0.$$

2. (a) $$\begin{vmatrix} 2 & -1 & 2 \\ 1 & 2 & -4 \\ 3 & 1 & 2 \end{vmatrix} \underset{C3+(2)C2}{=} \begin{vmatrix} 2 & -1 & 0 \\ 1 & 2 & 0 \\ 3 & 1 & 4 \end{vmatrix} = -4\begin{vmatrix} 2 & -1 \\ 1 & 2 \end{vmatrix} = 20.$$

(c) $$\begin{vmatrix} 1 & -2 & 3 \\ -3 & 6 & -9 \\ 4 & 5 & 7 \end{vmatrix} \underset{R2+(3)R1}{=} \begin{vmatrix} 1 & -2 & 3 \\ 0 & 0 & 0 \\ 4 & 5 & 7 \end{vmatrix} = 0.$$

3. (a) The given matrix can be obtained from A by multiplying the third row by 2, so its determinant is $2|A| = -4$.

(c) The given matrix can be obtained from A by adding twice row 1 to row 2, so its determinant is $|A| = -2$.

4. (a) The given matrix can be obtained from A by interchanging columns 2 and 3, so its determinant is $-|A| = -5$.

   (c) The given matrix is the transpose of A, and $|A^t| = |A| = 5$.

5. The second answer is correct.

6. (a) Row 3 is all zeros. (c) Row 3 is -3 times row 1.

7. (a) Column 3 is all zeros. (c) Row 3 is 3 times row 1.

8. (a) $|2A| = (2)^2|A| = 12$. (c) $|A^2| = |A||A| = 9$. (e) $|(A^2)^t| = |A^2| = 9$.

9. (a) $|AB| = |A||B| = -6$. (c) $|A^tB| = |A^t||B| = |A||B| = -6$.

   (e) $|2AB^{-1}| = (2)^3|A||B^{-1}| = 8|A|/|B| = -12$.

10. (a) The given matrix can be obtained from A by interchanging rows 1 and 2 and then interchanging rows 2 and 3. Thus its determinant is $(-1)(-1)|A| = 3$.

    (c) The given matrix can be obtained from A by interchanging rows 1 and 2 and then interchanging columns 2 and 3 in the resulting matrix. Thus the determinant of the given matrix is $(-1)(-1)|A| = 3$.

12. Expand by row (or column) 1 at each stage.

$$\begin{vmatrix} a_{11} & 0 & 0 & \cdots & 0 \\ 0 & a_{22} & 0 & \cdots & 0 \\ 0 & 0 & a_{33} & \cdots & 0 \\ \vdots & \vdots & & & \vdots \\ 0 & 0 & 0 & \cdots & a_{nn} \end{vmatrix} = a_{11}\begin{vmatrix} a_{22} & 0 & \cdots & 0 \\ 0 & a_{33} & \cdots & 0 \\ \vdots & \vdots & & \vdots \\ 0 & 0 & \cdots & a_{nn} \end{vmatrix} = a_{11}a_{22}\begin{vmatrix} a_{33} & \cdots & 0 \\ \vdots & & \vdots \\ 0 & \cdots & a_{nn} \end{vmatrix}$$

$$= \ldots = a_{11}a_{22}a_{33}\ldots a_{nn}.$$

15. Suppose the sum of the elements in each column is zero.

$$|A| = \begin{vmatrix} a_{11} & \cdots & a_{1n} \\ a_{21} & \cdots & a_{2n} \\ a_{31} & \cdots & a_{3n} \\ \vdots & & \vdots \\ a_{n1} & \cdots & a_{nn} \end{vmatrix} \underset{R1+R2}{=} \begin{vmatrix} a_{11}+a_{21} & \cdots & a_{1n}+a_{2n} \\ a_{21} & \cdots & a_{2n} \\ a_{31} & \cdots & a_{3n} \\ \vdots & & \vdots \\ a_{n1} & \cdots & a_{nn} \end{vmatrix}$$

$$\underset{R1+R3}{=} \begin{vmatrix} a_{11}+a_{21}+a_{31} & \cdots & a_{1n}+a_{2n}+a_{3n} \\ a_{21} & \cdots & a_{2n} \\ a_{31} & \cdots & a_{3n} \\ \vdots & & \vdots \\ a_{n1} & \cdots & a_{nn} \end{vmatrix} \underset{R1+R4}{=} \cdots$$

$$\underset{R1+Rn}{=} \begin{vmatrix} a_{11}+a_{21}+a_{31}+\ldots+a_{n1} & \cdots & a_{1n}+a_{2n}+a_{3n}+\ldots+a_{nn} \\ a_{21} & \cdots & a_{2n} \\ a_{31} & \cdots & a_{3n} \\ \vdots & & \vdots \\ a_{n1} & \cdots & a_{nn} \end{vmatrix} = 0, \text{ because the}$$

first row is all zeros.

17. $|AB| = |A||B| = |B||A| = |BA|$.

Exercise Set 3.3, page 131

1. (a) $3 \times -1 \times 4 = -12$ (c) $9 \times 3 \times 1 = 27$

2. (a) $2 \times 3 \times 5 \times -2 = -60$ (c) $7 \times 6 \times 0 \times 8 = 0$

3. (a) $\begin{vmatrix} 1 & 0 & -1 \\ 2 & 1 & 2 \\ -1 & 1 & 1 \end{vmatrix} \underset{\substack{R2+(-2)R1 \\ R3+R1}}{=} \begin{vmatrix} 1 & 0 & -1 \\ 0 & 1 & 4 \\ 0 & 1 & 0 \end{vmatrix} \underset{R3+(-1)R2}{=} \begin{vmatrix} 1 & 0 & -1 \\ 0 & 1 & 4 \\ 0 & 0 & -4 \end{vmatrix} = -4.$

(c) $\begin{vmatrix} 1 & -2 & 3 \\ -1 & 2 & 1 \\ 2 & 1 & 3 \end{vmatrix} \underset{\substack{R2+R1 \\ R3+(-2)R1}}{=} \begin{vmatrix} 1 & -2 & 3 \\ 0 & 0 & 4 \\ 0 & 5 & -3 \end{vmatrix} \underset{R2 \leftrightarrow R3}{=} -\begin{vmatrix} 1 & -2 & 3 \\ 0 & 5 & -3 \\ 0 & 0 & 4 \end{vmatrix} = -20.$

4. (a) $\begin{vmatrix} 2 & 3 & 8 \\ -2 & -3 & 4 \\ 4 & 6 & -2 \end{vmatrix} \underset{\substack{R2+R1 \\ R3+(-2)R1}}{=} \begin{vmatrix} 2 & 3 & 8 \\ 0 & 0 & 12 \\ 0 & 0 & -18 \end{vmatrix} = 0.$

(c) $\begin{vmatrix} 2 & -1 & 4 \\ -2 & 1 & -3 \\ 0 & 3 & -2 \end{vmatrix} \underset{R2+R1}{=} \begin{vmatrix} 2 & -1 & 4 \\ 0 & 0 & 1 \\ 0 & 3 & -2 \end{vmatrix} \underset{R2 \leftrightarrow R3}{=} -\begin{vmatrix} 2 & -1 & 4 \\ 0 & 3 & -2 \\ 0 & 0 & 1 \end{vmatrix} = -6.$

5. (a) $\begin{vmatrix} 2 & 4 & 1 \\ 0 & 0 & 3 \\ 0 & 1 & 2 \end{vmatrix} \underset{R2 \leftrightarrow R3}{=} -\begin{vmatrix} 2 & 4 & 1 \\ 0 & 1 & 2 \\ 0 & 0 & 3 \end{vmatrix} = -6.$

(b) $\begin{vmatrix} 0 & 4 & 1 \\ 1 & 2 & 3 \\ 4 & 1 & 5 \end{vmatrix} \underset{R1 \leftrightarrow R2}{=} -\begin{vmatrix} 1 & 2 & 3 \\ 0 & 4 & 1 \\ 4 & 1 & 5 \end{vmatrix} \underset{R3+(-4)R1}{=} -\begin{vmatrix} 1 & 2 & 3 \\ 0 & 4 & 1 \\ 0 & -7 & -7 \end{vmatrix}$

$\underset{R3+(7/4)R2}{=} -\begin{vmatrix} 1 & 2 & 3 \\ 0 & 4 & 1 \\ 0 & 0 & -21/4 \end{vmatrix} = 21.$

6. (a) $\begin{vmatrix} -1 & 2 & 0 & 1 \\ 1 & 1 & -1 & 0 \\ 2 & 1 & 1 & 0 \\ -1 & -1 & 0 & 1 \end{vmatrix} \underset{\substack{R2+R1 \\ R3+(2)R1 \\ R4+(-1)R1}}{=} \begin{vmatrix} -1 & 2 & 0 & 1 \\ 0 & 3 & -1 & 1 \\ 0 & 5 & 1 & 2 \\ 0 & -3 & 0 & 0 \end{vmatrix}$

$$\underset{\substack{R3+(-5/3)R2\\ R4+R2}}{=} \begin{vmatrix} -1 & 2 & 0 & 1 \\ 0 & 3 & -1 & 1 \\ 0 & 0 & 8/3 & 1/3 \\ 0 & 0 & -1 & 1 \end{vmatrix} \underset{R4+(3/8)R3}{=} \begin{vmatrix} -1 & 2 & 0 & 1 \\ 0 & 3 & -1 & 1 \\ 0 & 0 & 8/3 & 1/3 \\ 0 & 0 & 0 & 9/8 \end{vmatrix} = -9.$$

(c) $$\begin{vmatrix} -1 & 1 & 2 & 1 \\ 1 & -1 & 3 & -1 \\ 2 & -2 & 3 & 1 \\ 1 & -1 & 0 & 1 \end{vmatrix} \underset{\substack{R2+R1\\ R3+(2)R1\\ R4+R1}}{=} \begin{vmatrix} -1 & 1 & 2 & 1 \\ 0 & 0 & 5 & 0 \\ 0 & 0 & 7 & 3 \\ 0 & 0 & 2 & 2 \end{vmatrix} = 0.$$

7. (a) $$\begin{vmatrix} 1 & -1 & 0 & 2 \\ -1 & 1 & 0 & 0 \\ 2 & -2 & 0 & 1 \\ 3 & 1 & 5 & -1 \end{vmatrix} \underset{\substack{R2+R1\\ R3+(-2)R1\\ R4+(-3)R1}}{=} \begin{vmatrix} 1 & -1 & 0 & 2 \\ 0 & 0 & 0 & 2 \\ 0 & 0 & 0 & -3 \\ 0 & 4 & 5 & -7 \end{vmatrix} \underset{R2\leftrightarrow R4}{=} -\begin{vmatrix} 1 & -1 & 0 & 2 \\ 0 & 4 & 5 & -7 \\ 0 & 0 & 0 & -3 \\ 0 & 0 & 0 & 2 \end{vmatrix} = 0.$$

(c) $$\begin{vmatrix} 2 & 1 & 3 & 1 \\ -2 & 3 & -1 & 2 \\ 2 & 1 & 2 & 3 \\ -4 & -2 & 0 & -1 \end{vmatrix} \underset{\substack{R2+R1\\ R3+(-1)R1\\ R4+(2)R1}}{=} \begin{vmatrix} 2 & 1 & 3 & 1 \\ 0 & 4 & 2 & 3 \\ 0 & 0 & -1 & 2 \\ 0 & 0 & 6 & 1 \end{vmatrix} \underset{R4+(6)R3}{=} \begin{vmatrix} 2 & 1 & 3 & 1 \\ 0 & 4 & 2 & 3 \\ 0 & 0 & -1 & 2 \\ 0 & 0 & 0 & 13 \end{vmatrix}$$

$$= -104.$$

12. (a) $|A^{-1}| = 1/|A|$, but if $A = A^{-1}$ then $|A| = |A^{-1}|$, so $|A| = 1/|A|$ and $|A|$ must be ± 1.

14. If $AB = I_n$ then $|A||B| = |AB| = |I_n| = 1$. If the product of two real numbers is not zero, then neither number is zero, so $|A| \neq 0$ and $|B| \neq 0$.

Exercise Set 3.4, page 139

1. (a) The determinant is 5. The matrix is invertible.

   (c) The determinant is zero. The matrix is singular. The inverse does not exist.

2. (a) The determinant is –6. The matrix is invertible.

(c) The determinant is 7. The matrix is invertible.

3. (a) The determinant is 18. The matrix is invertible.

(c) The determinant is –105. The matrix is invertible.

4. (a) The determinant is zero. The matrix is singular. The inverse does not exist.

(c) The determinant is –27. The matrix is invertible.

5. (a) The determinant is –10. $\begin{bmatrix} 1 & 4 \\ 3 & 2 \end{bmatrix}^{-1} = \frac{-1}{10}\begin{bmatrix} 2 & -4 \\ -3 & 1 \end{bmatrix}.$

(c) The determinant is zero. The inverse does not exist.

6. (a) The determinant is –3.

$$\begin{bmatrix} 1 & 2 & 3 \\ 0 & 1 & 2 \\ 4 & 5 & 3 \end{bmatrix}^{-1} = \frac{-1}{3}\begin{bmatrix} \begin{vmatrix} 1 & 2 \\ 5 & 3 \end{vmatrix} & -\begin{vmatrix} 2 & 3 \\ 5 & 3 \end{vmatrix} & \begin{vmatrix} 2 & 3 \\ 1 & 2 \end{vmatrix} \\ -\begin{vmatrix} 0 & 2 \\ 4 & 3 \end{vmatrix} & \begin{vmatrix} 1 & 3 \\ 4 & 3 \end{vmatrix} & -\begin{vmatrix} 1 & 3 \\ 0 & 2 \end{vmatrix} \\ \begin{vmatrix} 0 & 1 \\ 4 & 5 \end{vmatrix} & -\begin{vmatrix} 1 & 2 \\ 4 & 5 \end{vmatrix} & \begin{vmatrix} 1 & 2 \\ 0 & 1 \end{vmatrix} \end{bmatrix} = \frac{-1}{3}\begin{bmatrix} -7 & 9 & 1 \\ 8 & -9 & -2 \\ -4 & 3 & 1 \end{bmatrix}.$$

(c) The determinant is –4.

$$\begin{bmatrix} 1 & 2 & -1 \\ 2 & 4 & -3 \\ 1 & -2 & 0 \end{bmatrix}^{-1} = \frac{-1}{4}\begin{bmatrix} -6 & 2 & -2 \\ -3 & 1 & 1 \\ -8 & 4 & 0 \end{bmatrix}.$$

7. (a) The determinant is -1.

$$\begin{bmatrix} 5 & 2 & 4 \\ 2 & 1 & 2 \\ 4 & 2 & 3 \end{bmatrix}^{-1} = -\begin{bmatrix} \begin{vmatrix} 1 & 2 \\ 2 & 3 \end{vmatrix} & -\begin{vmatrix} 2 & 4 \\ 2 & 3 \end{vmatrix} & \begin{vmatrix} 2 & 4 \\ 1 & 2 \end{vmatrix} \\ -\begin{vmatrix} 2 & 2 \\ 4 & 3 \end{vmatrix} & \begin{vmatrix} 5 & 4 \\ 4 & 3 \end{vmatrix} & -\begin{vmatrix} 5 & 4 \\ 2 & 2 \end{vmatrix} \\ \begin{vmatrix} 2 & 1 \\ 4 & 2 \end{vmatrix} & -\begin{vmatrix} 5 & 2 \\ 4 & 2 \end{vmatrix} & \begin{vmatrix} 5 & 2 \\ 2 & 1 \end{vmatrix} \end{bmatrix} = -\begin{bmatrix} -1 & 2 & 0 \\ 2 & -1 & -2 \\ 0 & -2 & 1 \end{bmatrix}.$$

(c) The determinant is zero. The inverse does not exist.

8. (a) $x_1 = \dfrac{\begin{vmatrix} 8 & 2 \\ 19 & 5 \end{vmatrix}}{\begin{vmatrix} 1 & 2 \\ 2 & 5 \end{vmatrix}} = \dfrac{2}{1} = 2.$ $\quad x_2 = \dfrac{\begin{vmatrix} 1 & 8 \\ 2 & 19 \end{vmatrix}}{\begin{vmatrix} 1 & 2 \\ 2 & 5 \end{vmatrix}} = \dfrac{3}{1} = 3.$

(c) $x_1 = \dfrac{\begin{vmatrix} 11 & 3 \\ -1 & 1 \end{vmatrix}}{\begin{vmatrix} 1 & 3 \\ -2 & 1 \end{vmatrix}} = \dfrac{14}{7} = 2.$ $\quad x_2 = \dfrac{\begin{vmatrix} 1 & 11 \\ -2 & -1 \end{vmatrix}}{\begin{vmatrix} 1 & 3 \\ -2 & 1 \end{vmatrix}} = \dfrac{21}{7} = 3.$

9. (a) $x_1 = \dfrac{\begin{vmatrix} -1 & 1 \\ 3 & 1 \end{vmatrix}}{\begin{vmatrix} 3 & 1 \\ 1 & 1 \end{vmatrix}} = \dfrac{-4}{2} = -2.$ $\quad x_2 = \dfrac{\begin{vmatrix} 3 & -1 \\ 1 & 3 \end{vmatrix}}{\begin{vmatrix} 3 & 1 \\ 1 & 1 \end{vmatrix}} = \dfrac{10}{2} = 5.$

(b) $x_1 = \dfrac{\begin{vmatrix} 11 & 2 \\ 14 & 3 \end{vmatrix}}{\begin{vmatrix} 3 & 2 \\ 2 & 3 \end{vmatrix}} = \dfrac{5}{5} = 1.$ $\quad x_2 = \dfrac{\begin{vmatrix} 3 & 11 \\ 2 & 14 \end{vmatrix}}{\begin{vmatrix} 3 & 2 \\ 2 & 3 \end{vmatrix}} = \dfrac{20}{5} = 4.$

10. (a) $|A| = \begin{vmatrix} 1 & 3 & 4 \\ 2 & 6 & 9 \\ 3 & 1 & -2 \end{vmatrix} = 8,\ |A_1| = \begin{vmatrix} 3 & 3 & 4 \\ 5 & 6 & 9 \\ 7 & 1 & -2 \end{vmatrix} = 8,\ |A_2| = \begin{vmatrix} 1 & 3 & 4 \\ 2 & 5 & 9 \\ 3 & 7 & -2 \end{vmatrix} = 16,$

$$|A_3| = \begin{vmatrix} 1 & 3 & 3 \\ 2 & 6 & 5 \\ 3 & 1 & 7 \end{vmatrix} = -8, \text{ so } x_1 = \frac{8}{8} = 1, x_2 = \frac{16}{8} = 2, x_3 = \frac{-8}{8} = -1.$$

(c) $$|A| = \begin{vmatrix} 2 & 1 & 3 \\ 3 & -2 & 4 \\ 1 & 4 & -2 \end{vmatrix} = 28, |A_1| = \begin{vmatrix} 2 & 1 & 3 \\ 2 & -2 & 4 \\ 1 & 4 & -2 \end{vmatrix} = 14, |A_2| = \begin{vmatrix} 2 & 2 & 3 \\ 3 & 2 & 4 \\ 1 & 1 & -2 \end{vmatrix} = 7,$$

$$|A_3| = \begin{vmatrix} 2 & 1 & 2 \\ 3 & -2 & 2 \\ 1 & 4 & 1 \end{vmatrix} = 7, \text{ so } x_1 = \frac{14}{28} = \frac{1}{2}, x_2 = \frac{7}{28} = \frac{1}{4}, x_3 = \frac{7}{28} = \frac{1}{4}.$$

11. (a) $$|A| = \begin{vmatrix} 1 & 4 & 2 \\ 1 & 4 & -1 \\ 2 & 6 & 1 \end{vmatrix} = -6, |A_1| = \begin{vmatrix} 5 & 4 & 2 \\ 2 & 4 & -1 \\ 7 & 6 & 1 \end{vmatrix} = -18, |A_2| = \begin{vmatrix} 1 & 5 & 2 \\ 1 & 2 & -1 \\ 2 & 7 & 1 \end{vmatrix} = 0,$$

$$|A_3| = \begin{vmatrix} 1 & 4 & 5 \\ 1 & 4 & 2 \\ 2 & 6 & 7 \end{vmatrix} = -6, \text{ so } x_1 = \frac{-18}{-6} = 3, \ x_2 = \frac{0}{-6} = 0, \ x_3 = \frac{-6}{-6} = 1.$$

(c) $$|A| = \begin{vmatrix} 8 & -2 & 1 \\ 2 & -1 & 6 \\ 6 & 1 & 4 \end{vmatrix} = -128, |A_1| = \begin{vmatrix} 1 & -2 & 1 \\ 3 & -1 & 6 \\ 3 & 1 & 4 \end{vmatrix} = -16, |A_2| = \begin{vmatrix} 8 & 1 & 1 \\ 2 & 3 & 6 \\ 6 & 3 & 4 \end{vmatrix} = -32,$$

$$|A_3| = \begin{vmatrix} 8 & -2 & 1 \\ 2 & -1 & 3 \\ 6 & 1 & 3 \end{vmatrix} = -64, \text{ so } x_1 = \frac{-16}{-128} = \frac{1}{8}, x_2 = \frac{-32}{-128} = \frac{1}{4}, x_3 = \frac{-64}{-128} = \frac{1}{2}.$$

12. (a) $|A| = 0$ so this system of equations cannot be solved using Cramer's rule.

(c) $$|A| = \begin{vmatrix} 3 & 6 & -1 \\ 1 & -2 & 3 \\ 4 & -2 & 5 \end{vmatrix} = 24, |A_1| = \begin{vmatrix} 3 & 6 & -1 \\ 2 & -2 & 3 \\ 5 & -2 & 5 \end{vmatrix} = 12, |A_2| = \begin{vmatrix} 3 & 3 & -1 \\ 1 & 2 & 3 \\ 4 & 5 & 5 \end{vmatrix} = 9,$$

$$|A_3| = \begin{vmatrix} 3 & 6 & 3 \\ 1 & -2 & 2 \\ 4 & -2 & 5 \end{vmatrix} = 18, \text{ so } x_1 = \frac{12}{24} = \frac{1}{2}, x_2 = \frac{9}{24} = \frac{3}{8}, x_3 = \frac{18}{24} = \frac{3}{4}.$$

13. (a) The determinant of the coefficient matrix is zero, so there is not a unique solution.

    (c) The determinant of the coefficient matrix is zero, so there is not a unique solution.

14. (a) The determinant of the coefficient matrix is 42, so there is a unique solution.

    (c) The determinant of the coefficient matrix is zero, so there is not a unique solution.

15. The system of equations will have nontrivial solutions if the determinant of the coefficient matrix is zero, i.e., if

$$\begin{vmatrix} 1-\lambda & 6 \\ 5 & 2-\lambda \end{vmatrix} = 0.$$

$$\begin{vmatrix} 1-\lambda & 6 \\ 5 & 2-\lambda \end{vmatrix} = (1-\lambda)(2-\lambda) - 30 = 2 - 3\lambda + \lambda^2 - 30 = \lambda^2 - 3\lambda - 28 = 0, \text{ so}$$

$(\lambda-7)(\lambda+4) = 0$ and $\lambda = 7$ or $\lambda = -4$. Substituting $\lambda = 7$ in the given equations, one finds that the general solution is $x_1 = x_2 = r$. For $\lambda = -4$ the general solution is $x_1 = -6r/5$, $x_2 = r$.

17. The system of equations will have nontrivial solutions if the determinant of the coefficient matrix is zero, i.e., if

$$\begin{vmatrix} 5-\lambda & 4 & 2 \\ 4 & 5-\lambda & 2 \\ 2 & 2 & 2-\lambda \end{vmatrix} = 0.$$

$$\begin{vmatrix} 5-\lambda & 4 & 2 \\ 4 & 5-\lambda & 2 \\ 2 & 2 & 2-\lambda \end{vmatrix} = (5-\lambda)(5-\lambda)(2-\lambda) + 16 + 16 - 4(5-\lambda) - 4(5-\lambda) - 16(2-\lambda)$$

$= 10 - 21\lambda + 12\lambda^2 - \lambda^3 = (1-\lambda)(1-\lambda)(10-\lambda) = 0$, so $\lambda = 1$ or $\lambda = 10$. For $\lambda = 1$

the general solution is $x_1 = -s - r/2$, $x_2 = s$, $x_3 = r$. For $\lambda = 10$ the general solution is

$x_1 = x_2 = 2r$, $x_3 = r$.

18. $AX = \lambda X = \lambda I_n X$, so $AX - \lambda I_n X = 0$. Thus $(A - \lambda I_n)X = 0$, and this system of equations has a nontrivial solution if and only if $|A - \lambda I_n| = 0$.

22. $A = (A^{-1})^{-1} = \frac{1}{|A^{-1}|}$ adj(A^{-1}), so adj$(A^{-1}) = |A^{-1}|A = \frac{1}{|A|}A =$ [adj$(A)]^{-1}$ from exercise 21.

25. If $|A| = \pm 1$, then $A^{-1} = \frac{1}{|A|}$ adj(A) = ±adj(A), and since all elements of A are integers, all elements of adj(A) are integers (because adding and multiplying integers gives integer results).

27. $AX = B_2$ has a unique solution if and only if $|A| \neq 0$ if and only if $AX = B_1$ has a unique solution.

Chapter 3 Review Exercises, page 141

1. (a) 3x1 - 2x5 = -7. (b) -3x6 - 0x1 = -18. (c) 9x4 - 7x1 = 29.

2. (a) $M_{12} = \begin{vmatrix} -3 & 1 \\ 7 & 2 \end{vmatrix} = -3\times 2 - 1\times 7 = -13$. $C_{12} = (-1)^{1+2}M_{12} = 13$.

(b) $M_{31} = \begin{vmatrix} 1 & 0 \\ 4 & 1 \end{vmatrix} = 1 \times 1 - 0 \times 4 = 1.$ $\qquad C_{31} = (-1)^{3+1}M_{31} = 1.$

(c) $M_{22} = \begin{vmatrix} 2 & 0 \\ 7 & 2 \end{vmatrix} = 2 \times 2 - 0 \times 4 = 4.$ $\qquad C_{22} = (-1)^{2+2}M_{22} = 4.$

3. (a) using row 1:

$$\begin{vmatrix} 1 & 2 & -3 \\ 0 & 2 & 5 \\ 4 & 1 & 2 \end{vmatrix} = \begin{vmatrix} 2 & 5 \\ 1 & 2 \end{vmatrix} - 2\begin{vmatrix} 0 & 5 \\ 4 & 2 \end{vmatrix} + (-3)\begin{vmatrix} 0 & 2 \\ 4 & 1 \end{vmatrix} = -1 + 40 + 24 = 63.$$

using column 1:

$$= \begin{vmatrix} 2 & 5 \\ 1 & 2 \end{vmatrix} - 0\begin{vmatrix} 2 & -3 \\ 1 & 2 \end{vmatrix} + 4\begin{vmatrix} 2 & -3 \\ 2 & 5 \end{vmatrix} = -1 + 0 + 64 = 63.$$

(b) using row 3:

$$\begin{vmatrix} 0 & 5 & 3 \\ 2 & -3 & 1 \\ 2 & 7 & 3 \end{vmatrix} = 2\begin{vmatrix} 5 & 3 \\ -3 & 1 \end{vmatrix} - 7\begin{vmatrix} 0 & 3 \\ 2 & 1 \end{vmatrix} + 3\begin{vmatrix} 0 & 5 \\ 2 & -3 \end{vmatrix} = 28 + 42 - 30 = 40.$$

using column 2:

$$= -5\begin{vmatrix} 2 & 1 \\ 2 & 3 \end{vmatrix} + (-3)\begin{vmatrix} 0 & 3 \\ 2 & 3 \end{vmatrix} - 7\begin{vmatrix} 0 & 3 \\ 2 & 1 \end{vmatrix} = -20 + 18 + 42 = 40$$

4. $\begin{vmatrix} x & x \\ 2 & x-3 \end{vmatrix} = x(x-3) - 2x = x^2 - 5x = -6,\ x^2 - 5x + 6 = 0.$ Thus $(x-3)(x-2) = 0$,

so $x = 3$ or $x = 2$.

5. (a) $\begin{vmatrix} 1 & 2 & -1 \\ 3 & 1 & 1 \\ 2 & 4 & 1 \end{vmatrix} \underset{R3+(-2)R1}{=} \begin{vmatrix} 1 & 2 & -1 \\ 3 & 1 & 1 \\ 0 & 0 & 3 \end{vmatrix} = 3\begin{vmatrix} 1 & 2 \\ 3 & 1 \end{vmatrix} = -15.$

(b) $\begin{vmatrix} 5 & 3 & 4 \\ 4 & 6 & 1 \\ 2 & -3 & 7 \end{vmatrix} \underset{\substack{R2+(-2)R1 \\ R3+R1}}{=} \begin{vmatrix} 5 & 3 & 4 \\ -6 & 0 & -7 \\ 7 & 0 & 11 \end{vmatrix} = -3\begin{vmatrix} -6 & -7 \\ 7 & 11 \end{vmatrix} = 51.$

(c) $\begin{vmatrix} 1 & 4 & -2 \\ 2 & 3 & 1 \\ -1 & 5 & 6 \end{vmatrix} \underset{\substack{R2+(-2)R1 \\ R3+R1}}{=} \begin{vmatrix} 1 & 4 & -2 \\ 0 & -5 & 5 \\ 0 & 9 & 4 \end{vmatrix} = \begin{vmatrix} -5 & 5 \\ 9 & 4 \end{vmatrix} = -65.$

6. (a) This matrix can be obtained from A by multiplying row 2 by 3, so its determinant is $3|A| = 6$.

   (b) This matrix can be obtained from A by adding −2 times row 1 to row 2, so its determinant is $|A| = 2$.

   (c) This matrix can be obtained from A by multiplying row 1 by 2, row 2 by −1, and row 3 by 3, so its determinant is $2x{-}1x3x|A| = -12$.

7. (a) $\begin{vmatrix} 1 & 2 & 4 \\ -1 & 4 & 3 \\ 2 & 0 & 5 \end{vmatrix} \underset{R2+(-2)R1}{=} \begin{vmatrix} 1 & 2 & 4 \\ -3 & 0 & -5 \\ 2 & 0 & 5 \end{vmatrix} = -2\begin{vmatrix} -3 & -5 \\ 2 & 5 \end{vmatrix} = 10.$

   (b) $\begin{vmatrix} -1 & 3 & 2 \\ 0 & 5 & 2 \\ 1 & 7 & 6 \end{vmatrix} \underset{R3+R1}{=} \begin{vmatrix} -1 & 3 & 2 \\ 0 & 5 & 2 \\ 0 & 10 & 8 \end{vmatrix} = -1\begin{vmatrix} 5 & 2 \\ 10 & 8 \end{vmatrix} = -20.$

   (c) $\begin{vmatrix} 2 & -3 & 5 \\ 4 & 0 & 6 \\ 1 & 2 & 7 \end{vmatrix} \underset{R3+(2/3)R1}{=} \begin{vmatrix} 2 & -3 & 5 \\ 4 & 0 & 6 \\ 7/3 & 0 & 31/3 \end{vmatrix} = -(-3)\begin{vmatrix} 4 & 6 \\ 7/3 & 31/3 \end{vmatrix} = 82.$

8. (a) $|3A| = 3^3|A| = 27x{-}2 = -54$. (b) $|2AA^t| = 2^3|A||A^t| = 8|A||A| = 32$.

   (c) $|A^3| = |A|^3 = (-2)^3 = -8$.

   (d) $|(A^tA)^2| = (|A^tA|)^2 = (|A^t||A|)^2 = (|A||A|)^2 = |A|^4 = 16$.

   (e) $|(A^t)^3| = (|A^t|)^3 = (|A|)^3 = (-2)^3 = -8$.

   (f) $|2A^t(A^{-1})^2| = (2)^3\,|A^t||(A^{-1})^2| = 8|A^t||A^{-1}|^2 = 8|A|(1/|A|)^2 = 8/|A| = -4$.

9. If A is 2x2, then $cA = \begin{bmatrix} ca_{11} & ca_{12} \\ ca_{21} & ca_{22} \end{bmatrix}$, and $|cA| = c^2a_{11}a_{22} - c^2a_{12}a_{21} = c^2|A|$.

If A is 3x3, then $cA = \begin{bmatrix} ca_{11} & ca_{12} & ca_{13} \\ ca_{21} & ca_{22} & ca_{23} \\ ca_{31} & ca_{32} & ca_{33} \end{bmatrix}$. The (i,j)th cofactor is 2x2 and so from above

is c^2C_{ij}, where C_{ij} is the (i,j)th cofactor of A. Thus

$|cA| = ca_{11}c^2C_{11} + ca_{12}c^2C_{12} + ca_{13}c^2C_{13} = c^3(a_{11}C_{11} + a_{12}C_{12} + a_{13}C_{13}) = c^3|A|$.

Repeat this argument for each size A, until A is nxn and the (i,j)th cofactor is $c^{n-1}C_{ij}$, where C_{ij} is the (i,j)th cofactor of A. Then

$|cA| = ca_{11}c^{n-1}C_{11} + \ldots + ca_{1n}c^{n-1}C_{1n} = c^n(a_{11}C_{11} + \ldots + a_{1n}C_{1n}) = c^n|A|$.

10. $|B| \neq 0$ so B^{-1} exists, and A and B^{-1} can be multiplied. Let $C = AB^{-1}$. Then $CB = AB^{-1}B = A$.

11. $|CA| = |C||A| = |A||C| = |AC|$, so that $|C^{-1}||CA| = |C^{-1}||AC|$. But $|C^{-1}||CA| = |C^{-1}CA| = |A|$

and $|C^{-1}||AC| = |C^{-1}AC|$, so that $|A| = |C^{-1}AC|$.

12. If A is upper triangular then any element in A is zero if its row number is greater than its column number. If $i > j$ then for every k with $1 \leq k \leq n$, $a_{ik}a_{kj} = 0$ because either $i \geq k$ so that $a_{ik} = 0$, or $k \geq i > j$, so that $a_{kj} = 0$. Thus if $i > j$, the (i,j)th term $a_{i1}a_{1j} + a_{i2}a_{2j} + \ldots + a_{in}a_{nj}$ of A^2 is zero because each summand is zero. So A^2 is upper triangular. The proof is similar for lower triangular matrices.

13. If $A^2 = A$ then $|A||A| = |A|$, so $|A| = 1$ or zero. If A is also invertible then $|A| \neq 0$, so $|A| = 1$.

14. (a) $\begin{vmatrix} 3 & 5 \\ 1 & 2 \end{vmatrix} = 1$, so $\begin{bmatrix} 3 & 5 \\ 1 & 2 \end{bmatrix}^{-1} = \begin{bmatrix} 2 & -5 \\ -1 & 3 \end{bmatrix}$.

(b) $\begin{vmatrix} 3 & 2 \\ -1 & 5 \end{vmatrix} = 17$, so $\begin{bmatrix} 3 & 2 \\ -1 & 5 \end{bmatrix}^{-1} = \frac{1}{17}\begin{bmatrix} 5 & -2 \\ 1 & 3 \end{bmatrix}$.

(c) $\begin{vmatrix} 1 & 4 & -1 \\ 0 & 2 & 0 \\ 1 & 6 & -1 \end{vmatrix} = 0$, so the inverse does not exist.

(d) $\begin{vmatrix} 2 & 1 & 3 \\ 0 & 2 & 9 \\ 4 & 2 & 11 \end{vmatrix} = 20$, so $\begin{bmatrix} 2 & 1 & 3 \\ 0 & 2 & 9 \\ 4 & 2 & 11 \end{bmatrix}^{-1}$

$$= \frac{1}{20}\begin{bmatrix} \begin{vmatrix} 2 & 9 \\ 2 & 11 \end{vmatrix} & -\begin{vmatrix} 1 & 3 \\ 2 & 11 \end{vmatrix} & \begin{vmatrix} 1 & 3 \\ 2 & 9 \end{vmatrix} \\ -\begin{vmatrix} 0 & 9 \\ 4 & 11 \end{vmatrix} & \begin{vmatrix} 2 & 3 \\ 4 & 11 \end{vmatrix} & -\begin{vmatrix} 2 & 3 \\ 0 & 9 \end{vmatrix} \\ \begin{vmatrix} 0 & 2 \\ 4 & 2 \end{vmatrix} & -\begin{vmatrix} 2 & 1 \\ 4 & 2 \end{vmatrix} & \begin{vmatrix} 2 & 1 \\ 0 & 2 \end{vmatrix} \end{bmatrix} = \frac{1}{20}\begin{bmatrix} 4 & -5 & 3 \\ 36 & 10 & -18 \\ -8 & 0 & 4 \end{bmatrix}.$$

15. (a) $x_1 = \dfrac{\begin{vmatrix} -1 & 1 \\ 18 & -5 \end{vmatrix}}{\begin{vmatrix} 2 & 1 \\ 3 & -5 \end{vmatrix}} = \dfrac{-13}{-13} = 1$, $\quad x_2 = \dfrac{\begin{vmatrix} 2 & -1 \\ 3 & 18 \end{vmatrix}}{\begin{vmatrix} 2 & 1 \\ 3 & -5 \end{vmatrix}} = \dfrac{39}{-13} = -3$.

(b) $|A| = \begin{vmatrix} 1 & 1 & 1 \\ 2 & -1 & 3 \\ 4 & 5 & 1 \end{vmatrix} = 8$, $|A_1| = \begin{vmatrix} 1 & 1 & 1 \\ 5 & -1 & 3 \\ 3 & 5 & 1 \end{vmatrix} = 16$, $|A_2| = \begin{vmatrix} 1 & 1 & 1 \\ 2 & 5 & 3 \\ 4 & 3 & 1 \end{vmatrix} = -8$,

$|A_3| = \begin{vmatrix} 1 & 1 & 1 \\ 2 & -1 & 5 \\ 4 & 5 & 3 \end{vmatrix} = 0$, so $x_1 = \frac{16}{8} = 2$, $x_2 = \frac{-8}{8} = -1$, and $x_3 = \frac{0}{8} = 0$.

16. If A is not invertible, $|A| = 0$. The (i,j)th term of A[adj(A)] is $a_{i1}C_{j1} + a_{i2}C_{j2} + \ldots + a_{in}C_{jn}$. If $i = j$ this term is $|A| = 0$ and if $i \neq j$ it is the determinant of the matrix obtained from A by replacing row j with row i, i.e., it is the determinant of a matrix having two equal rows, and so it is zero. Thus A[adj(A)] is the zero matrix.

17. $|A|$ is the product of the diagonal elements, and since $|A| \neq 0$ all diagonal elements must be nonzero.

18. If $|A| = \pm 1$ then $A^{-1} = \pm \text{adj}(A)$, so $X = A^{-1}AX = A^{-1}B = \pm \text{adj}(A)B$. If all the elements of A and of B are integers, then all the elements of adj(A) are integers and all the elements of the product adj(A)B are integers, so X has all integer components.

Chapter 4

Exercise Set 4.1, page 151

1.

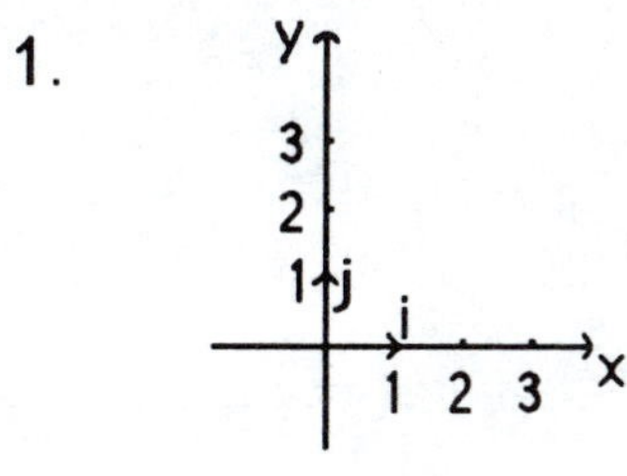

2.

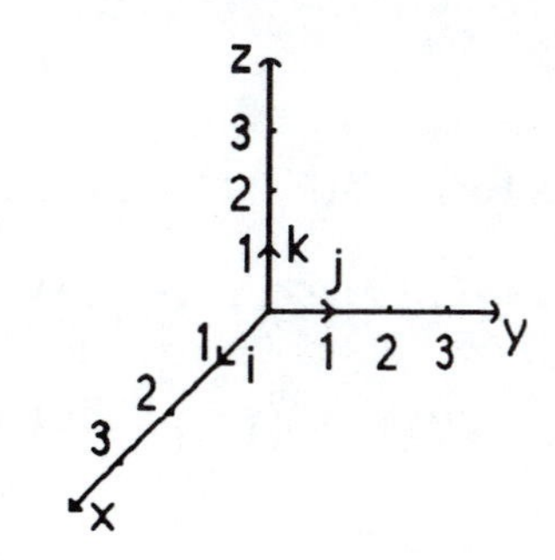

3. (a)

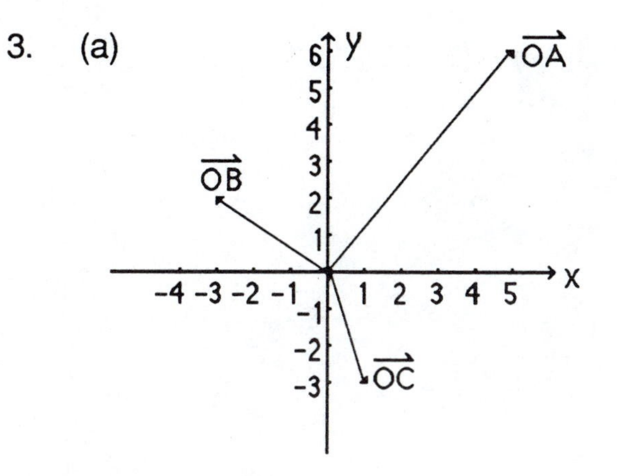

(c)

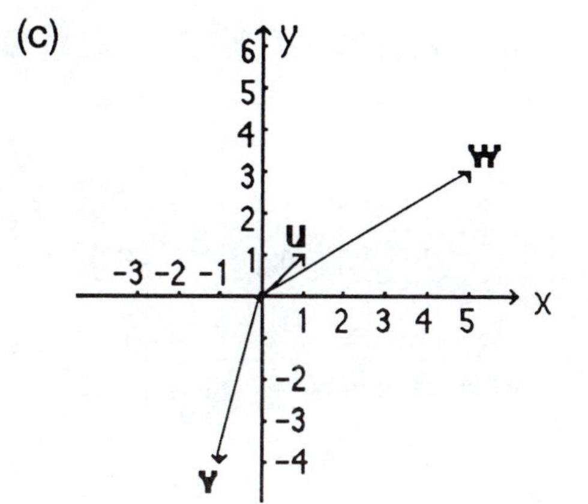

4. (a)

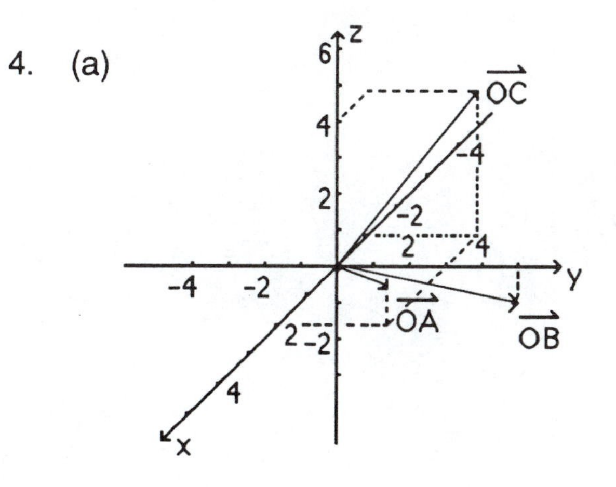

5. (a) $3(1,4) = (3,12)$. (c) $(1/2)(2,6) = (1,3)$. (d) $(-1/2)(2,4,2) = (-1,-2,-1)$.

(g) $-5(1,-4,3,-2,5) = (-5,20,-15,10,-25)$.

6. (b) $\mathbf{u} + 3\mathbf{v} = (1,2) + 3(4,-1) = (13,-1)$.

(d) $2\mathbf{u} + 3\mathbf{v} - \mathbf{w} = 2(1,2) + 3(4,-1) - (-3,5) = (17,-4)$.

(e) $-3\mathbf{u} + 4\mathbf{v} - 2\mathbf{w} = -3(1,2) + 4(4,-1) - 2(-3,5) = (19,-20)$.

7. (a) $\mathbf{u} + \mathbf{w} = (2,1,3) + (2,4,-2) = (4,5,1)$. (b) $2\mathbf{u} + \mathbf{v} = 2(2,1,3) + (-1,3,2) = (3,5,8)$.

(d) $5\mathbf{u} - 2\mathbf{v} + 6\mathbf{w} = 5(2,1,3) - 2(-1,3,2) + 6(2,4,-2) = (24,23,-1)$.

9. (b) $2\mathbf{v} - 3\mathbf{w} = 2\begin{bmatrix} -1 \\ -4 \end{bmatrix} - 3\begin{bmatrix} 4 \\ -6 \end{bmatrix} = \begin{bmatrix} -14 \\ 10 \end{bmatrix}$.

(d) $-3\mathbf{u} - 2\mathbf{v} + 4\mathbf{w} = -3\begin{bmatrix} 2 \\ 3 \end{bmatrix} - 2\begin{bmatrix} -1 \\ -4 \end{bmatrix} + 4\begin{bmatrix} 4 \\ -6 \end{bmatrix} = \begin{bmatrix} 12 \\ -25 \end{bmatrix}$.

10. (a) $\mathbf{u} + 2\mathbf{v} = \begin{bmatrix} 1 \\ 2 \\ -1 \end{bmatrix} + 2\begin{bmatrix} 3 \\ 0 \\ 1 \end{bmatrix} = \begin{bmatrix} 7 \\ 2 \\ 1 \end{bmatrix}$.

(c) $3\mathbf{u} - 2\mathbf{v} + 4\mathbf{w} = 3\begin{bmatrix} 1 \\ 2 \\ -1 \end{bmatrix} - 2\begin{bmatrix} 3 \\ 0 \\ 1 \end{bmatrix} + 4\begin{bmatrix} -1 \\ 0 \\ 5 \end{bmatrix} = \begin{bmatrix} -7 \\ 6 \\ 15 \end{bmatrix}$.

11. Additional solutions are $(5,-2,1,0) + (11,-4,1,1) = (16,-6,2,1)$,

$2(5,-2,1,0) = (10,-4,2,0)$, $-3(11,-4,1,1) = (-33,12,-3,-3)$,

$(5,-2,1,0) - (11,-4,1,1) = (-6,2,0,-1)$, and $2(5,-2,1,0) - 3(11,-4,1,1) = (-23,8,-1,-3)$.

13. (a) The resultant force is (3,2) + (5,−5) = (8,−3). (c) (9,−1) + (−3,2) = (6,1).

14. (a) The resultant force is (1,4,6) + (2,5,1) = (3,9,7).

15. The resultant force is (0,3) + (3,1) = (3,4).

(6,8) = (3,1) + (3,7), so to double the magnitude (0,3) must be replaced by (3,7).

16. The resultant force is (5,1) + (3,4) = (8,5).

(9,7) = (5,1) + (4,6), so (3,4) must be replaced by (4,6).

18. (a) temperature - scalar (b) acceleration - vector (c) pressure - scalar
(d) frequency - scalar (e) gravity - vector (f) position - vector
(g) time - scalar (h) sound - scalar (i) cost - scalar

Exercise Set 4.2, page 156

1. (b) (a,a) + (b,b) = (a+b,a+b) and c(a,a) = (ca,ca); thus the sum and scalar product of vectors in the set are also in the set, and so the set is a subspace of $\mathbf{R}^2$. The set is the line given by the equation x = y.

(d) (a,a+3b) + (d,d+3e) = (a+d,a+d+3(b+e)) and c(a,a+3b) = (ca,ca+3cb); thus the sum and scalar product of vectors in the set are also in the set. This set is all of $\mathbf{R}^2$.

2. (a) (a,b,0) + (d,e,0) = (a+d,b+e,0) and c(a,b,0) = (ca,cb,0); thus the sum and scalar product of vectors in the set are also in the set, and so the set is a subspace of $\mathbf{R}^3$. The set is the xy plane.

(c) (a,2a,b) + (d,2d,e) = (a+d,2(a+d),b+e) and c(a,2a,b) = (ca,2ca,cb); thus the sum and scalar product of vectors in the set are also in the set, and so the set is a subspace of $\mathbf{R}^3$. The set is the plane given by the equation y = 2x.

3. (a) (a,a,a) + (b,b,b) = (a+b,a+b,a+b) and c(a,a,a) = (ca,ca,ca); thus the sum and scalar product of vectors in the set are also in the set, and so the set is a subspace of $\mathbf{R}^3$. The set is the line given by x = y = z.

   (c) (a,a+b,3a) + (d,d+e,3d) = (a+d,a+d+b+e,3(a+d)) and c(a,a+b,3a) = (ca,ca+cb,3ca); thus the sum and scalar product of vectors in the set are also in the set, and so the set is a subspace of $\mathbf{R}^3$. The set is the plane given by the equation z = 3x.

5. A is a subspace of $\mathbf{R}^2$. (See exercise 1(c).)

   B is not a subspace of $\mathbf{R}^2$, because $c(a,a^2) = (ca,ca^2)$ is not of the form (a,a^2).

   C is not a subspace of $\mathbf{R}^2$, because $c(a,a^2+3) = (ca,ca^2+3c)$ is not of the form (a,a^2+3).

6. (a) (a,b,a+3)+(c,d,c+3) = (a+c, b+d, a+c+6), so the sum is not in the set. Thus the set is not a subspace of $\mathbf{R}^3$.

   (c) (a,b,2)+(c,d,2) = (a+c,b+d,4), so the sum is not in the set. Thus the set is not a subspace of $\mathbf{R}^3$.

7. (b) $(a,a^2,5a)+(b,b^2,5b) = (a+b, a^2+b^2, 5(a+b))$. a^2+b^2 is not the square of a+b, so the sum is not in the set and therefore the set is not a subspace of $\mathbf{R}^3$.

   (d) (a, b, 2a+3b+6)+(c, d, 2c+3d+6) = (a+c, b+d, 2(a+c)+3(b+d)+12), so the sum is not in the set. Thus the set is not a subspace of $\mathbf{R}^3$.

8. (a) Yes, this set is a subspace of $\mathbf{R}^3$. (a,b,c)+(d,e,f) = (a+d, b+e, c+f) and (a+d)+(b+e)+(c+f) = (a+b+c)+(d+e+f) = 0. Also k(a,b,c) = (ka,kb,kc) and ka+kb+kc = k(a+b+c) = 0.

   (c) No, this set is not a subspace of $\mathbf{R}^3$. The sum of two such vectors is not necessarily such a vector. (0,1,1)+(1,0,1) = (1,1,2), which is not in the set.

   (e) No, this set is not a subspace of $\mathbf{R}^3$. The sum of two such vectors is not necessarily in the set. (0,1,2)+(2,3,3) = (2,4,5), which is not in the set.

9. (a) No, this set is not a subspace of $\mathbf{R}^3$. If k = 1/2, then k(1,b,c) = (1/2,kb,kc) and the first component is not an integer.

   (c) No, this set is not a subspace of $\mathbf{R}^3$. If k is irrational, then k(1,b,c) = (k,kb,kc) and the first component is irrational.

10. (b) Yes, this set is a subspace of $\mathbf{R}^2$. $(a,b^3)+(d,e^3) = (a+d,b^3+e^3)$ and $c(a,b^3) = (ca,cb^3)$. b^3+e^3 is the cube of some real number and so is cb^3.

    (c) No, this set is not a subspace of $\mathbf{R}^2$. If k is negative, then k(a,b) = (ka,kb) and ka is negative.

    (e) No, this set is not a subspace of $\mathbf{R}^2$. If k is negative, then k(a,b) = (ka,kb) and ka is positive if a is negative and kb is negative if b is positive.

11. (a) The set of vectors (a,0,0), where a is nonnegative, is closed under addition but not under scalar multiplication. If k is negative, then k(a,0,0) = (ka,0,0) and ka is negative if a is positive.

    (b) The set of vectors (a,b,0), where $a \neq b$ unless a = b = 0, is closed under scalar multiplication but not under addition. (1,2,0)+(2,1,0) = (3,3,0).

12. (a) If (a,a+1,b) = (0,0,0) then a = a+1 = b = 0. a = a+1 is impossible. Therefore (0,0,0) is not in the set.

    (c) If (a,b,a+b−4) = (0,0,0) then a = b = a+b−4 = 0, but if a = b = 0 then a+b−4 = −4, not zero, so (0,0,0) is not in the set.

13. (a) The set of vectors of the form (a,a^2) contains the zero vector but is not a subspace of $\mathbf{R}^2$. (See exercise 5.)

14. (a,2a,b) = (p,2p,q) if and only if a = p and b = q , so U and V consist of the same vectors. (a,2a,b)+(d,2d,e) = (a+d, 2(a+d), b+e) and c(a,2a,b) = (ca,2ca,cb), so the set is a subspace of $\mathbf{R}^3$.

16. Let a = p and b+2 = r^3. (Every real number is the cube of some real number.) U = V. (a,3a,b+2)+(d,3d,e+2) = (a+d, 3(a+d), b+2+e+2). Thus the second term is three times the first and the third term is two more than some number, so the sum is in the set. c(a,3a,b+2) = (ca,3ca,cb+2c). Thus again the second term is three times the first and the third term can be written h+2, where h = cb+2c−2, so U = V is a subspace of $\mathbf{R}^3$.

Exercise Set 4.3, page 163

1. (a) To find a and b such that (−1,7) = a(1,−1) + b(2,4) we solve the equations −1 = a + 2b and 7 = −a + 4b and find that a = −3 and b = 1, so that (−1,7) = −3(1,−1) + (2,4). Thus (−1,7) is a linear combination of (1,−1) and (2,4).

   (c) If (−1,15) = a(−1,4) + b(2,−8) then −1 = −a + 2b and 15 = 4a − 8b. This system of equations has no solution, so (−1,15) is not a linear combination of(−1,4) and (2,−8).

2. (a) (7,5) = −3(1,−1) + 2(5,1).   (d) (4,0) = −(−1,2) + (3,2) + 0(6,4).

3. (a) (−3,3,7) = 2(1,−1,2) − (2,1,0) + 3(−1,2,1).

   (d) The system of equations 0 = −a + b + c, 10 = 2a + 3b + 8c, 8 = 3a + b + 5c has many solutions. The general solution is a = 2 − c, b = 2 − 2c, c = any real number. Thus (0,10,8) = (2−c)(−1,2,3) + (2−2c)(1,3,1) + c(1,8,5), where c is any real number.

4. (a) $(4,-4,6) = (1,2,-3) + \frac{c+3}{2}(2,-4,6) + c(-1,2,-3)$.

   (b) The system of equations 4 = −a + 2b, 3 = b + c, 8 = a + 3b + 5c has no solution. Therefore (4,3,8) is not a linear combination of (−1,0,1), (2,1,3), and (0,1,5).

5. (a) To show that the vectors (1,1) and (1,−1) span $\mathbf{R}^2$ one must show that any vector (x_1,x_2) in $\mathbf{R}^2$ can be written as a linear combination of (1,1) and (1,−1). $(x_1,x_2) = a(1,1) + b(1,-1)$ if and only if $x_1 = a + b$ and $x_2 = a - b$. The determinant of the coefficient matrix for this system of equations is $\begin{vmatrix} 1 & 1 \\ 1 & -1 \end{vmatrix} = -2$, so the system

has a unique solution and we have shown that the given vectors span $\mathbf{R}^2$. The solution of this system of equations is $a = \frac{x_1 + x_2}{2}$, $b = \frac{x_1 - x_2}{2}$, so that

$$(x_1,x_2) = \frac{x_1 + x_2}{2}(1,1) + \frac{x_1 - x_2}{2}(1,-1) \text{ and } (3,5) = 4(1,1) + (-1)(1,-1).$$

(c) $(x_1,x_2) = a(1,3) + b(3,10)$ if and only if $x_1 = a + 3b$ and $x_2 = 3a + 10b$. The unique solution of this system of equations is $a = 10x_1 - 3x_2$, $b = -3x_1 + x_2$, so the given vectors span $\mathbf{R}^2$ and $(x_1,x_2) = (10x_1 - 3x_2)(1,3) + (-3x_1 + x_2)(3,10)$. $(3,5) = 15(1,3) + (-4)(3,10)$.

6. (a) $(x_1,x_2,x_3) = a(1,2,1) + b(-1,-1,0) + c(2,5,4)$ if and only if $x_1 = a - b + 2c$, $x_2 = 2a - b + 5c$, and $x_3 = a + 4c$. The determinant of the coefficient matrix for this system of equations is $\begin{vmatrix} 1 & -1 & 2 \\ 2 & -1 & 5 \\ 3 & 0 & 4 \end{vmatrix} = -5$, so the system of equations has a unique solution and the given vectors span $\mathbf{R}^3$. Using Cramer's rule, the solution is

$$a = \frac{1}{-5}\begin{vmatrix} x_1 & -1 & 2 \\ x_2 & -1 & 5 \\ x_3 & 0 & 4 \end{vmatrix} = \frac{1}{-5}(-4x_1 + 4x_2 - 3x_3),$$

$$b = \frac{1}{-5}\begin{vmatrix} 1 & x_1 & 2 \\ 2 & x_2 & 5 \\ 3 & x_3 & 4 \end{vmatrix} = \frac{1}{-5}(7x_1 - 2x_2 - x_3),$$

$$c = \frac{1}{-5}\begin{vmatrix} 1 & -1 & x_1 \\ 2 & -1 & x_2 \\ 3 & 0 & x_3 \end{vmatrix} = \frac{1}{-5}(3x_1 - 3x_2 + x_3).$$

$$(1,3,-2) = \frac{14}{-5}(1,2,3) + \frac{3}{-5}(-1,-1,0) + \frac{8}{5}(2,5,4).$$

(c) $(x_1,x_2,x_3) = a(5,1,3) + b(2,0,1) + c(-2,-3,-1)$ if and only if $x_1 = 5a + 2b - 2c$, $x_2 = a - 3c$, $x_3 = 3a + b - c$. The determinant of the coefficient matrix for this system of equations is $\begin{vmatrix} 5 & 2 & -2 \\ 1 & 0 & -3 \\ 3 & 1 & -1 \end{vmatrix} = -3$, so the system of equations has a unique solution and the given vectors span $\mathbf{R}^3$. The solution of the system is $a = \frac{1}{-3}(3x_1 - 6x_3)$, $b = \frac{1}{-3}(-8x_1 + x_2 + 13x_3)$, $c = \frac{1}{-3}(x_1 + x_2 - 2x_3)$, and $(1,3,-2) = -5(5,1,3) + \frac{31}{3}(2,0,1) + \frac{8}{-3}(-2,-3,-1)$.

7. (a) $(x_1,x_2) = a(1,-3) + b(2,-5)$ if and only if $x_1 = a + 2b$ and $x_2 = -3a - 5b$. The determinant of the coefficient matrix for this system of equations is $\begin{vmatrix} 1 & 2 \\ -3 & -5 \end{vmatrix} \neq 0$, so the system has a unique solution and the given vectors span $\mathbf{R}^2$.

(c) $(3,-1) = -(-3,1)$, so all the vectors spanned by these two vectors are just the multiples of $(-3,1)$. These vectors do not span all of $\mathbf{R}^2$.

(d) Any one of the three given vectors is a linear combination of the other two, so we'll be clever and make one of the coefficients zero. $(x_1,x_2) = a(3,2) + 0(1,1) + c(1,0)$ if and only if $x_1 = 3a + c$ and $x_2 = 2a$. This system of equations has the unique solution $a = x_2/2$ and $c = x_1 - (3/2)x_2$, so the given vectors span $\mathbf{R}^2$. (We could find a more general solution by not making one of the coefficients zero, but we don't need it.)

8. (a) $(x_1,x_2,x_3) = a(2,1,0) + b(-1,3,1) + c(4,5,0)$ if and only if $x_1 = 2a - b + 4c$, $x_2 = a + 3b + 5c$, $x_3 = b$. The determinant of the coefficient matrix for this system of equations is $\begin{vmatrix} 2 & -1 & 4 \\ 1 & 3 & 5 \\ 0 & 1 & 0 \end{vmatrix} \neq 0$, so the system of equations has a unique solution and the given vectors span $\mathbf{R}^3$.

(c) $(x_1,x_2,x_3) = a(1,2,1) + b(-1,3,0) + c(0,5,1)$ if and only if $x_1 = a - b$, $x_2 = 2a + 3b + 5c$, $x_3 = a + c$. The determinant of the coefficient matrix for this system of equations is $\begin{vmatrix} 1 & -1 & 0 \\ 2 & 3 & 5 \\ 1 & 0 & 1 \end{vmatrix} = 0$, so the system of equations does not have a unique solution. It has either many solutions or no solutions, depending on the values of x_1, x_2, and x_3. If the system is solved using Gaussian elimination, the last row of the coefficient matrix becomes all zeros. Thus if $(x_1,x_2,x_3) = (0,0,1)$ the last row gives $0 = 1$, and there is no solution. $(0,0,1)$ cannot be expressed as a linear combination of $(1,2,1)$, $(-1,3,0)$, and $(0,5,1)$, so these vectors do not span $\mathbf{R}^3$.

9. $(1,2,3) + (1,2,0) = (2,4,3)$, $(1,2,3) - (1,2,0) = (0,0,3)$, $2(1,2,3) = (2,4,6)$.

11. $-(1,2,3) = (-1,-2,-3)$, $2(1,2,3) = (2,4,6)$, $(1/2)(1,2,3) = (1/2,1,3/2)$.

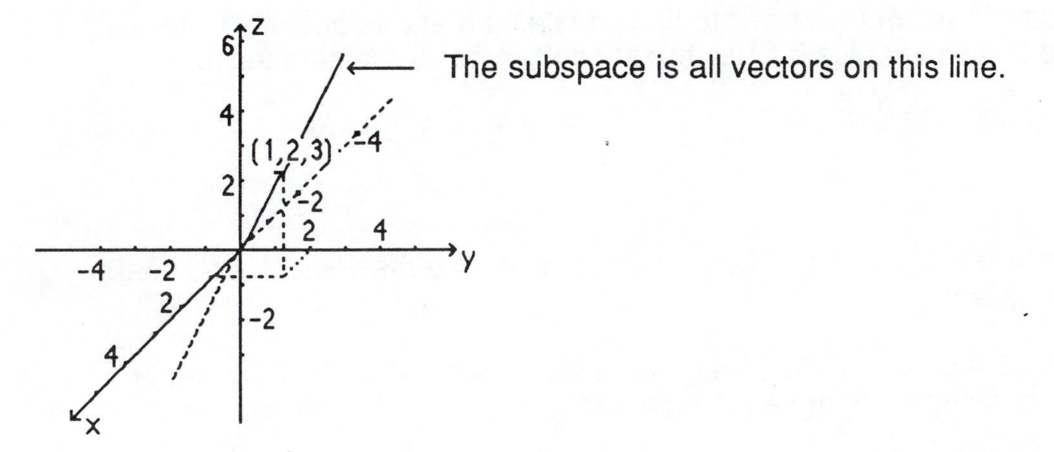

14. $-(1,2,-1,3) = (-1,-2,1,-3)$, $2(1,2,-1,3) = (2,4,-2,6)$, $(.1)(1,2,-1,3) = (.1,.2,-.1,.3)$.

15. $-(2,1,-3,4) = (-2,-1,3,-4)$, $(2,1,-3,4) + (-3,0,1,5) = (-1,1,-2,9)$, $5(4,1,2,0) = (20,5,10,0)$.

16. U is the subspace of all vectors $a(-1,3)$ and V is the subspace of all vectors $b(-2,6)$. $a(-1,3) = b(-2,6)$ if and only if $2a = b$. Thus every vector in U is also in V and every vector in V is also in U, so U = V.

20. By definition the subspace generated by $\mathbf{u}$ is the set of all vectors $a\mathbf{u}$, where a is a real number. This set is the set of all vectors on the line through the origin determined by $\mathbf{u}$.

21. If $\mathbf{v} = a\mathbf{v}_1 + b\mathbf{v}_2$ then $\mathbf{v} = \frac{a}{c_1} c_1\mathbf{v}_1 + \frac{b}{c_2} c_2\mathbf{v}_2$.

24. If $\mathbf{v}_1$ and $\mathbf{v}_2$ span V then any vector $\mathbf{v}$ in V can be written as a linear combination of $\mathbf{v}_1$ and $\mathbf{v}_2$: $\mathbf{v} = a_1\mathbf{v}_1 + a_2\mathbf{v}_2$. In particular $\mathbf{v}_3 = b_1\mathbf{v}_1 + b_2\mathbf{v}_2$. Thus $\mathbf{v} = a_1\mathbf{v}_1 + a_2\mathbf{v}_2 = (a_1 - b_1)\mathbf{v}_1 + (a_2 - b_1)\mathbf{v}_2 + \mathbf{v}_3$, so that $\mathbf{v}_1$, $\mathbf{v}_2$, and $\mathbf{v}_3$ span V. (It is also trivially true that $\mathbf{v} = a_1\mathbf{v}_1 + a_2\mathbf{v}_2 + 0\mathbf{v}_3$.)

Exercise Set 4.4, page 169

1. For each pair of vectors we need to find constants a and b, not both zero, such that a times the first vector plus b times the second vector is the zero vector.

   (a) $2(-1,2) + (2,-4) = \mathbf{0}$.

   (c) $3(-2,3) + (6,-9) = \mathbf{0}$.

2. (a) If $(0,0) = a(1,0) + b(0,1) = (a,b)$, then $a = b = 0$, so the vectors are linearly independent.

   (c) We must show that if $(0,0) = a(-1,3) + b(2,5) = (-a+2b, 3a+5b)$, then $a = b = 0$. The system of homogeneous equations $-a+2b = 0$, $3a+5b = 0$ has coefficient matrix $\begin{bmatrix} -1 & 2 \\ 3 & 5 \end{bmatrix}$. The determinant of this matrix is not zero, so the system of equations has the unique solution $a = b = 0$. Thus the vectors are linearly independent.

3. For each set of vectors $\mathbf{u},\mathbf{v},\mathbf{w}$ we need to find constants a, b, and c, not all zero, such that $a\mathbf{u} + b\mathbf{v} + c\mathbf{w} = \mathbf{0}$.

   (a) $-2(1,-2,3) - 3(-2,4,1) + (-4,8,9) = \mathbf{0}$, so the vectors are linearly dependent.

   $(-4,8,9) = 2(1,-2,3) + 3(-2,4,1)$.

   (c) $(3,4,1) + 3(2,1,0) - (9,7,1) = \mathbf{0}$, so the vectors are linearly dependent.

   $(3,4,1) = -3(2,1,0) + (9,7,1)$.

4. (a) We must show that if $(0,0,0) = a(1,2,5) + b(1,-2,1) + c(2,1,4)$ $= (a+b+2c, 2a-2b+c, 5a+b+4c)$, then $a = b = c = 0$. The system of homogeneous equations $a+b+2c = 0$, $2a-2b+c = 0$, $5a+b+4c = 0$ has coefficient matrix $\begin{bmatrix} 1 & 1 & 2 \\ 2 & -2 & 1 \\ 5 & 1 & 4 \end{bmatrix}$. The determinant of this matrix is not zero, so the system of equations has the unique solution $a = b = c = 0$. Thus the vectors are linearly independent.

(c) We must show that if $(0,0,0) = a(1,3,-4) + b(3,-1,4) + c(1,0,-2)$ $= (a+3b+c, 3a-b, -4a+4b-2c)$, then $a = b = c = 0$. The system of homogeneous equations $a+3b+c = 0$, $3a-b = 0$, $-4a+4b-2c = 0$ has coefficient matrix $\begin{bmatrix} 1 & 3 & 1 \\ 3 & -1 & 0 \\ -4 & 4 & -2 \end{bmatrix}$. The determinant of this matrix is not zero, so the system of equations has the unique solution $a = b = c = 0$. Thus the vectors are linearly independent.

5. (a) $2(2,-1,3) + (-4,2,-6) = \mathbf{0}$, so the vectors $(2,-1,3)$ and $(-4,2,-6)$ are linearly dependent and any set of vectors containing these vectors is linearly dependent.

(c) $(3,0,4) + (-3,0,-4) = \mathbf{0}$, so the vectors $(3,0,4)$ and $(-3,0,-4)$ are linearly dependent and any set of vectors containing these vectors is linearly dependent.

6. (a) $2(1,2) + 3(-1,4) - (-1,16) = \mathbf{0}$, so the vectors are linearly dependent.

(d) If $(0,0,0) = a(1,2,8) + b(1,-1,-1) + c(1,0,3) = (a+b+c, 2a-b, 8a-b+3c)$, then $a+b+c = 0$, $2a-b = 0$, and $8a-b+3c = 0$. This system of homogeneous equations has coefficient matrix $\begin{bmatrix} 1 & 1 & 1 \\ 2 & -1 & 0 \\ 8 & -1 & 3 \end{bmatrix}$. The determinant of this matrix is not zero, so the system of equations has the unique solution $a = b = c = 0$. Thus the vectors are linearly independent.

(e) If $(0,0,0,0) = a(1,0,0,0) + b(0,1,0,0) + c(0,0,1,0) + d(0,0,0,1) = (a,b,c,d)$, then $a = b = c = d = 0$, and so the vectors are linearly independent.

7. (a) $-4 = -2 \times 2$, so $t = -2 \times -1 = 2$.

   (c) The vectors are linearly dependent if $(2t+6, 4t) = c(2, -t)$, i.e., if $2c = 2t+6$ and $-tc = 4t$. From the second equation, $c = -4$, and therefore, from the first equation, $t = -7$.

10. Let $c_1\mathbf{v}_1 + c_2\mathbf{v}_2 = \mathbf{0}$, where c_1 and c_2 are not both zero. Any linear combination of $\mathbf{v}_1 + \mathbf{v}_2$ and $\mathbf{v}_1 - \mathbf{v}_2$ is a linear combination of $\mathbf{v}_1$ and $\mathbf{v}_2$: $a_1(\mathbf{v}_1 + \mathbf{v}_2) + a_2(\mathbf{v}_1 - \mathbf{v}_2)$ $= (a_1 + a_2)\mathbf{v}_1 + (a_1 - a_2)\mathbf{v}_2$. Thus if $a_1 + a_2 = c_1$ and $a_1 - a_2 = c_2$, i.e., $a_1 = \dfrac{c_1+c_2}{2}$ and $a_2 = \dfrac{c_1-c_2}{2}$, then at least one of a_1 and a_2 is not zero and $a_1(\mathbf{v}_1 + \mathbf{v}_2) + a_2(\mathbf{v}_1 - \mathbf{v}_2) = \mathbf{0}$.

11. (a) Let $c_1\mathbf{v}_1 + c_2\mathbf{v}_2 + c_3\mathbf{v}_3 = \mathbf{0}$, where c_1, c_2, and c_3 are not all zero. $a_1\mathbf{v}_1 + a_2(\mathbf{v}_1 + \mathbf{v}_2) + a_3\mathbf{v}_3 = (a_1 + a_2)\mathbf{v}_1 + a_2\mathbf{v}_2 + a_3\mathbf{v}_3 = \mathbf{0}$ if $a_1 + a_2 = c_1$, $a_2 = c_2$, and $a_3 = c_3$; i.e., $a_1 = c_1 - c_2$, $a_2 = c_2$, and $a_3 = c_3$. At least one of a_1, a_2, and a_3 is not zero, so the set $\{\mathbf{v}_1, \mathbf{v}_1 + \mathbf{v}_2, \mathbf{v}_3\}$ is linearly dependent.

12. (a) $a_1\mathbf{v}_1 + a_2(\mathbf{v}_1 + \mathbf{v}_2) + a_3\mathbf{v}_3 = (a_1 + a_2)\mathbf{v}_1 + a_2\mathbf{v}_2 + a_3\mathbf{v}_3 = \mathbf{0}$ only if $a_1 + a_2 = 0$, $a_2 = 0$, and $a_3 = 0$ (because $\mathbf{v}_1$, $\mathbf{v}_2$, and $\mathbf{v}_3$ are linearly independent), i.e. only if $a_1 = a_2 = a_3 = 0$. Thus the set $\{\mathbf{v}_1, \mathbf{v}_1 + \mathbf{v}_2, \mathbf{v}_3\}$ is linearly independent.

16. It is more likely that the vectors will be independent. For them to be dependent, one has to be a multiple of the other.

18. The computer is more likely to state that the vectors are linearly independent when they are linearly dependent, because round-off error causes zero to look to the computer like a nonzero number close to zero. Thinking geometrically, a small move by one or more of the vectors (round-off error) is not likely to place them in the same plane if they are not already in the same plane, but is more likely to move one of the vectors out of the plane of the other two if they are in fact in the same plane.

Exercise Set 4.5, page 176

1. For two vectors to be linearly dependent one must be a multiple of the other. In each case below neither vector is a multiple of the other, so the vectors are linearly independent. We show that each set spans $\mathbf{R}^2$.

   (a) $(x_1,x_2) = \dfrac{3x_2-x_1}{5}(1,2) + \dfrac{2x_1-x_2}{5}(3,1).$

   (c) $(x_1,x_2) = \dfrac{x_1+x_2}{2}(1,1) + \dfrac{-x_1+x_2}{2}(-1,1).$

2. (a) and (c) It is necessary to show either that the set of vectors is linearly independent or that the set spans $\mathbf{R}^2$. For two vectors to be linearly dependent one must be a multiple of the other. In each case neither vector is a multiple of the other, so the vectors are linearly independent and therefore a basis for $\mathbf{R}^2$.

3. (a) These two vectors are linearly independent since neither is a multiple of the other. Thus they are a basis for $\mathbf{R}^2$.

   (b) $(-2,6) = -2(1,-3)$, so these vectors are not linearly independent and therefore not a basis for $\mathbf{R}^2$.

4. (a) $a(1,1,1) + b(0,1,2) + c(3,0,1) = (0,0,0)$ if and only if $a+3c = 0$, $a+b = 0$, and $a+2b+c = 0$. The coefficient matrix for this homogeneous system of equations is $\begin{bmatrix} 1 & 0 & 3 \\ 1 & 1 & 0 \\ 1 & 2 & 1 \end{bmatrix}$. Its determinant is nonzero, so the system has the unique solution $a = b = c = 0$, and the three vectors are linearly independent. Thus they are a basis for $\mathbf{R}^3$.

5. (a) $a(1,-1,2) + b(2,0,1) + c(3,0,0) = (0,0,0)$ if and only if $a+2b+3c = 0$, $-a = 0$, and $2a+b = 0$. The coefficient matrix for this homogeneous system of equations is $\begin{bmatrix} 1 & 2 & 3 \\ -1 & 0 & 0 \\ 2 & 1 & 0 \end{bmatrix}$. Its determinant is nonzero, so the system has the unique solution

$a = b = c = 0$, and the three vectors are linearly independent. Thus they are a basis for $\mathbf{R}^3$.

(b) $2(2,1,0) + (-1,1,1) = (3,3,1)$, so the vectors are linearly dependent and therefore not a basis for $\mathbf{R}^3$.

6. (a) These vectors are linearly dependent.

(b) A basis for $\mathbf{R}^2$ can contain only two vectors.

(d) The third vector is a multiple of the second, so the set is not linearly independent.

(f) $(1,4)$ is not in $\mathbf{R}^3$.

7. $(1,4,3) = 3(-1,2,1) + 2(2,-1,0)$, and $(-1,2,1)$ and $(2,-1,0)$ are linearly independent, so the dimension is 2 and $(-1,2,1)$ and $(2,-1,0)$ are a basis for the subspace.

8. $(1,2,-1) = 2(1,3,1) - (1,4,3)$.

10. $(-3,3,-6) = -\frac{3}{2}(2,-2,4)$.

12. The set $\{(1,2),(0,1)\}$ is a basis for $\mathbf{R}^2$.

13. The set $\{(1,1,1),(1,0,-2),(1,0,0)\}$ is a basis for $\mathbf{R}^3$.

15. (a) $(a,a,b) = a(1,1,0) + b(0,0,1)$, so the linearly independent set $\{(1,1,0),(0,0,1)\}$ spans the subspace of vectors of the form (a,a,b) and is therefore a basis. The dimension of the space is 2 since there are 2 vectors in the basis.

(c) $(a,b,a+b) = a(1,0,1) + b(0,1,1)$, so the linearly independent set $\{(1,0,1),(0,1,1)\}$ spans the isubspace of vectors of the form $(a,b,a+b)$ and is therefore a basis. The dimension of the space is 2 since there are 2 vectors in the basis.

(e) $(a,b,c) = (a, b, -a-b) = a(1,0,-1) + b(0,1,-1)$, so the linearly independent set $\{(1,0,-1),(0,1,-1)\}$ spans the subspace of vectors of the form (a,b,c) where $a+b+c = 0$. Thus the set $\{(1,0,-1),(0,1,-1)\}$ is a basis for the subspace. The dimension of the space is 2 since there are 2 vectors in the basis.

16. (a) (a, b, a+b, a−b) = a(1,0,1,1) + b(0,1,1,−1), so the linearly independent set {(1,0,1,1),(0,1,1,−1)} spans the subspace of vectors of the form (a, b, a+b, a−b) and is therefore a basis. The dimension of the space is 2 since there are 2 vectors in the basis.

    (c) (2a, b, a+3b, c) = a(2,0,1,0) + b(0,1,3,0)+c(0,0,0,1), so the linearly independent set {(2,0,1,0), (0,1,3,0),(0,0,0,1)} spans the subspace of vectors of the form (2a,b,a+3b,0) and is therefore a basis. The dimension of the space is 3 since there are 3 vectors in the basis.

17. Since dim(V) = 2 any basis for V must contain 2 vectors. In Example 3 in Section 4.4 it was shown that the vectors $\mathbf{u}_1 = \mathbf{v}_1 + \mathbf{v}_2$ and $\mathbf{u}_2 = \mathbf{v}_1 - \mathbf{v}_2$ are linearly independent. Thus from Theorem 4.11 the set $\{\mathbf{u}_1, \mathbf{u}_2\}$ is a basis for V.

19. Since V is n-dimensional any basis for V must contain n vectors. We need only show that the vectors $c\mathbf{v}_1, c\mathbf{v}_2, \ldots, c\mathbf{v}_n$ are linearly independent.
If $a_1c\mathbf{v}_1 + a_2c\mathbf{v}_2 + \ldots + a_nc\mathbf{v}_n = \mathbf{0}$, then $a_1c = a_2c = \ldots = a_nc = 0$ and since $c \neq 0$, $a_1 = a_2 = \ldots = a_n = 0$. Thus the vectors $c\mathbf{v}_1, c\mathbf{v}_2, \ldots, c\mathbf{v}_n$ are linearly independent and a basis for V.

20. (a) This statement is false. Any set of vectors containing the zero vector is linearly dependent.

    (c) true

22. If m > n, then any basis of W is linearly dependent by Theorem 4.7. But a basis is linearly independent, so $m \leq n$.

23. (a) False. The vectors are linearly dependent.

    (c) True. (1,2,3) + (0,1,4) = (1,3,7), and (1,2,3) and (0,1,4) are linearly independent, so the dimension is 2.

24. (b) False. If the two vectors are linearly dependent you can't get a basis no matter what vector you add to them.

    (d) True. The number of vectors in any set of linearly independent vectors is less than or equal to the dimension of the vector space.

Exercise Set 4.6, page 184

1. (a) The row vectors are linearly independent, so the rank of the matrix is 2.

   (c) The row vectors are linearly independent, so the rank of the matrix is 2.

2. (a) The first row is in the space spanned by the other two linearly independent rows. The rank of the matrix is 2.

   (c) Rows 2 and 3 are multiples of row 1, so the rank of the matrix is 1.

3. (a) $\begin{bmatrix} 1 & 2 & -1 \\ 2 & 5 & 2 \\ 0 & 2 & 9 \end{bmatrix} \approx \begin{bmatrix} 1 & 2 & -1 \\ 0 & 1 & 4 \\ 0 & 2 & 9 \end{bmatrix} \approx \begin{bmatrix} 1 & 2 & -1 \\ 0 & 1 & 4 \\ 0 & 0 & 1 \end{bmatrix}$, so the vectors (1,2,−1), (0,1,4),

   and (0,0,1) are a basis for the row space and the rank of the matrix is 3.

   (c) $\begin{bmatrix} 1 & -3 & 2 \\ -2 & 6 & -4 \\ -1 & 3 & -2 \end{bmatrix} \approx \begin{bmatrix} 1 & -3 & 2 \\ 0 & 0 & 0 \\ 0 & 0 & 0 \end{bmatrix}$, so the vector (1,−3,2) is a basis for the row space

   and the rank of the matrix is 1.

4. (a) $\begin{bmatrix} 1 & 4 & 0 \\ -1 & -3 & 3 \\ 2 & 9 & 5 \end{bmatrix} \approx \begin{bmatrix} 1 & 4 & 0 \\ 0 & 1 & 3 \\ 0 & 1 & 5 \end{bmatrix} \approx \begin{bmatrix} 1 & 4 & 0 \\ 0 & 1 & 3 \\ 0 & 0 & 1 \end{bmatrix}$, so the vectors (1,4,0), (0,1,3), and

   (0,0,1) are a basis for the row space and the rank of the matrix is 3.

   (c) $\begin{bmatrix} 1 & 2 & 3 \\ 0 & -1 & -1 \\ 3 & 4 & 7 \end{bmatrix} \approx \begin{bmatrix} 1 & 2 & 3 \\ 0 & 1 & 1 \\ 0 & -2 & -2 \end{bmatrix} \approx \begin{bmatrix} 1 & 2 & 3 \\ 0 & 1 & 1 \\ 0 & 0 & 0 \end{bmatrix}$, so the vectors (1,2,3) and (0,1,1)

   are a basis for the row space and the rank of the matrix is 2.

5. (a) $\begin{bmatrix} 1 & 2 & 3 & 4 \\ -1 & 2 & 0 & 1 \\ 0 & 1 & 0 & 2 \end{bmatrix} \approx \begin{bmatrix} 1 & 2 & 3 & 4 \\ 0 & 4 & 3 & 5 \\ 0 & 1 & 0 & 2 \end{bmatrix} \approx \begin{bmatrix} 1 & 2 & 3 & 4 \\ 0 & 1 & 0 & 2 \\ 0 & 4 & 3 & 5 \end{bmatrix} \approx \begin{bmatrix} 1 & 2 & 3 & 4 \\ 0 & 1 & 0 & 2 \\ 0 & 0 & 1 & -1 \end{bmatrix}$, so the

vectors (1,2,3,4), (0,1,0,2), and (0,0,1,−1) are a basis for the row space and the rank of the matrix is 3.

(b) $\begin{bmatrix} 1 & 2 & -1 & 4 \\ 0 & 1 & -2 & 3 \\ -1 & 0 & -3 & 2 \end{bmatrix} \approx \begin{bmatrix} 1 & 2 & -1 & 4 \\ 0 & 1 & -2 & 3 \\ 0 & 2 & -4 & 6 \end{bmatrix} \approx \begin{bmatrix} 1 & 2 & -1 & 4 \\ 0 & 1 & -2 & 3 \\ 0 & 0 & 0 & 0 \end{bmatrix}$, so the vectors (1,2,−1,4)

and (0,1,−2,3) are a basis for the row space and the rank of the matrix is 2.

6. (a) The given vectors are linearly independent, so they are a basis for the space they

span, which is $\mathbf{R}^3$. Also $\begin{bmatrix} 1 & 3 & 2 \\ 0 & 1 & 4 \\ 1 & 4 & 9 \end{bmatrix} \approx \begin{bmatrix} 1 & 3 & 2 \\ 0 & 1 & 4 \\ 0 & 1 & 7 \end{bmatrix} \approx \begin{bmatrix} 1 & 3 & 2 \\ 0 & 1 & 4 \\ 0 & 0 & 1 \end{bmatrix}$, so the row

vectors of any of these matrices are a basis for the same vector space.

(c) $\begin{bmatrix} 1 & -1 & 3 \\ 1 & 0 & 1 \\ -2 & 1 & -4 \end{bmatrix} \approx \begin{bmatrix} 1 & -1 & 3 \\ 0 & 1 & -2 \\ 0 & -1 & 2 \end{bmatrix} \approx \begin{bmatrix} 1 & -1 & 3 \\ 0 & 1 & -2 \\ 0 & 0 & 0 \end{bmatrix}$, so the vectors (1,−1,3) and

(0,1,−2) are a basis for the space spanned by the given vectors.

7. (a) $\begin{bmatrix} 1 & 3 & -1 & 4 \\ 1 & 3 & 0 & 6 \\ -1 & -3 & 0 & -8 \end{bmatrix} \approx \begin{bmatrix} 1 & 3 & -1 & 4 \\ 0 & 0 & 1 & 2 \\ 0 & 0 & -1 & -4 \end{bmatrix} \approx \begin{bmatrix} 1 & 3 & -1 & 4 \\ 0 & 0 & 1 & 2 \\ 0 & 0 & 0 & 1 \end{bmatrix}$, so the vectors

(1,3,−1,4), (0,0,1,2), and (0,0,0,1) are a basis for the subspace. The given vectors are linearly independent, so they are a basis also.

(c) $\begin{bmatrix} 1 & 2 & 3 & 4 \\ 0 & -1 & 2 & 3 \\ 2 & 3 & 8 & 11 \\ 2 & 3 & 6 & 8 \end{bmatrix} \approx \begin{bmatrix} 1 & 2 & 3 & 4 \\ 0 & 1 & -2 & -3 \\ 0 & -1 & 2 & 3 \\ 0 & -1 & 0 & 0 \end{bmatrix} \approx \begin{bmatrix} 1 & 2 & 3 & 4 \\ 0 & 1 & -2 & -3 \\ 0 & 0 & 0 & 0 \\ 0 & 0 & -2 & -3 \end{bmatrix}$

$$\approx \begin{bmatrix} 1 & 2 & 3 & 4 \\ 0 & 1 & -2 & -3 \\ 0 & 0 & 1 & 3/2 \\ 0 & 0 & 0 & 0 \end{bmatrix}$$, so the vectors (1,2,3,4), (0,1,−2,−3), and (0,0,1,3/2) are a basis for the subspace.

8. $$A = \begin{bmatrix} 1 & 2 & -1 \\ 0 & 1 & 3 \\ 1 & 4 & 6 \end{bmatrix} \approx \begin{bmatrix} 1 & 2 & -1 \\ 0 & 1 & 3 \\ 0 & 2 & 7 \end{bmatrix} \approx \begin{bmatrix} 1 & 2 & -1 \\ 0 & 1 & 3 \\ 0 & 0 & 1 \end{bmatrix}$$, so the vectors (1,2,−1), (0,1,3), and (0,0,1) are a basis for the row space of A.

$$A^t = \begin{bmatrix} 1 & 0 & 1 \\ 2 & 1 & 4 \\ -1 & 3 & 6 \end{bmatrix} \approx \begin{bmatrix} 1 & 0 & 1 \\ 0 & 1 & 2 \\ 0 & 3 & 7 \end{bmatrix} \approx \begin{bmatrix} 1 & 0 & 1 \\ 0 & 1 & 2 \\ 0 & 0 & 1 \end{bmatrix}$$, so the vectors (1,0,1), (0,1,2), and (0,0,1) are a basis for the row space of A^t. Therefore the column vectors

$\begin{bmatrix} 1 \\ 0 \\ 1 \end{bmatrix}$, $\begin{bmatrix} 0 \\ 1 \\ 2 \end{bmatrix}$, and $\begin{bmatrix} 0 \\ 0 \\ 1 \end{bmatrix}$ are a basis for the column space of A. Both the row space and the column space of A have dimension 3.

10. (b) The largest possible rank is the smaller of the two numbers m and n.

11. dim(column space of A) = dim(row space of A) $\leq$ number of rows of A = m < n

12. (a) The rank of A cannot be greater than 3, so a basis for the column space of A can contain no more than three vectors. The four column vectors in A must therefore be linearly dependent by Theorem 4.7.

14. If the n columns of A are linearly independent, they are a basis for the column space of A and dim(column space of A) is n, so rank(A) = n.

If rank(A) = n, then the n column vectors of A are basis for the column space of A because they span the column space of A (Theorem 4.11), but this means they are linearly independent.

Chapter 4 Review Exercises, page 185

1. (a) $\mathbf{u} + \mathbf{w} = (3,-1,5) + (0,1,-3) = (3,0,2)$.

   (b) $3\mathbf{u} + \mathbf{v} = 3(3,-1,5) + (2,3,7) = (11,0,22)$.

   (c) $\mathbf{u} - 2\mathbf{w} = (3,-1,5) - 2(0,1,-3) = (3,-3,11)$.

   (d) $4\mathbf{u} - 2\mathbf{v} + 3\mathbf{w} = 4(3,-1,5) - 2(2,3,7) + 3(0,1,-3) = (8,-7,-3)$.

   (e) $2\mathbf{u} - 5\mathbf{v} - \mathbf{w} = 2(3,-1,5) - 5(2,3,7) - (0,1,-3) = (-4,-18,-22)$.

2. (a) $(a, b, a-2) + (c, d, c-2) = (a+c, b+d, a+c-4)$, so the sum of two such vectors is not in the set. Thus the set is not a subspace of $\mathbf{R}^3$.

   (b) $(a,-2a,3a) + (b,-2b,3b) = (a+b, -2(a+b), 3(a+b))$ and $c(a, -2a, 3a) = (ca, -2ca, 3ca)$, so the sum and scalar product of vectors in the set is in the set. Thus the set is a subspace of $\mathbf{R}^3$.

   (c) $(a, b, 2a-3b) + (e, f, 2e-3f) = (a+e, b+f, 2(a+e)-3(b+f))$ and $c(a, b, 2a-3b) = (ca, cb, 2ca-3cb)$, so the sum and scalar product of vectors in the set is in the set. Thus the set is a subspace of $\mathbf{R}^3$.

   (d) $(a,2,b) + (c,2,d) = (a+c, 4, b+d)$, so the sum of vectors in the set is not in the set. Thus the set is not a subspace of $\mathbf{R}^3$.

3. Only the subset (a) is a subspace of $\mathbf{R}^3$. None of the other subsets is closed under scalar multiplication.

4. (a) $(3,15,-4) = 2(1,2,-1) + 3(2,4,0) - (5,1,2)$.

   (b) $(-3,-4,7) = a(5,-1,3) + b(2,0,3) + c(4,1,0)$ if and only if $-3 = 5a + 2b + 4c$, $-4 = -a + c$, and $7 = 3a + 3b$. This system of equations has the unique solution $a = 25/21$, $b = 24/21$, $c = -59/21$, so $(-3,-4,7)$ is a linear combination of $(5,-1,3)$, $(2,0,3)$, and $(4,1,0)$.

5. (a) $(x_1,x_2,x_3) = a(1,2,3) + b(-2,4,1) + c(0,6,4)$ if and only if $x_1 = a - 2b$, $x_2 = 2a + 4b + 6c$, and $x_3 = 3a + b + 4c$. The coefficient matrix for this system of equations has nonzero determinant, so the system has a unique solution. Thus each vector in $\mathbf{R}^3$ is a linear combination of the given vectors.

   (b) $(x_1,x_2,x_3) = a(-2,1,0) + b(1,2,-1) + c(-1,8,-3)$ if and only if $x_1 = -2a + b - c$, $x_2 = a + 2b + 8c$, $x_3 = -b - 3c$. The coefficient matrix for this system of equations has determinant zero, so the coefficient matrix is row equivalent to a 3x3 matrix having at least one row of zeros. Thus the augmented matrix is row equivalent to a 3x4 matrix having 0 0 0 N for its third row, where N is not zero for some values of x_1, x_2, and x_3. So for some values of x_1, x_2, and x_3 the system of equations has no solution. The given set of vectors does not span $\mathbf{R}^3$.

   A shorter argument is that the given vectors are not linearly independent and, in fact, span a two-dimensional subspace of $\mathbf{R}^3$, so they certainly do not span $\mathbf{R}^3$.

6. (a) $a(1,-2,0) + b(0,1,3) + c(2,0,12) = (0,0,0)$ if and only if $a + 2c = 0$, $-2a + b = 0$, and $3b + 12c = 0$. The coefficient matrix for this homogeneous system of equations has determinant zero, so there are solutions with a, b, and c not all zero. Thus these vectors are linearly dependent.

   (b) $a(-1,18,7) + b(-1,4,1) + c(1,3,2) = (0,0,0)$ if and only if $-a - b + c = 0$, $18a + 4b + 3c = 0$, and $7a + b + 2c = 0$. The coefficient matrix for this homogeneous system of equations has determinant zero, so there are solutions with a, b, and c not all zero. Thus these vectors are linearly dependent.

   (c) $a(5,-1,3) + b(2,1,0) + c(3,-2,2) = (0,0,0)$ if and only if $5a + 2b + 3c = 0$, $-a + b - 2c = 0$, and $3a + 2c = 0$. The coefficient matrix for this homogeneous system of equations has nonzero determinant, so the solution $a = b = c = 0$ is unique. The vectors are therefore linearly independent.

7. (a) and (b) In each case the set consists of two linearly independent vectors. By Theorem 4.11 they are therefore a basis for $\mathbf{R}^2$.

   (c) and (d) In each case the set consists of three linearly independent vectors. By Theorem 4.11 they are therefore a basis for $\mathbf{R}^3$.

8. $(10,9,8) = 2(-1,3,1) + 3(4,1,2)$.

9. Any of the sets {(1,−2,3), (4,1,−1), (1,0,0)}, {(1,−2,3), (4,1,−1), (0,1,0)}, {(1,−2,3), (4,1,−1), (0,0,1)} is a basis for $\mathbf{R}^3$.

10. (a,b,c,a−2b+3c) = a(1,0,0,1) + b(0,1,0,−2) + c(0,0,1,3). The linearly independent set {(1,0,0,1), (0,1,0,−2), (0,0,1,3)} is a basis for the subspace.

11. (a) $\begin{bmatrix} 1 & 2 & -1 \\ -1 & 3 & 4 \\ 0 & 5 & 3 \end{bmatrix} \approx \begin{bmatrix} 1 & 2 & -1 \\ 0 & 5 & 3 \\ 0 & 5 & 3 \end{bmatrix} \approx \begin{bmatrix} 1 & 2 & -1 \\ 0 & 1 & 3/5 \\ 0 & 0 & 0 \end{bmatrix}$, so the rank of the matrix is 2.

(b) $\begin{bmatrix} 2 & 1 & 4 \\ -2 & 0 & -1 \\ 3 & 2 & 7 \end{bmatrix} \approx \begin{bmatrix} 1 & 1/2 & 2 \\ 0 & 1 & 3 \\ 0 & 1/2 & 1 \end{bmatrix} \approx \begin{bmatrix} 1 & 1/2 & 2 \\ 0 & 1 & 3 \\ 0 & 0 & 1 \end{bmatrix}$, so the rank of the matrix is 3.

(c) $\begin{bmatrix} -2 & 4 & 8 \\ 1 & -2 & 4 \\ 4 & -8 & 16 \end{bmatrix} \approx \begin{bmatrix} 1 & -2 & -4 \\ 0 & 0 & 1 \\ 0 & 0 & 0 \end{bmatrix}$, so the rank of the matrix is 2.

12. $\begin{bmatrix} 1 & -2 & 3 & 4 \\ -1 & 3 & 1 & -2 \\ 2 & -3 & 10 & 10 \end{bmatrix} \approx \begin{bmatrix} 1 & -2 & 3 & 4 \\ 0 & 1 & 4 & 2 \\ 0 & 1 & 4 & 2 \end{bmatrix} \approx \begin{bmatrix} 1 & -2 & 3 & 4 \\ 0 & 1 & 4 & 2 \\ 0 & 0 & 0 & 0 \end{bmatrix}$, so the vectors

(1,−2,3,4) and (0,1,4,2) are a basis for the subspace.

13. $\mathbf{v} = a\mathbf{v}_1 + b\mathbf{v}_2 = a\mathbf{v}_1 + b\mathbf{v}_2 + 0\mathbf{v}_3$.

14. If $a\mathbf{v}_1 + b\mathbf{v}_2 = \mathbf{0}$ then $a\mathbf{v}_1 + b\mathbf{v}_2 + 0\mathbf{v}_3 = \mathbf{0}$ and since the set $\{\mathbf{v}_1, \mathbf{v}_2, \mathbf{v}_3\}$ is linearly independent this means a = b = 0, so $\{\mathbf{v}_1, \mathbf{v}_2\}$ must be linearly independent.

If $a\mathbf{v}_1 + c\mathbf{v}_3 = \mathbf{0}$ then $a\mathbf{v}_1 + 0\mathbf{v}_2 + c\mathbf{v}_3 = \mathbf{0}$ and since the set $\{\mathbf{v}_1, \mathbf{v}_2, \mathbf{v}_3\}$ is linearly independent this means a = c = 0, so $\{\mathbf{v}_1, \mathbf{v}_3\}$ must be linearly independent.

If $b\mathbf{v}_2 + c\mathbf{v}_3 = \mathbf{0}$ then $0\mathbf{v}_1 + b\mathbf{v}_2 + c\mathbf{v}_3 = \mathbf{0}$ and since the set $\{\mathbf{v}_1, \mathbf{v}_2, \mathbf{v}_3\}$ is linearly independent this means b = c = 0, so $\{\mathbf{v}_2, \mathbf{v}_3\}$ must be linearly independent.

Since $\{\mathbf{v}_1, \mathbf{v}_2, \mathbf{v}_3\}$ is linearly independent, none of the three vectors can be the zero vector. Thus $a\mathbf{v}_1 = \mathbf{0}$ (or $b\mathbf{v}_2 = \mathbf{0}$ or $c\mathbf{v}_3 = \mathbf{0}$) only if a = 0 (or b = 0 or c = 0).

15. $a(\mathbf{v}_1 + 2\mathbf{v}_2) + b(3\mathbf{v}_1 - \mathbf{v}_2) = (a + 3b)\mathbf{v}_1 + (2a - b)\mathbf{v}_2$. There are scalars c and d, not both zero, with $c\mathbf{v}_1 + d\mathbf{v}_2 = \mathbf{0}$. Solve the system c = a + 3b, d = 2a − b: $a = \frac{3d+c}{7}$, $b = \frac{2c-d}{7}$. 3d+c and 2c−d cannot both be zero unless both c and d are zero, so at least one of a and b is nonzero and $a(\mathbf{v}_1 + 2\mathbf{v}_2) + b(3\mathbf{v}_1 - \mathbf{v}_2) = c\mathbf{v}_1 + d\mathbf{v}_2 = \mathbf{0}$. So $\mathbf{v}_1 + 2\mathbf{v}_2$ and $3\mathbf{v}_1 - \mathbf{v}_2$ are linearly dependent.

16. (a) True. The vectors are both in the subspace which has dimension 2, and they are linearly independent.

    (b) False. The dimension of $\mathbf{R}^2$ is 2. No set of more than two vectors can be linearly independent.

    (c) True. This is a direct result of Theorem 4.11 and the definitions.

    (d) True. If two vectors in $\mathbf{R}^2$ are not a basis they are linearly dependent and therefore collinear.

    (e) False. For the vectors to be linearly dependent all three would have to lie on the same line or in the same plane. It is much more likely that the first two vectors would not lie on the same line and that the third vector would not lie in the plane of the first two. Reasoning in terms of matrices, if you write the vectors as rows of a matrix, the determinant of the matrix would have to be zero for the vectors to be linearly dependent. Zero is only one of an infinite number of possible values for the determinant of a 3x3 matrix.

17. (a) false (b) true (c) true (d) true (e) false

18. If rank(A) = n, the row space of A is $\mathbf{R}^n$, so the reduced echelon form of A must have n linearly independent rows; i.e., it must be I_n. If A is row equivalent to I_n, the rows of I_n are linear combinations of the rows of A. But the rows of I_n span $\mathbf{R}^n$, so the rows of A span $\mathbf{R}^n$ and rank(A) = n.

Chapter 5

Exercise Set 5.1, page 198

1. (a) $(2,1)\cdot(3,4) = 2x3 + 1x4 = 6 + 4 = 10$ (c) $(2,0)\cdot(0,-1) = 2x0 + 0x{-1} = 0$

2. (a) $(1,2,3)\cdot(4,1,0) = 1x4 + 2x1 + 3x0 = 4 + 2 + 0 = 6$

   (c) $(7,1,-2)\cdot(3,-5,8) = 7x3 + 1x{-5} + {-2}x8 = 21 - 5 - 16 = 0$

3. (a) $(5,1)\cdot(2,-3) = 5x2 + 1x{-3} = 10 - 3 = 7$

   (c) $(7,1,2,-4)\cdot(3,0,-1,5) = 7x3 + 1x0 + 2x{-1} + {-4}x5 = 21 + 0 - 2 - 20 = -1$

   (e) $(1,2,3,0,0,0)\cdot(0,0,0,-2,-4,9) = 1x0 + 2x0 + 3x0 + 0x{-2} + 0x{-4} + 0x9 = 0$

4. (a) $||(1,2)|| = \sqrt{(1,2)\cdot(1,2)} = \sqrt{1x1+2x2} = \sqrt{5}$

   (c) $||(4,0)|| = \sqrt{(4,0)\cdot(4,0)} = \sqrt{4x4+0x0} = \sqrt{16} = 4$

   (d) $||(-3,1)|| = \sqrt{(-3,1)\cdot(-3,1)} = \sqrt{-3x{-3}+1x1} = \sqrt{10}$

5. (a) $||(1,3,-1)|| = \sqrt{(1,3,-1)\cdot(1,3,-1)} = \sqrt{1x1+3x3+{-1}x{-1}} = \sqrt{11}$

   (c) $||(5,1,1)|| = \sqrt{(5,1,1)\cdot(5,1,1)} = \sqrt{5x5+1x1+1x1} = \sqrt{27} = 3\sqrt{3}$

   (e) $||(7,-2,-3)|| = \sqrt{(7,-2,-3)\cdot(7,-2,-3)} = \sqrt{7x7+{-2}x{-2}+{-3}x{-3}} = \sqrt{62}$

6. (a) $||(5,2)|| = \sqrt{(5,2)\cdot(5,2)} = \sqrt{5x5+2x2} = \sqrt{29}$

   (b) $||(-4,2,3)|| = \sqrt{(-4,2,3)\cdot(-4,2,3)} = \sqrt{-4x{-4}+2x2+3x3} = \sqrt{29}$

   (e) $||(-3,0,1,4,2)|| = \sqrt{(-3,0,1,4,2)\cdot(-3,0,1,4,2)} = \sqrt{-3x{-3}+0x0+1x1+4x4+2x2} = \sqrt{30}$

7. (a) $\frac{(1,3)}{\|(1,3)\|} = \left(\frac{1}{\sqrt{10}}, \frac{3}{\sqrt{10}}\right)$ (c) $\frac{(1,2,3)}{\|(1,2,3)\|} = \left(\frac{1}{\sqrt{14}}, \frac{2}{\sqrt{14}}, \frac{3}{\sqrt{14}}\right)$

(d) $\frac{(-2,4,0)}{\|(-2,4,0)\|} = \left(\frac{-2}{\sqrt{20}}, \frac{4}{\sqrt{20}}, 0\right) = \left(\frac{-1}{\sqrt{5}}, \frac{2}{\sqrt{5}}, 0\right)$

8. (a) $\frac{(4,2)}{\|(4,2)\|} = \left(\frac{4}{2\sqrt{5}}, \frac{2}{2\sqrt{5}}\right) = \left(\frac{2}{\sqrt{5}}, \frac{1}{\sqrt{5}}\right)$ (c) $\frac{(7,2,0,1)}{\|(7,2,0,1)\|} = \left(\frac{7}{3\sqrt{6}}, \frac{2}{3\sqrt{6}}, 0, \frac{1}{3\sqrt{6}}\right)$

(e) $\frac{(0,0,0,7,0,0)}{\|(0,0,0,7,0,0)\|} = (0,0,0,1,0,0)$

9. (a) $\cos\theta = \frac{(-1,1)\cdot(0,1)}{\|(-1,1)\|\,\|(0,1)\|} = \frac{1}{\sqrt{2}}$, so $\theta = \frac{\pi}{4} = 45°$

(b) $\cos\theta = \frac{(2,0)\cdot(1,\sqrt{3})}{\|(2,0)\|\,\|(1,\sqrt{3})\|} = \frac{2}{4} = \frac{1}{2}$, so $\theta = \frac{\pi}{3} = 60°$

10. (a) $\cos\theta = \frac{(4,-1)\cdot(2,3)}{\|(4,-1)\|\,\|(2,3)\|} = \frac{5}{\sqrt{17}\sqrt{13}}$

(b) $\cos\theta = \frac{(3,-1,2)\cdot(4,1,1)}{\|(3,-1,2)\|\,\|(4,1,1)\|} = \frac{13}{\sqrt{14}\sqrt{18}} = \frac{13}{6\sqrt{7}}$

(d) $\cos\theta = \frac{(7,1,0,0)\cdot(3,2,1,0)}{\|(7,1,0,0)\|\,\|(3,2,1,0)\|} = \frac{23}{\sqrt{50}\sqrt{14}} = \frac{23}{10\sqrt{7}}$

11. (a) $(1,3)\cdot(3,-1) = 1\text{x}3 + 3\text{x}-1 = 0$, so the vectors are orthogonal.

12. (c) $(7,1,0)\cdot(2,-14,3) = 7\text{x}2 + 1\text{x}-14 + 0\text{x}3 = 0$, so the vectors are orthogonal.

13. (a) If (a,b) is orthogonal to $(1,3)$ then $(a,b)\cdot(1,3) = a + 3b = 0$, so $a = -3b$. Thus any vector of the form $(-3b,b)$ is orthogonal to $(1,3)$.

(c) If (a,b) is orthogonal to $(-4,-1)$ then $(a,b)\cdot(-4,-1) = -4a - b = 0$, so $b = -4a$. Thus any vector of the form $(a,-4a)$ is orthogonal to $(-4,-1)$.

14. (b) If (a,b,c) is orthogonal to $(1,-2,3)$ then $(a,b,c)\cdot(1,-2,3) = a - 2b + 3c = 0$, so $a = 2b-3c$. Thus any vector of the form $(2b-3c,b,c)$ is orthogonal to $(1,-2,3)$.

(d) If (a,b,c,d) is orthogonal to $(5,0,1,1)$ then $(a,b,c,d)\cdot(5,0,1,1) = 5a + c + d = 0$, so $d = -5a-c$. Thus any vector of the form $(a,b,c,-5a-c)$ is orthogonal to $(5,0,1,1)$.

(f) If (a,b,c,d,e) is orthogonal to $(0,-2,3,1,5)$ then $(a,b,c,d,e)\cdot(0,-2,3,1,5) = -2b + 3c + d + 5e = 0$, so $d = 2b-3c-5e$. Thus any vector of the form $(a,b,c,2b-3c-5e,e)$ is orthogonal to $(0,-2,3,1,5)$.

15. If (a,b,c) is orthogonal to both $(1,2,-1)$ and $(3,1,0)$ then $(a,b,c)\cdot(1,2,-1) = a + 2b - c = 0$ and $(a,b,c)\cdot(3,1,0) = 3a + b = 0$. These equations yield the solution $b = -3a$ and $c = -5a$, so any vector of the form $(a,-3a,-5a)$ is orthogonal to both $(1,2,-1)$ and $(3,1,0)$.

16. (a) $d = \sqrt{(6-2)^2+(5-2)^2} = 5.$ (c) $d = \sqrt{(7-2)^2+(-3-2)^2} = \sqrt{50} = 5\sqrt{2}.$

17. (b) $d = \sqrt{(1-2)^2+(2-1)^2+(3-0)^2} = \sqrt{11}.$

(d) $d^2 = (5-2)^2+(1-0)^2+(0-1)^2+(0-3)^2 = 20$, so $d = \sqrt{20} = 2\sqrt{5}.$

(e) $d^2 = (-3-2)^2+(1-1)^2+(1-4)^2+(0-1)^2+(2+1)^2 = 44$, so $d = \sqrt{44} = 2\sqrt{11}.$

19. $\mathbf{u}$ is a scalar multiple of $\mathbf{v}$, so it has the same direction as $\mathbf{v}$. The magnitude of $\mathbf{u}$ is

$$||\mathbf{u}|| = \frac{1}{||\mathbf{v}||}\sqrt{(v_1)^2+(v_2)^2+\ldots+(v_n)^2} = \frac{||\mathbf{v}||}{||\mathbf{v}||} = 1,$$ so $\mathbf{u}$ is a unit vector.

21. If $\mathbf{u}\cdot\mathbf{v} = \mathbf{u}\cdot\mathbf{w}$ then $\mathbf{u}\cdot(\mathbf{v}-\mathbf{w}) = 0$ for all vectors $\mathbf{u}$ in U. Since $\mathbf{v}-\mathbf{w}$ is a vector in U this means that $(\mathbf{v}-\mathbf{w})\cdot(\mathbf{v}-\mathbf{w}) = 0$. Therefore $\mathbf{v}-\mathbf{w} = \mathbf{0}$, so $\mathbf{v} = \mathbf{w}$.

23. (a) vector (c) makes no sense (f) scalar (h) scalar

24. $||c(3,0,4)|| = \sqrt{3c\text{x}3c+4c\text{x}4c} = |c|\sqrt{9+16} = 5|c| = 15$, so $|c| = 3$ and $c = \pm 3$.

26. $(a,b)\cdot(-b,a) = ax-b + bxa = 0$, so $(-b,a)$ is orthogonal to (a,b).

27. Suppose (a,b) is orthogonal to (u_1,u_2). Then $(a,b)\cdot(u_1,u_2) = au_1 + bu_2 = 0$, so $au_1 = -bu_2$. Thus if $u_1 \neq 0$ (a,b) is orthogonal to (u_1,u_2) if and only if (a,b) is of the form $\left(\frac{-bu_2}{u_1}, b\right) = \frac{b}{u_1}(-u_2,u_1)$; i.e., (a,b) is orthogonal to (u_1,u_2) if and only if (a,b) is in the vector space with basis $(-u_2,u_1)$. If $u_1 = 0$ then $b = 0$ and (a,b) is orthogonal to (u_1,u_2) if and only if (a,b) is of the form $a(1,0)$; i.e., (a,b) is in the vector space with basis $(1,0)$.

30. $(\mathbf{u} + \mathbf{v})\cdot(\mathbf{u} - \mathbf{v}) = (u_1 + v_1)(u_1 - v_1) + (u_2 + v_2)(u_2 - v_2) + \ldots + (u_n + v_n)(u_n - v_n)$

$$= u_1^2 - v_1^2 + u_2^2 - v_2^2 + \ldots + u_n^2 - v_n^2$$

$$= u_1^2 + u_2^2 + \ldots + u_n^2 - v_1^2 - v_2^2 - \ldots - v_n^2 = ||\mathbf{u}|| - ||\mathbf{v}||, \text{ so } ||\mathbf{u}|| = ||\mathbf{v}||$$

if and only if $||\mathbf{u}|| - ||\mathbf{v}|| = 0$ if and only if $(\mathbf{u} + \mathbf{v})\cdot(\mathbf{u} - \mathbf{v}) = 0$ if and only if $\mathbf{u} + \mathbf{v}$ and $\mathbf{u} - \mathbf{v}$ are orthogonal.

31. The two-dimensional subspace with basis (1,0,0) and (0,1,0) is orthogonal to the one-dimensional subspace with basis (0,0,1).

34. (b) $||\mathbf{u}|| = \max_{i=1,\ldots,n} |u_i| \geq 0$ since the absolute value of any number is equal to or greater than zero.

$\max_{i=1,\ldots,n} |u_i| = 0$ if and only if all $|u_i| = 0$.

$||c\mathbf{u}|| = \max_{i=1,\ldots,n} |cu_i| = |c| \max_{i=1,\ldots,n} |u_i| = |c|\ ||\mathbf{u}||$.

$||(1,2)|| = |2| = 2$, $||(-3,4)|| = |4| = 4$, $||(1,2,-5)|| = |-5| = 5$, and $||(0,-2,7)|| = |7| = 7$.

35. (c) $d(\mathbf{x},\mathbf{z}) = ||\mathbf{x}-\mathbf{y}+\mathbf{y}-\mathbf{z}|| \leq ||\mathbf{x}-\mathbf{y}|| + ||\mathbf{y}-\mathbf{z}|| = d(\mathbf{x},\mathbf{y}) + d(\mathbf{y},\mathbf{z})$, from the triangle inequality.

Exercise Set 5.2, page 211

1. (a) $(1,2)\cdot(2,-1) = 0$, so the vectors are orthogonal.

   (b) $(3,-1)\cdot(0,5) = -5$, so the vectors are not orthogonal.

   (d) $(4,1)\cdot(2,-3) = 5$, so the vectors are not orthogonal.

2. (a) $(1,2,1)\cdot(4,-2,0) = 0$, $(1,2,1)\cdot(2,4,-10) = 0$, and $(4,-2,0)\cdot(2,4,-10) = 0$, so the set of vectors is orthogonal.

   (b) $(3,-1,1)\cdot(2,0,1) = 7$, so the set of vectors is not orthogonal.

3. (a) $\left(\frac{1}{3},\frac{2}{3},\frac{2}{3}\right)\cdot\left(\frac{2}{3},-\frac{2}{3},\frac{1}{3}\right)=0$, $\left(\frac{1}{3},\frac{2}{3},\frac{2}{3}\right)\cdot\left(\frac{2}{3},\frac{1}{3},-\frac{2}{3}\right)=0$, and

   $\left(\frac{2}{3},-\frac{2}{3},\frac{1}{3}\right)\cdot\left(\frac{2}{3},\frac{1}{3},-\frac{2}{3}\right) = 0$, so the set of vectors is orthogonal.

   $\left(\frac{1}{3},\frac{2}{3},\frac{2}{3}\right)\cdot\left(\frac{1}{3},\frac{2}{3},\frac{2}{3}\right)=1$, $\left(\frac{2}{3},-\frac{2}{3},\frac{1}{3}\right)\cdot\left(\frac{2}{3},-\frac{2}{3},\frac{1}{3}\right) = 1$, and

   $\left(\frac{2}{3},\frac{1}{3},-\frac{2}{3}\right)\cdot\left(\frac{2}{3},\frac{1}{3},-\frac{2}{3}\right) = 1$, so all the vectors are unit vectors. Thus the vectors are an orthonormal set.

   (c) $\left(\frac{1}{\sqrt{2}},0,\frac{1}{\sqrt{2}}\right)\cdot\left(\frac{1}{\sqrt{2}},0,\frac{-1}{\sqrt{2}}\right)=0$, $\left(\frac{1}{\sqrt{2}},0,\frac{1}{\sqrt{2}}\right)\cdot(0,1,0) = 0$, and

   $\left(\frac{1}{\sqrt{2}},0,\frac{-1}{\sqrt{2}}\right)\cdot(0,1,0) = 0$, so the set of vectors is orthogonal.

   $\left(\frac{1}{\sqrt{2}},0,\frac{1}{\sqrt{2}}\right)\cdot\left(\frac{1}{\sqrt{2}},0,\frac{1}{\sqrt{2}}\right) = 1$, $\left(\frac{1}{\sqrt{2}},0,\frac{-1}{\sqrt{2}}\right)\cdot\left(\frac{1}{\sqrt{2}},0,\frac{-1}{\sqrt{2}}\right) = 1$, and

   $(0,1,0)\cdot(0,1,0) = 1$, so the vectors are unit vectors. Thus the vectors are an orthonormal set.

   (e) $\left(\frac{1}{\sqrt{32}},\frac{-2}{\sqrt{32}},\frac{5}{\sqrt{32}}\right)\cdot\left(\frac{1}{\sqrt{32}},\frac{-2}{\sqrt{32}},\frac{5}{\sqrt{32}}\right)=\frac{30}{32} \neq 1$, so $\left(\frac{1}{\sqrt{32}},\frac{-2}{\sqrt{32}},\frac{5}{\sqrt{32}}\right)$ is not a unit vector. Therefore the set is not orthonormal.

4. $\mathbf{v} = (\mathbf{v}\cdot\mathbf{u}_1)\mathbf{u}_1 + (\mathbf{v}\cdot\mathbf{u}_2)\mathbf{u}_2 + (\mathbf{v}\cdot\mathbf{u}_3)\mathbf{u}_3$.

$\mathbf{v}\cdot\mathbf{u}_1 = (2,-3,1)\cdot(0,-1,0) = 3$, $\mathbf{v}\cdot\mathbf{u}_2 = (2,-3,1)\cdot\left(\frac{3}{5},0,\frac{-4}{5}\right) = \frac{2}{5}$, and

$\mathbf{v}\cdot\mathbf{u}_3 = (2,-3,1)\cdot\left(\frac{4}{5},0,\frac{3}{5}\right) = \frac{11}{5}$, so $\mathbf{v} = 3\mathbf{u}_1 + \frac{2}{5}\mathbf{u}_2 + \frac{11}{5}\mathbf{u}_3$.

6. $\text{proj}_{\mathbf{u}}\mathbf{v} = \dfrac{\mathbf{v}\cdot\mathbf{u}}{\mathbf{u}\cdot\mathbf{u}}\mathbf{u}$.

(a) $\text{proj}_{\mathbf{u}}\mathbf{v} = \dfrac{(7,4)\cdot(1,2)}{(1,2)\cdot(1,2)}(1,2) = 3(1,2) = (3,6)$.

(c) $\text{proj}_{\mathbf{u}}\mathbf{v} = \dfrac{(4,6,4)\cdot(1,2,3)}{(1,2,3)\cdot(1,2,3)}(1,2,3) = 2(1,2,3) = (2,4,6)$.

(e) $\text{proj}_{\mathbf{u}}\mathbf{v} = \dfrac{(1,2,3,0)\cdot(1,-1,2,3)}{(1,-1,2,3)\cdot(1,-1,2,3)}(1,-1,2,3) = \frac{1}{3}(1,-1,2,3) = \left(\frac{1}{3},\frac{-1}{3},\frac{2}{3},1\right)$.

7. (a) $\text{proj}_{\mathbf{u}}\mathbf{v} = \dfrac{(1,2)\cdot(2,5)}{(2,5)\cdot(2,5)}(2,5) = \frac{12}{29}(2,5) = \left(\frac{24}{29},\frac{60}{29}\right)$.

(c) $\text{proj}_{\mathbf{u}}\mathbf{v} = \dfrac{(1,2,3)\cdot(1,2,0)}{(1,2,0)\cdot(1,2,0)}(1,2,0) = (1,2,0)$.

(e) $\text{proj}_{\mathbf{u}}\mathbf{v} = \dfrac{(2,-1,3,1)\cdot(-1,2,1,3)}{(-1,2,1,3)\cdot(-1,2,1,3)}(-1,2,1,3) = \frac{2}{15}(-1,2,1,3) = \left(\frac{-2}{15},\frac{4}{15},\frac{2}{15},\frac{2}{5}\right)$.

8. (a) $\mathbf{u}_1 = (1,2)$, $\mathbf{u}_2 = (-1,3) - \dfrac{(-1,3)\cdot(1,2)}{(1,2)\cdot(1,2)}(1,2) = (-1,3) - (1,2) = (-2,1)$ is an orthogonal

basis for $\mathbf{R}^2$. $||\mathbf{u}_1|| = \sqrt{5}$ and $||\mathbf{u}_2|| = \sqrt{5}$, so the set $\left\{\left(\frac{1}{\sqrt{5}},\frac{2}{\sqrt{5}}\right),\left(\frac{-2}{\sqrt{5}},\frac{1}{\sqrt{5}}\right)\right\}$

is an orthonormal basis for $\mathbf{R}^2$.

(c) $\mathbf{u}_1 = (1,-1)$, $\mathbf{u}_2 = (4,-2) - \dfrac{(4,-2)\cdot(1,-1)}{(1,-1)\cdot(1,-1)}(1,-1) = (1,1)$ is an orthogonal

basis for $\mathbf{R}^2$. $||\mathbf{u}_1|| = \sqrt{2}$ and $||\mathbf{u}_2|| = \sqrt{2}$, so the set $\left\{\left(\frac{1}{\sqrt{2}},\frac{-1}{\sqrt{2}}\right),\left(\frac{1}{\sqrt{2}},\frac{1}{\sqrt{2}}\right)\right\}$

is an orthonormal basis for $\mathbf{R}^2$.

9. (a) $\mathbf{u}_1 = (1,1,1)$, $\mathbf{u}_2 = (2,0,1) - \dfrac{(2,0,1)\cdot(1,1,1)}{(1,1,1)\cdot(1,1,1)}(1,1,1) = (1,-1,0)$,

$$\mathbf{u}_3 = (2,4,5) - \frac{(2,4,5)\cdot(1,1,1)}{(1,1,1)\cdot(1,1,1)}(1,1,1) - \frac{(2,4,5)\cdot(1,-1,0)}{(1,-1,0)\cdot(1,-1,0)}(1,-1,0)$$

$= (2,4,5) - \dfrac{11}{3}(1,1,1) - (-1)(1,-1,0) = \left(\dfrac{-2}{3},\dfrac{-2}{3},\dfrac{4}{3}\right)$ is an orthogonal basis for

$\mathbf{R}^3$. $\|\mathbf{u}_1\| = \sqrt{3}$, $\|\mathbf{u}_2\| = \sqrt{2}$, and $\|\mathbf{u}_3\| = \dfrac{\sqrt{24}}{3} = \dfrac{2\sqrt{6}}{3}$, so the set

$\left\{\left(\dfrac{1}{\sqrt{3}}, \dfrac{1}{\sqrt{3}}, \dfrac{1}{\sqrt{3}}\right), \left(\dfrac{1}{\sqrt{2}}, \dfrac{-1}{\sqrt{2}}, 0\right), \left(\dfrac{-1}{\sqrt{6}}, \dfrac{-1}{\sqrt{6}}, \dfrac{2}{\sqrt{6}}\right)\right\}$ is an orthonormal

basis for $\mathbf{R}^3$.

10. (a) $\mathbf{u}_1 = (1,0,2)$, $\mathbf{u}_2 = (-1,0,1) - \dfrac{(-1,0,1)\cdot(1,0,2)}{(1,0,2)\cdot(1,0,2)}(1,0,2) = (-1,0,1) - \dfrac{1}{5}(1,0,2)$

$= \left(\dfrac{-6}{5}, 0, \dfrac{3}{5}\right)$ is an orthogonal basis for the subspace of $\mathbf{R}^3$. $\|\mathbf{u}_1\| = \sqrt{5}$ and

$\|\mathbf{u}_2\| = \dfrac{3}{5}\sqrt{5}$, so the set $\left\{\left(\dfrac{1}{\sqrt{5}}, 0, \dfrac{2}{\sqrt{5}}\right), \left(\dfrac{-2}{\sqrt{5}}, 0, \dfrac{1}{\sqrt{5}}\right)\right\}$ is an orthonormal basis

for the subspace.

11. (b) $\mathbf{u}_1 = (3,0,0,0)$, $\mathbf{u}_2 = (0,1,2,1)$ (because $\mathbf{u}_1\cdot\mathbf{u}_2 = 0$),

$\mathbf{u}_3 = (0,-1,3,2) - 0(3,0,0,0) - \dfrac{(0,-1,3,2)\cdot(0,1,2,1)}{(0,1,2,1)\cdot(0,1,2,1)}(0,1,2,1) = \left(0, \dfrac{-13}{6}, \dfrac{4}{6}, \dfrac{5}{6}\right)$ is an

orthogonal basis for the subspace of $\mathbf{R}^4$. $\|\mathbf{u}_1\| = 3$, $\|\mathbf{u}_2\| = \sqrt{6}$, and $\|\mathbf{u}_3\| = \dfrac{1}{6}\sqrt{210}$,

so the set $\left\{(1,0,0,0), \left(0, \dfrac{1}{\sqrt{6}}, \dfrac{2}{\sqrt{6}}, \dfrac{1}{\sqrt{6}}\right), \left(0, \dfrac{-13}{\sqrt{210}}, \dfrac{4}{\sqrt{210}}, \dfrac{5}{\sqrt{210}}\right)\right\}$

is an orthonormal basis for the subspace.

12. To actually construct such a vector start with any vector that is not a multiple of the given vector (1,2,−1,−1) and use the Gram-Schmidt process to find a vector orthogonal to (1,2,−1,−1). The vector (−2,1,0,0) is orthogonal to (1,2,−1,−1).

14. First find an orthonormal basis for W. $\mathbf{u}_1 = (0,-1,3)$,

$$\mathbf{u}_2 = (1,1,2) - \frac{(1,1,2)\cdot(0,-1,3)}{(0,-1,3)\cdot(0,-1,3)}(0,-1,3) = (1,1,2) - \frac{1}{2}(0,-1,3) = \left(1, \frac{3}{2}, \frac{1}{2}\right)$$ is an

orthogonal basis for the subspace of $\mathbf{R}^3$. $||\mathbf{u}_1|| = \sqrt{10}$ and $||\mathbf{u}_2|| = \frac{1}{2}\sqrt{14}$, so the set

$\left\{\left(0, \frac{-1}{\sqrt{10}}, \frac{3}{\sqrt{10}}\right), \left(\frac{2}{\sqrt{14}}, \frac{3}{\sqrt{14}}, \frac{1}{\sqrt{14}}\right)\right\}$ is an orthonormal basis for W.

(a) $\text{proj}_W(3,-1,2) = \left((3,-1,2)\cdot\left(0,\frac{-1}{\sqrt{10}},\frac{3}{\sqrt{10}}\right)\right)\left(0,\frac{-1}{\sqrt{10}},\frac{3}{\sqrt{10}}\right)$

$+ \left((3,-1,2)\cdot\left(\frac{2}{\sqrt{14}},\frac{3}{\sqrt{14}},\frac{1}{\sqrt{14}}\right)\right)\left(\frac{2}{\sqrt{14}},\frac{3}{\sqrt{14}},\frac{1}{\sqrt{14}}\right) = \frac{7}{10}(0,-1,3) + \frac{5}{14}(2,3,1)$

$= \left(\frac{5}{7}, \frac{13}{35}, \frac{86}{35}\right).$

(c) $\text{proj}_W(4,2,1) = \left((4,2,1)\cdot\left(0,\frac{-1}{\sqrt{10}},\frac{3}{\sqrt{10}}\right)\right)\left(0,\frac{-1}{\sqrt{10}},\frac{3}{\sqrt{10}}\right)$

$+ \left((4,2,1)\cdot\left(\frac{2}{\sqrt{14}},\frac{3}{\sqrt{14}},\frac{1}{\sqrt{14}}\right)\right)\left(\frac{2}{\sqrt{14}},\frac{3}{\sqrt{14}},\frac{1}{\sqrt{14}}\right) = \frac{1}{10}(0,-1,3) + \frac{15}{14}(2,3,1)$

$= \left(\frac{15}{7}, \frac{109}{35}, \frac{48}{35}\right).$

15. First find an orthonormal basis for V. $\mathbf{u}_1 = (-1,0,2,1)$,

$$\mathbf{u}_2 = (1,-1,0,3) - \frac{(1,-1,0,3)\cdot(-1,0,2,1)}{(-1,0,2,1)\cdot(-1,0,2,1)}(-1,0,2,1) = (1,-1,0,3) - \frac{1}{3}(-1,0,2,1)$$

$= \left(\frac{4}{3}, -1, \frac{-2}{3}, \frac{8}{3}\right)$ is an orthogonal basis for the subspace of $\mathbf{R}^3$. $||\mathbf{u}_1|| = \sqrt{6}$ and

$||\mathbf{u}_2|| = \frac{1}{3}\sqrt{93}$, so the set $\left\{\left(\frac{-1}{\sqrt{6}}, 0, \frac{2}{\sqrt{6}}, \frac{1}{\sqrt{6}}\right), \left(\frac{4}{\sqrt{93}}, \frac{-3}{\sqrt{93}}, \frac{-2}{\sqrt{93}}, \frac{8}{\sqrt{93}}\right)\right\}$

is an orthonormal basis for V.

(a) $\text{proj}_V(1,-1,1,-1) = \left((1,-1,1,-1)\cdot\left(\frac{-1}{\sqrt{6}},0,\frac{2}{\sqrt{6}},\frac{1}{\sqrt{6}}\right)\right)\left(\frac{-1}{\sqrt{6}},0,\frac{2}{\sqrt{6}},\frac{1}{\sqrt{6}}\right)$

$$+\left((1,-1,1,-1)\cdot\left(\frac{4}{\sqrt{93}},\frac{-3}{\sqrt{93}},\frac{-2}{\sqrt{93}},\frac{8}{\sqrt{93}}\right)\right)\left(\frac{4}{\sqrt{93}},\frac{-3}{\sqrt{93}},\frac{-2}{\sqrt{93}},\frac{8}{\sqrt{93}}\right)$$

$$= 0(-1,0,2,1)+\frac{-1}{31}(4,-3,-2,8)=\left(\frac{-4}{31},\frac{3}{31},\frac{2}{31},\frac{-8}{31}\right).$$

(c) $$\text{proj}_V(3,2,1,0)=\left((3,2,1,0)\cdot\left(\frac{-1}{\sqrt{6}},0,\frac{2}{\sqrt{6}},\frac{1}{\sqrt{6}}\right)\right)\left(\frac{-1}{\sqrt{6}},0,\frac{2}{\sqrt{6}},\frac{1}{\sqrt{6}}\right)$$

$$+\left((3,2,1,0)\cdot\left(\frac{4}{\sqrt{93}},\frac{-3}{\sqrt{93}},\frac{-2}{\sqrt{93}},\frac{8}{\sqrt{93}}\right)\right)\left(\frac{4}{\sqrt{93}},\frac{-3}{\sqrt{93}},\frac{-2}{\sqrt{93}},\frac{8}{\sqrt{93}}\right)$$

$$=\frac{-1}{6}(-1,0,2,1)+\frac{4}{93}(4,-3,-2,8)=\left(\frac{21}{62},\frac{-4}{31},\frac{-13}{31},\frac{11}{62}\right).$$

16. $V=\{a(1,1,0)+b(0,0,1)\}$. The set $\{(1,1,0),(0,0,1)\}$ is an orthogonal basis. The vectors $\left(\frac{1}{\sqrt{2}},\frac{1}{\sqrt{2}},0\right)$ and $(0,0,1)$ are therefore an orthonormal basis. $\mathbf{v}=\mathbf{w}+\mathbf{w}_\perp$, where

$$\mathbf{w}=\text{proj}_V\mathbf{v}=\left((1,2,-1)\cdot\left(\frac{1}{\sqrt{2}},\frac{1}{\sqrt{2}},0\right)\right)\left(\frac{1}{\sqrt{2}},\frac{1}{\sqrt{2}},0\right)+((1,2,-1)\cdot(0,0,1))(0,0,1)$$

$$=\frac{3}{2}(1,1,0)+(-1)(0,0,1)=\left(\frac{3}{2},\frac{3}{2},-1\right)\text{ and}$$

$$\mathbf{w}_\perp=\mathbf{v}-\text{proj}_V\mathbf{v}=(1,2,-1)-\left(\frac{3}{2},\frac{3}{2},-1\right)=\left(\frac{-1}{2},\frac{1}{2},0\right).$$

18. $W=\{a(1,0,1)+b(0,1,1)\}$. We must find an orthogonal basis. $\mathbf{u}_1=(1,0,1)$ and

$$\mathbf{u}_2=(0,1,1)-\frac{(0,1,1)\cdot(1,0,1)}{(1,0,1)\cdot(1,0,1)}(1,0,1)=(0,1,1)-\frac{1}{2}(1,0,1)=\left(\frac{-1}{2},1,\frac{1}{2}\right)\text{ are an}$$

orthogonal basis. The vectors $\left(\frac{1}{\sqrt{2}},0,\frac{1}{\sqrt{2}}\right)$ and $\left(\frac{-1}{\sqrt{6}},\frac{2}{\sqrt{6}},\frac{1}{\sqrt{6}}\right)$ are therefore an orthonormal basis. $\mathbf{v}=\mathbf{w}+\mathbf{w}_\perp$, where

$$\mathbf{w}=\text{proj}_W\mathbf{v}=\left((3,2,1)\cdot\left(\frac{1}{\sqrt{2}},0,\frac{1}{\sqrt{2}}\right)\right)\left(\frac{1}{\sqrt{2}},0,\frac{1}{\sqrt{2}}\right)$$

$$+\left((3,2,1)\cdot\left(\frac{-1}{\sqrt{6}},\frac{2}{\sqrt{6}},\frac{1}{\sqrt{6}}\right)\right)\left(\frac{-1}{\sqrt{6}},\frac{2}{\sqrt{6}},\frac{1}{\sqrt{6}}\right)=2(1,0,1)+\frac{1}{3}(-1,2,1)=\left(\frac{5}{3},\frac{2}{3},\frac{7}{3}\right)$$

and $\mathbf{w}_\perp = \mathbf{v} - \text{proj}_W\mathbf{v} = (3,2,1) - \left(\frac{5}{3}, \frac{2}{3}, \frac{7}{3}\right) = \left(\frac{4}{3}, \frac{4}{3}, \frac{-4}{3}\right)$.

19. $W = \{a(1,1,0) + b(0,0,1)\}$. The set $\{(1,1,0), (0,0,1)\}$ is an orthogonal basis. The vectors $\left(\frac{1}{\sqrt{2}}, \frac{1}{\sqrt{2}}, 0\right)$ and $(0,0,1)$ are therefore an orthonormal basis.

$$\text{proj}_W\mathbf{x} = \left((1,3,-2)\cdot\left(\frac{1}{\sqrt{2}}, \frac{1}{\sqrt{2}}, 0\right)\right)\left(\frac{1}{\sqrt{2}}, \frac{1}{\sqrt{2}}, 0\right) + ((1,3,-2)\cdot(0,0,1))(0,0,1)$$

$$= 2(1,1,0) + (-2)(0,0,1) = (2,2,-2), \text{ so}$$

$$d(\mathbf{x},W) = \|\mathbf{x} - \text{proj}_W\mathbf{x}\| = \|(1,3,-2) - (2,2,-2)\| = \|(-1,1,0)\| = \sqrt{2}.$$

21. $W = \{a(1,2,3)\}$. The vector $\left(\frac{1}{\sqrt{14}}, \frac{2}{\sqrt{14}}, \frac{3}{\sqrt{14}}\right)$ is an orthonormal basis for W.

$$\text{proj}_W\mathbf{x} = \left((1,3,-2)\cdot\left(\frac{1}{\sqrt{14}}, \frac{2}{\sqrt{14}}, \frac{3}{\sqrt{14}}\right)\right)\left(\frac{1}{\sqrt{14}}, \frac{2}{\sqrt{14}}, \frac{3}{\sqrt{14}}\right) = \left(\frac{1}{14}, \frac{2}{14}, \frac{3}{14}\right), \text{ so}$$

$$d(\mathbf{x},W) = \|\mathbf{x} - \text{proj}_W\mathbf{x}\| = \left\|(1,3,-2) - \left(\frac{1}{14}, \frac{2}{14}, \frac{3}{14}\right)\right\| = \left\|\left(\frac{13}{14}, \frac{40}{14}, \frac{-31}{14}\right)\right\| = \frac{\sqrt{2730}}{14}.$$

22. $(1,2,-2)$ and $(6,1,4)$ are orthogonal. If the vector (a,b,c) is orthogonal to both, then $(1,2,-2)\cdot(a,b,c) = a + 2b - 2c = 0$ and $(6,1,4)\cdot(a,b,c) = 6a + b + 4c = 0$. One solution to this system of equations is $a = -10$, $b = 16$, and $c = 11$. The set $\{(1,2,-2), (6,1,4), (-10,16,11)\}$ is therefore an orthogonal basis for $\mathbf{R}^3$.

24. $\{\mathbf{u}_1, \mathbf{u}_2, \ldots, \mathbf{u}_n\}$ is a set of n linearly independent vectors in a vector space with dimension n. Therefore by Theorem 4.11, $\{\mathbf{u}_1, \mathbf{u}_2, \ldots, \mathbf{u}_n\}$ is a basis for the vector space.

25. $\text{proj}_\mathbf{u}\mathbf{v} = \frac{\mathbf{v}\cdot\mathbf{u}}{\mathbf{u}\cdot\mathbf{u}}\mathbf{u}$, so $(\mathbf{v} - \text{proj}_\mathbf{u}\mathbf{v})\cdot\mathbf{u} = (\mathbf{v} - \frac{\mathbf{v}\cdot\mathbf{u}}{\mathbf{u}\cdot\mathbf{u}}\mathbf{u})\cdot\mathbf{u} = \mathbf{v}\cdot\mathbf{u} - \frac{\mathbf{v}\cdot\mathbf{u}}{\mathbf{u}\cdot\mathbf{u}}\mathbf{u}\cdot\mathbf{u} = 0$.

26. We show that the column vectors are unit vectors and that they are mutually orthogonal.

(a) $\mathbf{a}_1 = \begin{bmatrix} 1 \\ 0 \end{bmatrix}$ and $\mathbf{a}_2 = \begin{bmatrix} 0 \\ 1 \end{bmatrix}$. $||\mathbf{a}_1|| = ||\mathbf{a}_2|| = 1$ and $\mathbf{a}_1 \cdot \mathbf{a}_2 = 0$.

(c) $\mathbf{a}_1 = \begin{bmatrix} \frac{\sqrt{3}}{2} \\ \frac{-1}{2} \end{bmatrix}$ and $\mathbf{a}_2 = \begin{bmatrix} \frac{1}{2} \\ \frac{\sqrt{3}}{2} \end{bmatrix}$. $||\mathbf{a}_1||^2 = \frac{3}{4} + \frac{1}{4} = 1$ and $||\mathbf{a}_2||^2 = \frac{1}{4} + \frac{3}{4} = 1$,

so $||\mathbf{a}_1|| = ||\mathbf{a}_2|| = 1$, and $\mathbf{a}_1 \cdot \mathbf{a}_2 = \frac{\sqrt{3}}{2} \times \frac{1}{2} - \frac{1}{2} \times \frac{\sqrt{3}}{2} = 0$.

30. It is proved in the text that if A is orthogonal then $A^{-1} = A^t$, so we need to show that if $A^{-1} = A^t$ then A is orthogonal. The (i,j)th element of $AA^{-1} = AA^t = I$ is $a_{1i}\, a_{1j} + a_{2i}\, a_{2j} + \dots + a_{ni}\, a_{nj}$, the dot product of the ith and jth columns of A. If $i \neq j$ this sum is zero because the nondiagonal elements of I are zero. If $i = j$ this sum is 1 because the diagonal elements of I are 1. Thus we have proved that the columns of A are mutually orthogonal and that they are all unit vectors, so A is an orthogonal matrix.

31. To show that A^{-1} is unitary it is necessary to show that $(A^{-1})^{-1} = (\overline{A^{-1}})^t$.

Since A is unitary $A^{-1} = \overline{A}^t$. Thus $(A^{-1})^t = (\overline{A}^t)^t = \overline{A}$, so that $(\overline{A^{-1}})^t = A = (A^{-1})^{-1}$.

Exercise Set 5.3, page 221

We use the symbol x to denote cross product of matrices.

1. (a) $\mathbf{u} \times \mathbf{v} = (2\text{x}4 - 3\text{x}0, 3\text{x}{-1} - 1\text{x}4, 1\text{x}0 - 2\text{x}{-1}) = (8,-7,2)$.

(c) $\mathbf{u} \times \mathbf{w} = (2\text{x}{-1} - 3\text{x}2, 3\text{x}1 - 1\text{x}{-1}, 1\text{x}2 - 2\text{x}1) = (-8,4,0)$.

(e) $(\mathbf{u} \times \mathbf{v}) \times \mathbf{w} = (8,-7,2) \times \mathbf{w} = (-7\text{x}{-1} - 2\text{x}2, 2\text{x}1 - 8\text{x}{-1}, 8\text{x}2 - (-7)\text{x}1) = (3,10,23)$.

2. (a) $\mathbf{u} \times \mathbf{v} = \begin{vmatrix} \mathbf{i} & \mathbf{j} & \mathbf{k} \\ -2 & 2 & 4 \\ 3 & 0 & 5 \end{vmatrix} = (10,22,-6).$ (c) $\mathbf{w} \times \mathbf{v} = \begin{vmatrix} \mathbf{i} & \mathbf{j} & \mathbf{k} \\ 4 & -2 & 1 \\ 3 & 0 & 5 \end{vmatrix} = (-10,-17,6).$

(e) $(\mathbf{w} \times \mathbf{v}) \times \mathbf{u} = \begin{vmatrix} \mathbf{i} & \mathbf{j} & \mathbf{k} \\ -10 & -17 & 6 \\ -2 & 2 & 4 \end{vmatrix} = (-80,28,-54).$

3. (a) $\mathbf{u} \times \mathbf{v} = \begin{vmatrix} \mathbf{i} & \mathbf{j} & \mathbf{k} \\ 2 & 3 & 1 \\ -1 & 2 & 4 \end{vmatrix}$ (c) $\mathbf{w} \times \mathbf{u} = \begin{vmatrix} \mathbf{i} & \mathbf{j} & \mathbf{k} \\ 3 & 0 & -7 \\ 2 & 3 & 1 \end{vmatrix}$

$= 10\mathbf{i} - 9\mathbf{j} + 7\mathbf{k}.$ $= 21\mathbf{i} - 17\mathbf{j} + 9\mathbf{k}.$

(e) $(\mathbf{w} \times \mathbf{u}) \times \mathbf{v} = \begin{vmatrix} \mathbf{i} & \mathbf{j} & \mathbf{k} \\ 21 & -17 & 9 \\ -1 & 2 & 4 \end{vmatrix} = -86\mathbf{i} - 93\mathbf{j} + 25\mathbf{k}.$

4. (a) $\mathbf{u} \times \mathbf{v} = \begin{vmatrix} \mathbf{i} & \mathbf{j} & \mathbf{k} \\ 3 & 1 & -2 \\ 4 & -1 & 2 \end{vmatrix} = (0,-14,-7).$

(c) $(\mathbf{w} \times \mathbf{u}) \cdot \mathbf{v} = \begin{vmatrix} \mathbf{i} & \mathbf{j} & \mathbf{k} \\ 0 & 3 & -2 \\ 3 & 1 & -2 \end{vmatrix} \cdot \mathbf{v} = (-4,-6,-9) \cdot (4,-1,2) = -28.$

(d) $(\mathbf{w}+2\mathbf{u}) \times \mathbf{v} = \begin{vmatrix} \mathbf{i} & \mathbf{j} & \mathbf{k} \\ 6 & 5 & -6 \\ 4 & -1 & 2 \end{vmatrix} = (4,-36,-26).$

(f) $(\mathbf{v} \times \mathbf{u}) \cdot (\mathbf{w} \times \mathbf{v}) = (0,14,7) \cdot (4,-8,-12) = -196.$

5. (a) $\overrightarrow{AB} = (-3,4,6) - (1,2,1) = (-4,2,5)$ and $\overrightarrow{AC} = (1,8,3) - (1,2,1) = (0,6,2)$.

$$\overrightarrow{AB} \times \overrightarrow{AC} = \begin{vmatrix} \mathbf{i} & \mathbf{j} & \mathbf{k} \\ -4 & 2 & 5 \\ 0 & 6 & 2 \end{vmatrix} = (-26,8,-24), \text{ so area} = \frac{1}{2}\|(-26,8,-24)\| = \sqrt{329}.$$

(c) $\overrightarrow{AB} = (0,5,2) - (1,0,0) = (-1,5,2)$ and $\overrightarrow{AC} = (3,-4,8) - (1,0,0) = (2,-4,8)$.

$$\overrightarrow{AB} \times \overrightarrow{AC} = \begin{vmatrix} \mathbf{i} & \mathbf{j} & \mathbf{k} \\ -1 & 5 & 2 \\ 2 & -4 & 8 \end{vmatrix} = (48,12,-6), \text{ so area} = \frac{1}{2}\|(48,12,-6)\| = 3\sqrt{69}.$$

6. (a) $\mathbf{u} = \overrightarrow{AB} = (4,8,1) - (1,2,5) = (3,6,-4)$, $\mathbf{v} = \overrightarrow{AC} = (-3,2,3) - (1,2,5) = (-4,0,-2)$, and

$\mathbf{w} = \overrightarrow{AD} = (0,3,9) - (1,2,5) = (-1,1,4)$.

$$\mathbf{w}\cdot(\mathbf{u} \times \mathbf{v}) = \begin{vmatrix} 3 & 6 & -4 \\ -4 & 0 & -2 \\ -1 & 1 & 4 \end{vmatrix} = 130, \text{ so the volume} = |130| = 130.$$

(c) $\mathbf{u} = \overrightarrow{AB} = (-3,1,4) - (0,1,2) = (-3,0,2)$, $\mathbf{v} = \overrightarrow{AC} = (5,2,3) - (0,1,2) = (5,1,1)$,

and $\mathbf{w} = \overrightarrow{AD} = (-3,-2,1) - (0,1,2) = (-3,-3,-1)$.

$$\mathbf{w}\cdot(\mathbf{u} \times \mathbf{v}) = \begin{vmatrix} -3 & 0 & 2 \\ 5 & 1 & 1 \\ -3 & -3 & -1 \end{vmatrix} = -30, \text{ so the volume} = |-30| = 30.$$

8. $(\mathbf{i} \times \mathbf{j})\cdot\mathbf{k} = (0\text{x}0 - 0\text{x}1, 0\text{x}0 - 1\text{x}0, 1\text{x}1 - 0\text{x}0)\cdot(0,0,1) = (0,0,1)\cdot(0,0,1) = 0+0+1 = 1.$

12. $c(\mathbf{u} \times \mathbf{v}) = c(u_2v_3 - u_3v_2, u_3v_1 - u_1v_3, u_1v_2 - u_2v_1)$
$= (cu_2v_3 - cu_3v_2, cu_3v_1 - cu_1v_3, cu_1v_2 - cu_2v_1) = c\mathbf{u} \times \mathbf{v}$
$= (u_2cv_3 - u_3cv_2, u_3cv_1 - u_1cv_3, u_1cv_2 - u_2cv_1) = \mathbf{u} \times c\mathbf{v}.$

14. $\mathbf{u} = (u_1,u_2,u_3)$, $\mathbf{v} = (v_1,v_2,v_3)$, and $\mathbf{w} = (w_1,w_2,w_3)$. From exercise 11, $\mathbf{u} \times (\mathbf{v} \times \mathbf{w}) = (\mathbf{u}\cdot\mathbf{w})\mathbf{v} - (\mathbf{u}\cdot\mathbf{v})\mathbf{w}$. Therefore $\mathbf{v} \times (\mathbf{w} \times \mathbf{u}) = (\mathbf{v}\cdot\mathbf{u})\mathbf{w} - (\mathbf{v}\cdot\mathbf{w})\mathbf{u}$ and $\mathbf{w} \times (\mathbf{u} \times \mathbf{v}) = (\mathbf{w}\cdot\mathbf{v})\mathbf{u} - (\mathbf{w}\cdot\mathbf{u})\mathbf{v}$. Adding gives $\mathbf{u} \times (\mathbf{v} \times \mathbf{w}) + \mathbf{v} \times (\mathbf{w} \times \mathbf{u}) + \mathbf{w} \times (\mathbf{u} \times \mathbf{v})$ $= (\mathbf{u}\cdot\mathbf{w})\mathbf{v} - (\mathbf{u}\cdot\mathbf{v})\mathbf{w} + (\mathbf{v}\cdot\mathbf{u})\mathbf{w} - (\mathbf{v}\cdot\mathbf{w})\mathbf{u} + (\mathbf{w}\cdot\mathbf{v})\mathbf{u} - (\mathbf{w}\cdot\mathbf{u})\mathbf{v} = \mathbf{0}$.

15. $\mathbf{u} = (u_1,u_2,u_3)$, $\mathbf{v} = (v_1,v_2,v_3)$, and $\mathbf{w} = (w_1,w_2,w_3)$.

$$\mathbf{u}\cdot(\mathbf{v} \times \mathbf{w}) = \begin{vmatrix} u_1 & u_2 & u_3 \\ v_1 & v_2 & v_3 \\ w_1 & w_2 & w_3 \end{vmatrix} = 0$$ if and only if the row vectors **u**, **v** and **w** are linearly dependent, i.e., if and only if one is a linear combination of the other two. So $\mathbf{u}\cdot(\mathbf{v} \times \mathbf{w}) = 0$ if and only if one vector lies in the plane of the other two.

Exercise Set 5.4, page 227

1. (a) $P_0 = (x_0, y_0, z_0) = (1,-2,4)$ and $(a,b,c) = (1,1,1)$.

   point-normal form: $(x - 1) + (y + 2) + (z - 4) = 0$
   general form: $x + y + z - 3 = 0$

   (c) $P_0 = (x_0, y_0, z_0) = (0,0,0)$ and $(a,b,c) = (1,2,3)$.

   point-normal form: $x + 2y + 3z = 0$
   general form: $x + 2y + 3z = 0$

2. (a) $\overrightarrow{P_1P_2} = (1,0,2) - (1,-2,3) = (0,2,-1)$ and $\overrightarrow{P_1P_3} = (-1,4,6) - (1,-2,3) = (-2,6,3)$.

$$\overrightarrow{P_1P_2} \times \overrightarrow{P_1P_3} = \begin{vmatrix} \mathbf{i} & \mathbf{j} & \mathbf{k} \\ 0 & 2 & -1 \\ -2 & 6 & 3 \end{vmatrix} = (12,2,4)$$ is normal to the plane.

   $P_0 = (x_0, y_0, z_0) = (1,-2,3)$ and $(a,b,c) = (12,2,4)$.

   point-normal form: $12(x - 1) + 2(y + 2) + 4(z - 3) = 0$
   general form: $12x + 2y + 4z - 20 = 0$ or $6x + y + 2z - 10 = 0$

(c) $\overrightarrow{P_1P_2} = (3,5,4) - (-1,-1,2) = (4,6,2)$ and $\overrightarrow{P_1P_3} = (1,2,5) - (-1,-1,2) = (2,3,3)$.

$$\overrightarrow{P_1P_2} \times \overrightarrow{P_1P_3} = \begin{vmatrix} \mathbf{i} & \mathbf{j} & \mathbf{k} \\ 4 & 6 & 2 \\ 2 & 3 & 3 \end{vmatrix} = (12,-8,0) \text{ is normal to the plane.}$$

$P_0 = (x_0, y_0, z_0) = (-1,-1,2)$ and $(a,b,c) = (12,-8,0)$.

point-normal form: $12(x + 1) - 8(y + 1) = 0$
general form: $12x - 8y + 4 = 0$ or $3x - 2y + 1 = 0$

3. $(3,-2,4)$ is normal to the plane $3x - 2y + 4z - 3 = 0$ and $(-6,4,-8)$ is normal to the plane $-6x + 4y - 8z + 7 = 0$. $(-6,4,-8) = -2(3,-2,4)$, so the normals are parallel and therefore the planes are parallel.

5. $(2,-3,1)$ is normal to the plane $2x - 3y + z + 4 = 0$ and therefore to the parallel plane passing through $(1,2,-3)$.

   $P_0 = (x_0, y_0, z_0) = (1,2,-3)$ and $(a,b,c) = (2,-3,1)$.

   point-normal form: $2(x - 1) - 3(y - 2) + (z + 3) = 0$
   general form: $2x - 3y + z + 7 = 0$

6. (a) $P_0 = (x_0, y_0, z_0) = (1,2,3)$ and $(a,b,c) = (-1,2,4)$.

   parametric equations: $x = 1 - t,\ y = 2 + 2t,\ z = 3 + 4t,\ -\infty < t < \infty$

   symmetric equations: $-x + 1 = \frac{y-2}{2} = \frac{z-3}{4}$

   (c) $P_0 = (x_0, y_0, z_0) = (0,0,0)$ and $(a,b,c) = (-2,-3,5)$.

   parametric equations: $x = -2t,\ y = -3t,\ z = 5t,\ -\infty < t < \infty$

   symmetric equations: $-\frac{x}{2} = -\frac{y}{3} = \frac{z}{5}$

7. $P_0 = (x_0, y_0, z_0) = (1,2,-4)$ and $(a,b,c) = (2,3,1)$. $(x - 1, y - 2, z + 4) = t(2,3,1)$; thus the parametric equations are $x = 1 + 2t, y = 2 + 3t, z = -4 + t, -\infty < t < \infty$.

9. $P_0 = (x_0, y_0, z_0) = (4,-1,3)$ and $(a,b,c) = (2,-1,4)$. $(x - 4, y + 1, z - 3) = t(2,-1,4)$; thus the parametric equations are $x = 4 + 2t,\ y = -1 - t,\ z = 3 + 4t, -\infty < t < \infty$.

11. $\overrightarrow{P_1P_2} = (-1,1,3) - (3,-5,5) = (-4,6,-2)$ and $\overrightarrow{P_1P_3} = (5,-8,6) - (3,-5,5) = (2,-3,1)$.

$\overrightarrow{P_1P_2} = -2\overrightarrow{P_1P_3}$, so the points P_1, P_2, and P_3 are collinear. There are many planes through the line containing these three points.

12. $(-1,2,-4)$ is normal to the plane perpendicular to the given line.

$P_0 = (x_0, y_0, z_0) = (4,-1,5)$ and $(a,b,c) = (-1,2,-4)$.

point-normal form: $-(x - 4) + 2(y + 1) - 4(z - 5) = 0$
general form: $-x + 2y - 4z + 26 = 0$

13. The points on the line are of the form $(1 + t, 14 - t, 2 - t)$. If the line lies in the plane the points will satisfy the equation of the plane. $2(1 + t) - (14 - t) + 3(2 - t) + 6$ $= 2 - 14 + 6 + 6 + 2t + t - 3t = 0$, so the line lies in the plane.

16. The directions of the two lines are $(-4,4,8)$ and $(3,-1,2)$. $(-4,4,8)\cdot(3,-1,2) = 0$, so the lines are orthogonal. To find the point of intersection equate the expressions for x, y, and z: $-1 - 4t = 4 + 3h$, $4 + 4t = 1 - h$, $7 + 8t = 5 + 2h$. This system of equations has the unique solution $t = -1/2, h = -1$, so the point of intersection is $x = 1, y = 2, z = 3$.

17. The directions of the two lines are $(4,5,3)$ and $(1,-3,-2)$. $(4,5,3) \neq c(1,-3,-2)$, so the lines are not parallel. If the lines intersect there will be a solution to the system of equations $1 + 4t = 2 + h$, $2 + 5t = 1 - 3h$, $3 + 3t = -1 - 2h$. However the first equation gives $h = 4t - 1$, and substituting this into the other two equations gives two different values of t. Thus the system of equations has no solution and the lines do not intersect.

18. (a) Let (e,f,g) and (r,s,t) be two points on the plane and let h be any scalar. Then $2e + 3f - 4g = 0$ and $2r + 3s - 4t = 0$, so that $2(e+r) + 3(f+s) - 4(g+t) = 0$ and $2he + 3hf - 4hg = 0$; i.e., $(e,f,g) + (r,s,t) = (e+r, f+s, g+t)$ and $h(e,f,g) = (he,hf,hg)$ are also on the plane. Thus the set of all points on the plane is a subspace of $\mathbf{R}^3$.

A basis for the space must consist of two linearly independent vectors that are orthogonal to (2,3,−4). Two such vectors are (0,4,3) and (2,0,1).

(b) The point (0,0,0) is not on the plane, so the set of all points on the plane is not a subspace of $\mathbf{R}^3$.

Chapter 5 Review Exercises, page 228

1. (a) $(1,2)\cdot(3,-4) = 1\text{x}3 + 2\text{x}-4 = 3 - 8 = -5$.

(b) $(1,-2,3)\cdot(4,2,-7) = 1\text{x}4 + -2\text{x}2 + 3\text{x}-7 = 4 - 4 - 21 = -21$.

(c) $(2,2,-5)\cdot(3,2,-1) = 2\text{x}3 + 2\text{x}2 + -5\text{x}-1 = 6 + 4 + 5 = 15$.

2. (a) $||(1,-4)|| = \sqrt{(1,-4)\cdot(1,-4)} = \sqrt{1\text{x}1+-4\text{x}-4} = \sqrt{17}$.

(b) $||(-2,1,3)|| = \sqrt{(-2,1,3)\cdot(-2,1,3)} = \sqrt{-2\text{x}-2+1\text{x}1+3\text{x}3} = \sqrt{14}$.

(c) $||(1,-2,3,4)|| = \sqrt{(1,-2,3,4)\cdot(1,-2,3,4)} = \sqrt{1\text{x}1+-2\text{x}-2+3\text{x}3+4\text{x}4} = \sqrt{30}$.

3. (a) $\cos\theta = \dfrac{(-1,1)\cdot(2,3)}{||(-1,1)||\;||(2,3)||} = \dfrac{1}{\sqrt{2}\sqrt{13}} = \dfrac{1}{\sqrt{26}}$.

(b) $\cos\theta = \dfrac{(1,2,-3)\cdot(4,1,2)}{||(1,2,-3)||\;||(4,1,2)||} = 0$.

4. The vector (1,2,0) is orthogonal to (−2,1,5).

5. (a) $d = \sqrt{(1-5)^2+(-2-3)^2} = \sqrt{16+25} = \sqrt{41}$.

(b) $d = \sqrt{(3-7)^2+(2-1)^2+(1-2)^2} = \sqrt{18} = 3\sqrt{2}$.

(c) $d^2 = (3-4)^2+(1-1)^2+(-1-6)^2+(2-2)^2 = 50$, so $d = \sqrt{50} = 5\sqrt{2}$.

6. $\|c(1,2,3)\| = |c|\ \|(1,2,3)\| = |c|\sqrt{(1,2,3)\cdot(1,2,3)} = |c|\sqrt{1\text{x}1+2\text{x}2+3\text{x}3} = |c|\sqrt{14}$, so if

$\|c(1,2,3)\| = 196$ then $c = \pm\frac{196}{\sqrt{14}} = \pm 14\sqrt{14}$.

7. $\text{proj}_{\mathbf{u}}\mathbf{v} = \frac{\mathbf{v}\cdot\mathbf{u}}{\mathbf{u}\cdot\mathbf{u}}\mathbf{u}$.

(a) $\text{proj}_{\mathbf{u}}\mathbf{v} = \frac{(1,3)\cdot(2,4)}{(2,4)\cdot(2,4)}(2,4) = \frac{7}{10}(2,4) = \left(\frac{7}{5}, \frac{14}{5}\right)$.

(b) $\text{proj}_{\mathbf{u}}\mathbf{v} = \frac{(-1,3,4)\cdot(-1,2,4)}{(-1,2,4)\cdot(-1,2,4)}(-1,2,4) = \frac{23}{21}(-1,2,4) = \left(-\frac{23}{21}, \frac{46}{21}, \frac{92}{21}\right)$.

8. $\mathbf{u}_1 = (1,2,3,-1)$, $\mathbf{u}_2 = (2,0,-1,1) - \frac{(2,0,-1,1)\cdot(1,2,3,-1)}{(1,2,3,-1)\cdot(1,2,3,-1)}(1,2,3,-1)$

$= (2,0,-1,1) - \frac{-2}{15}(1,2,3,-1) = \left(\frac{32}{15}, \frac{4}{15}, \frac{-9}{15}, \frac{13}{15}\right)$, and

$\mathbf{u}_3 = (3,2,0,1) - \frac{(3,2,0,1)\cdot(1,2,3,-1)}{(1,2,3,-1)\cdot(1,2,3,-1)}(1,2,3,-1) - \frac{(3,2,0,1)\cdot(32,4,-9,13)}{(32,4,-9,13)\cdot(32,4,-9,13)}(32,4,-9,13)$

$= (3,2,0,1) - \frac{2}{5}(1,2,3,-1) - \frac{39}{430}(32,4,-9,13)$

$= \left(\frac{13}{5}, \frac{6}{5}, \frac{-6}{5}, \frac{7}{5}\right) - \frac{39}{86}\left(\frac{32}{5}, \frac{4}{5}, \frac{-9}{5}, \frac{13}{5}\right) = \left(\frac{-26}{86}, \frac{72}{86}, \frac{-33}{86}, \frac{19}{86}\right)$

are an orthogonal basis for the subspace of $\mathbf{R}^4$. $\|\mathbf{u}_1\| = \sqrt{15}$, $\|\mathbf{u}_2\| = \frac{1}{15}\sqrt{1290}$,

and $\|\mathbf{u}_3\| = \frac{1}{86}\sqrt{7310}$, so the set $\left\{\left(\frac{1}{\sqrt{15}}, \frac{2}{\sqrt{15}}, \frac{3}{\sqrt{15}}, \frac{-1}{\sqrt{15}}\right),\right.$

$\left.\left(\frac{32}{\sqrt{1290}}, \frac{4}{\sqrt{1290}}, \frac{-9}{\sqrt{1290}}, \frac{13}{\sqrt{1290}}\right), \left(\frac{-26}{\sqrt{7310}}, \frac{72}{\sqrt{7310}}, \frac{-33}{\sqrt{7310}}, \frac{19}{\sqrt{7310}}\right)\right\}$

is an orthonormal basis for the subspace.

9. $(x,y,x+2y) = x(1,0,1) + y(0,1,2)$. The vectors $(1,0,1)$ and $(0,1,2)$ span the subspace and are linearly independent so they are a basis. $\mathbf{u}_1 = (1,0,1)$ and

$\mathbf{u}_2 = (0,1,2) - \frac{(0,1,2)\cdot(1,0,1)}{(1,0,1)\cdot(1,0,1)}(1,0,1) = (0,1,2) - (1,0,1) = (-1,1,1)$ are an orthogonal basis, and $||\mathbf{u}_1|| = \sqrt{2}$ and $||\mathbf{u}_2|| = \sqrt{3}$, so the set $\left\{\left(\frac{1}{\sqrt{2}}, 0, \frac{1}{\sqrt{2}}\right), \left(\frac{-1}{\sqrt{3}}, \frac{1}{\sqrt{3}}, \frac{1}{\sqrt{3}}\right)\right\}$

is an orthonormal basis for the subspace.

10. $\mathbf{u}_1 = (2,1,1)$ and $\mathbf{u}_2 = (1,-1,3) - \frac{(1,-1,3)\cdot(2,1,1)}{(2,1,1)\cdot(2,1,1)}(2,1,1) = (1,-1,3) - \frac{2}{3}(2,1,1)$

$= \left(\frac{-1}{3}, \frac{-5}{3}, \frac{7}{3}\right)$ are an orthogonal basis for W. $||\mathbf{u}_1|| = \sqrt{6}$ and $||\mathbf{u}_2|| = \frac{5}{3}\sqrt{3}$,

so $\left(\frac{2}{\sqrt{6}}, \frac{1}{\sqrt{6}}, \frac{1}{\sqrt{6}}\right)$ and $\left(\frac{-1}{5\sqrt{3}}, \frac{-5}{5\sqrt{3}}, \frac{7}{5\sqrt{3}}\right)$ are an orthonormal basis.

$$\text{proj}_W(3,1,-2) = \left((3,1,-2)\cdot\left(\frac{2}{\sqrt{6}}, \frac{1}{\sqrt{6}}, \frac{1}{\sqrt{6}}\right)\right)\left(\frac{2}{\sqrt{6}}, \frac{1}{\sqrt{6}}, \frac{1}{\sqrt{6}}\right)$$

$$+ \left((3,1,-2)\cdot\left(\frac{-1}{5\sqrt{3}}, \frac{-5}{5\sqrt{3}}, \frac{7}{5\sqrt{3}}\right)\right)\left(\frac{-1}{5\sqrt{3}}, \frac{-5}{5\sqrt{3}}, \frac{7}{5\sqrt{3}}\right)$$

$$= \frac{5}{6}(2,1,1) + \frac{-22}{75}(-1,-5,7) = \left(\frac{294}{150}, \frac{345}{150}, \frac{-183}{150}\right).$$

11. $W = \{a(1,3,0) + b(0,0,1)\}$. $(1,3,0)$ and $(0,0,1)$ are an orthogonal basis for the subspace,

so $\left(\frac{1}{\sqrt{10}}, \frac{3}{\sqrt{10}}, 0\right)$ and $(0,0,1)$ are an orthonormal basis. Let $\mathbf{x} = (1,2,-4)$.

$$\text{proj}_W\mathbf{x} = \left((1,2,-4)\cdot\left(\frac{1}{\sqrt{10}}, \frac{3}{\sqrt{10}}, 0\right)\right)\left(\frac{1}{\sqrt{10}}, \frac{3}{\sqrt{10}}, 0\right) + ((1,2,-4)\cdot(0,0,1))(0,0,1)$$

$$= \frac{7}{10}(1,3,0) - 4(0,0,1) = \left(\frac{7}{10}, \frac{21}{10}, -4\right).$$ Thus

$$d(\mathbf{x},W) = ||\mathbf{x} - \text{proj}_W\mathbf{x}|| = \left\|(1,2,-4) - \left(\frac{7}{10}, \frac{21}{10}, -4\right)\right\| = \left\|\left(\frac{3}{10}, \frac{-1}{10}, 0\right)\right\| = \frac{1}{10}\sqrt{10}.$$

12. If A is orthogonal the rows of A form an orthonormal set. The rows of A are the columns of A^t, so from the definition of orthogonal matrix A^t is orthogonal. Interchange A and A^t in the argument above to show that if A^t is orthogonal then A is orthogonal.

13. $W = \{a(1,0,1) + b(0,1,-2)\}$. $\mathbf{u}_1 = (1,0,1)$ and $\mathbf{u}_2 = (0,1,-2) - \dfrac{(0,1,-2)\cdot(1,0,1)}{(1,0,1)\cdot(1,0,1)}(1,0,1)$

$= (0,1,-2) - (-1)(1,0,1) = (1,1,-1)$ are an orthogonal basis. The vectors

$\left(\frac{1}{\sqrt{2}}, 0, \frac{1}{\sqrt{2}}\right)$ and $\left(\frac{1}{\sqrt{3}}, \frac{1}{\sqrt{3}}, \frac{-1}{\sqrt{3}}\right)$ are therefore an orthonormal basis.

$\mathbf{v} = (1,3,-1) = \mathbf{w} + \mathbf{w}_\perp$, where

$$\mathbf{w} = \text{proj}_W\mathbf{v} = \left((1,3,-1)\cdot\left(\frac{1}{\sqrt{2}}, 0, \frac{1}{\sqrt{2}}\right)\right)\left(\frac{1}{\sqrt{2}}, 0, \frac{1}{\sqrt{2}}\right)$$

$$+\left((1,3,-1)\cdot\left(\frac{1}{\sqrt{3}}, \frac{1}{\sqrt{3}}, \frac{-1}{\sqrt{3}}\right)\right)\left(\frac{1}{\sqrt{3}}, \frac{1}{\sqrt{3}}, \frac{-1}{\sqrt{3}}\right) = \frac{5}{3}(1,1,-1) = \left(\frac{5}{3}, \frac{5}{3}, \frac{-5}{3}\right) \text{ and}$$

$$\mathbf{w}_\perp = \mathbf{v} - \text{proj}_W\mathbf{v} = (1,3,-1) - \left(\frac{5}{3}, \frac{5}{3}, \frac{-5}{3}\right) = \left(\frac{-2}{3}, \frac{4}{3}, \frac{2}{3}\right).$$

14. Suppose $\mathbf{u}$ and $\mathbf{v}$ are orthogonal and that $a\mathbf{u} + b\mathbf{v} = 0$, so that $a\mathbf{u} = -b\mathbf{v}$. Then $a\mathbf{u}\cdot a\mathbf{u} = a\mathbf{u}\cdot(-b\mathbf{v}) = (-ab)\mathbf{u}\cdot\mathbf{v} = 0$, so $a\mathbf{u} = \mathbf{0}$. Thus $a = 0$ since $\mathbf{u} \neq \mathbf{0}$. In the same way $b = 0$, and so $\mathbf{u}$ and $\mathbf{v}$ are linearly independent.

15. If $\mathbf{u}\cdot\mathbf{v} = 0$ and $\mathbf{u}\cdot\mathbf{w} = 0$ then $\mathbf{u}\cdot(a\mathbf{v} + b\mathbf{w}) = \mathbf{u}\cdot(a\mathbf{v}) + \mathbf{u}\cdot(b\mathbf{w}) = a(\mathbf{u}\cdot\mathbf{v}) + b(\mathbf{u}\cdot\mathbf{w}) = 0$.

16. (a) $\mathbf{u} \times \mathbf{v} = \begin{vmatrix} \mathbf{i} & \mathbf{j} & \mathbf{k} \\ 1 & 1 & -3 \\ 2 & -1 & 4 \end{vmatrix} = (1,-10,-3).$

(b) $2\mathbf{v} \times 3\mathbf{w} = \begin{vmatrix} \mathbf{i} & \mathbf{j} & \mathbf{k} \\ 4 & -2 & 8 \\ 6 & 0 & -6 \end{vmatrix} = (12,72,12).$

(c) $(\mathbf{w} \times \mathbf{u})\cdot\mathbf{v} = \begin{vmatrix} \mathbf{i} & \mathbf{j} & \mathbf{k} \\ 2 & 0 & -2 \\ 1 & 1 & -3 \end{vmatrix}\cdot\mathbf{v} = (2,4,2)\cdot(2,-1,4) = 8.$

(d) $(\mathbf{w}-3\mathbf{u}) \times \mathbf{v} = \begin{vmatrix} \mathbf{i} & \mathbf{j} & \mathbf{k} \\ -1 & -3 & 7 \\ 2 & -1 & 4 \end{vmatrix} = (-5,18,7).$

(e) $\mathbf{u}\cdot(\mathbf{v} \times \mathbf{w}) = \mathbf{u}\cdot\begin{vmatrix} \mathbf{i} & \mathbf{j} & \mathbf{k} \\ 2 & -1 & 4 \\ 2 & 0 & -2 \end{vmatrix} = (1,1,-3)\cdot(2,12,2) = 8.$

(f) $(\mathbf{v} \times \mathbf{u})\cdot(\mathbf{w} \times \mathbf{v}) = (-1,10,3)\cdot(-2,-12,-2) = -124.$

17. $\mathbf{u} = (4,1,0) - (1,-2,3) = (3,3,-3)$ and $\mathbf{v} = (5,-2,1) - (1,-2,3) = (4,0,-2)$.

$$\mathbf{u} \times \mathbf{v} = \begin{vmatrix} \mathbf{i} & \mathbf{j} & \mathbf{k} \\ 3 & 3 & -3 \\ 4 & 0 & -2 \end{vmatrix} = (-6,-6,-12), \text{ so area} = \frac{1}{2}||(-6,-6,-12)|| = \sqrt{54} = 3\sqrt{6}.$$

18. Let $\mathbf{u} = (u_1,u_2,u_3)$, $\mathbf{v} = (v_1,v_2,v_3)$, $\mathbf{w} = (w_1,w_2,w_3)$. $(\mathbf{u} \times \mathbf{v}) + (\mathbf{u} \times \mathbf{w})$

$= (u_2v_3 - u_3v_2, u_3v_1 - u_1v_3, u_1v_2 - u_2v_1) + (u_2w_3 - u_3w_2, u_3w_1 - u_1w_3, u_1w_2 - u_2w_1)$

$= (u_2(v_3+w_3) - u_3(v_2+w_2), u_3(v_1+w_1) - u_1(v_3+w_3), u_1(v_2+w_2) - u_2(v_1+w_1)) = \mathbf{u} \times (\mathbf{v} + \mathbf{w})$.

19. The coordinates of $\mathbf{u} \times \mathbf{v}$ are the determinants $\begin{vmatrix} u_2 & u_3 \\ v_2 & v_3 \end{vmatrix}, \begin{vmatrix} u_3 & u_1 \\ v_3 & v_1 \end{vmatrix}, \begin{vmatrix} u_1 & u_2 \\ v_1 & v_2 \end{vmatrix}$

and the coordinates of $\mathbf{v} \times \mathbf{u}$ are $\begin{vmatrix} v_2 & v_3 \\ u_2 & u_3 \end{vmatrix}, \begin{vmatrix} v_3 & v_1 \\ u_3 & u_1 \end{vmatrix}, \begin{vmatrix} v_1 & v_2 \\ u_1 & u_2 \end{vmatrix}$. Each coordinate of

$\mathbf{u} \times \mathbf{v}$ is therefore the negative of the corresponding coordinate of $\mathbf{v} \times \mathbf{u}$. Thus $\mathbf{u} \times \mathbf{v} = \mathbf{v} \times \mathbf{u}$ if and only if each coordinate of $\mathbf{u} \times \mathbf{v}$ is equal to its negative, i.e. if and only if all the coordinates of $\mathbf{u} \times \mathbf{v}$ are zero.

20. $(\mathbf{u}+\mathbf{v}) \times (\mathbf{u}-\mathbf{v}) = \mathbf{u} \times (\mathbf{u}-\mathbf{v}) + \mathbf{v} \times (\mathbf{u}-\mathbf{v}) = \mathbf{u} \times \mathbf{u} - \mathbf{u} \times \mathbf{v} + \mathbf{v} \times \mathbf{u} - \mathbf{v} \times \mathbf{v}$
$= -\mathbf{u} \times \mathbf{v} + \mathbf{v} \times \mathbf{u} = \mathbf{v} \times \mathbf{u} + \mathbf{v} \times \mathbf{u} = 2(\mathbf{v} \times \mathbf{u}) = 2\mathbf{v} \times \mathbf{u}$.

21. $P_0 = (x_0, y_0, z_0) = (1,-3,2)$ and $(a,b,c) = (1,-1,3)$.

point-normal form: $(x-1) - (y+3) + 3(z-2) = 0$
general form: $x - y + 3z - 10 = 0$

22. $(2,1,4) - (1,-3,1) = (1,4,3)$ and $(-1,4,2) - (1,-3,1) = (-2,7,1)$.

$$\begin{vmatrix} \mathbf{i} & \mathbf{j} & \mathbf{k} \\ 1 & 4 & 3 \\ -2 & 7 & 1 \end{vmatrix} = (-17,-7,15) \text{ is normal to the plane.}$$

$P_0 = (x_0, y_0, z_0) = (1,-3,1)$ and $(a,b,c) = (-17,-7,15)$.

point-normal form: $-17(x-1) - 7(y+3) + 15(z-1) = 0$
general form: $-17x - 7y + 15z - 19 = 0$

23. $P_0 = (x_0, y_0, z_0) = (1,-2,3)$ and $(a,b,c) = (2,-1,4)$.

parametric equations: $x = 1 + 2t,\ y = -2 - t,\ z = 3 + 4t,\ -\infty < t < \infty$

symmetric equations: $\frac{x-1}{2} = -y - 2 = \frac{z-3}{4}$

24. $P_0 = (x_0, y_0, z_0) = (-2,1,3)$ and $(a,b,c) = (1,-3,2)$.

parametric equations: $x = -2 + t,\ y = 1 - 3t,\ z = 3 + 2t,\ -\infty < t < \infty$

symmetric equations: $x + 2 = \frac{-y+1}{3} = \frac{z-3}{2}$

25. Substitute the coordinates of the points on the line into the left side of the equation for the plane: $2(2+t) - (4-3t) + (2-5t) = 4 + 2t - 4 + 3t + 2 - 5t = 2$, and we see that the points on the line lie in the plane, so the line lies in the plane.

Chapter 6

Exercise Set 6.1, page 240

1. axiom 9 $(kl)\mathbf{u} = (kl)\begin{bmatrix} a & b \\ c & d \end{bmatrix} = \begin{bmatrix} (kl)a & (kl)b \\ (kl)c & (kl)d \end{bmatrix} = \begin{bmatrix} k(la) & k(lb) \\ k(lc) & k(ld) \end{bmatrix} = k\begin{bmatrix} la & lb \\ lc & ld \end{bmatrix} = k(l\mathbf{u})$.

2. (a) $(f + g)(x) = f(x) + g(x) = x + 2 + x^2 - 1 = x^2 + x + 1$,

   $(2f)(x) = 2(f(x)) = 2(x + 2) = 2x + 4$, and $(3g)(x) = 3(g(x)) = 3(x^2 - 1) = 3x^2 - 3$.

3. f, g, and h are functions and c and d are scalars.

   axiom 7 $(c(f + g))(x) = c(f(x) + g(x)) = c(f(x)) + c(g(x)) = (cf)(x) + (cg)(x) = (cf + cg)(x)$,

   so $c(f + g) = cf + cg$.

4. (a) $\mathbf{u} + \mathbf{v} = (2 - i, 3 + 4i) + (5, 1 + 3i) = (7 - i, 4 + 7i)$ and

   $c\mathbf{u} = (3 - 2i)(2 - i, 3 + 4i) = (6 - 2 - 3i - 4i, 9 + 8 + 12i - 6i) = (4 - 7i, 17 + 6i)$.

5. (b) This set is not closed under addition. For example if $f(x) = 2$ for $0 \le x \le 1/2$ and $f(x) = 3$ for $1/2 < x \le 1$, and $g(x) = 3$ for $0 \le x \le 1/2$ and $g(x) = 2$ for $1/2 < x \le 1$ then both f and g are discontinuous functions on [0,1], but f + g is the constant function $(f + g)(x) = 5$, which is continuous.

6. (a) This set is a subspace. $\begin{bmatrix} 0 & a \\ b & 0 \end{bmatrix} + \begin{bmatrix} 0 & c \\ d & 0 \end{bmatrix} = \begin{bmatrix} 0 & a+c \\ b+d & 0 \end{bmatrix}$ and

   $k\begin{bmatrix} 0 & a \\ b & 0 \end{bmatrix} = \begin{bmatrix} 0 & ka \\ kb & 0 \end{bmatrix}$.

   (c) For $k \ne 0$ or 1, $k\begin{bmatrix} a & a^2 \\ b & b^2 \end{bmatrix} = \begin{bmatrix} ka & ka^2 \\ kb & kb^2 \end{bmatrix} \ne \begin{bmatrix} ka & (ka)^2 \\ kb & (kb)^2 \end{bmatrix}$, so the set is not a subspace.

(f) The matrices $\begin{bmatrix} 1 & 1 \\ 0 & 1 \end{bmatrix}$ and $\begin{bmatrix} -1 & 0 \\ 0 & 2 \end{bmatrix}$ are invertible but their sum $\begin{bmatrix} 0 & 1 \\ 0 & 3 \end{bmatrix}$ is not, so the set is not a subspace.

7. (a) $\begin{bmatrix} a & b & 0 \\ c & d & 0 \end{bmatrix} + \begin{bmatrix} e & f & 0 \\ g & h & 0 \end{bmatrix} = \begin{bmatrix} a+e & b+f & 0 \\ c+g & d+h & 0 \end{bmatrix}$ and $k\begin{bmatrix} a & b & 0 \\ c & d & 0 \end{bmatrix} = \begin{bmatrix} ka & kb & 0 \\ kc & kd & 0 \end{bmatrix}$,

so the set is closed under addition and scalar multiplication and is therefore a subspace.

8. Every element of P_2 is an element of P_3. Both P_2 and P_3 are vector spaces with the same operations and the same set of scalars, so P_2 is a subspace of P_3.

10. If $f(x) = ax^2 + bx + 3$ then $2f(x) = 2ax^2 + 2bx + 6$, which is not in S, so S is not closed under scalar multiplication and therefore is not a subspace of P_2.

11. $\mathbf{R}^n$ is a subset of $\mathbf{C}^n$, but the scalars for $\mathbf{C}^n$ are the complex numbers. If a nonzero element of $\mathbf{R}^n$ is multiplied by a complex number such as $1 + 2i$, the result is an element of $\mathbf{C}^n$ which is not an element of $\mathbf{R}^n$. Thus $\mathbf{R}^n$ is not closed under multiplication by the scalars in $\mathbf{C}^n$.

12. A subspace must be closed under scalar multiplication, so for any vector **v** in the subspace $0\mathbf{v} = \mathbf{0}$ must be in the subspace.

(a) $\begin{bmatrix} 0 & 0 \\ 0 & 0 \end{bmatrix} \neq \begin{bmatrix} a & 1 \\ b & c \end{bmatrix}$, so the zero vector $\begin{bmatrix} 0 & 0 \\ 0 & 0 \end{bmatrix}$ is not in the subset.

13. (a) $(f+g)(x) = f(x) + g(x)$ so $(f+g)(0) = 0 + 0 = 0$, and $(cf)(x) = c(f(x))$, so $(cf)(0) = c0 = 0$. Thus this subset is a subspace.

(b) $(f+g)(0) = f(0) + g(0) = 3 + 3 = 6$, so the subset is not closed under addition, and the subset is not a subspace.

14. (a) $c\mathbf{0} \underset{\text{axiom5}}{=} c\mathbf{0} + \mathbf{0} \underset{\text{axiom6}}{=} c\mathbf{0} + (c\mathbf{0} + -(c\mathbf{0})) \underset{\text{axiom4}}{=} (c\mathbf{0} + c\mathbf{0}) + -(c\mathbf{0})$

$$\underset{\text{axiom7}}{=} c(\mathbf{0} + \mathbf{0}) + -(c\mathbf{0}) \underset{\text{axiom5}}{=} c\mathbf{0} + -(c\mathbf{0}) \underset{\text{axiom6}}{=} \mathbf{0}.$$

15. (a) $\begin{bmatrix} 5 & 7 \\ 5 & -10 \end{bmatrix} = 2\begin{bmatrix} 1 & 2 \\ 3 & -4 \end{bmatrix} - \begin{bmatrix} 0 & 3 \\ 1 & 2 \end{bmatrix} + 3\begin{bmatrix} 1 & 2 \\ 0 & 0 \end{bmatrix}.$

(c) If $\begin{bmatrix} 4 & 1 \\ 7 & 10 \end{bmatrix} = a\begin{bmatrix} 1 & 1 \\ 1 & 1 \end{bmatrix} + b\begin{bmatrix} 3 & 1 \\ 0 & 0 \end{bmatrix} + c\begin{bmatrix} -1 & -1 \\ 2 & 3 \end{bmatrix}$, then $4 = a + 3b - c$,

$1 = a + b - c$, $7 = a + 2c$, and $10 = a + 3c$. This system of equations has no solution, so the first matrix is not a linear combination of the others.

16. (a) $3x^2 + 2x + 9 = 3(x^2 + 1) + 2(x + 3)$.

(b) If $2x^2 + x - 3 = a(x^2 - x + 1) + b(x^2 + 2x - 2)$, then $2 = a + b$, $1 = -a + 2b$, and $-3 = a - 2b$. This system of equations has no solution, so the first function is not a linear combination of the others.

17. (a) $1(2x^2 + 1) + (-1)(x^2 + 4x) + (-1)(x^2 - 4x + 1) = 0$, so the set is linearly dependent.

(c) If $a(x^2 + 3x - 1) + b(x + 3) + c(2x^2 - x + 1) = 0$, then $a + 2c = 0$, $3a + b - c = 0$, and $-a + 3b + c = 0$. This system of homogeneous equations has the unique solution $a = b = c = 0$, so the functions are linearly independent.

(e) If $a\begin{bmatrix} 1 & 2 \\ 3 & 1 \end{bmatrix} + b\begin{bmatrix} 1 & 1 \\ 1 & 1 \end{bmatrix} + c\begin{bmatrix} 2 & 1 \\ 4 & 2 \end{bmatrix} = \begin{bmatrix} 0 & 0 \\ 0 & 0 \end{bmatrix}$, then $a + b + 2c = 0$,

$2a + b + c = 0$, $3a + b + 4c = 0$, and $a + b + 2c = 0$. This system of equations has the unique solution $a = b = c = 0$, so the set is linearly independent.

18. (a) The set $\{x^3, x^2, x, 1\}$ is a basis. The dimension is 4.

(c) $\begin{bmatrix} 1 & 0 & 0 \\ 0 & 0 & 0 \end{bmatrix}, \begin{bmatrix} 0 & 1 & 0 \\ 0 & 0 & 0 \end{bmatrix}, \begin{bmatrix} 0 & 0 & 1 \\ 0 & 0 & 0 \end{bmatrix}, \begin{bmatrix} 0 & 0 & 0 \\ 1 & 0 & 0 \end{bmatrix}, \begin{bmatrix} 0 & 0 & 0 \\ 0 & 1 & 0 \end{bmatrix}$, and $\begin{bmatrix} 0 & 0 & 0 \\ 0 & 0 & 1 \end{bmatrix}$

are a basis. The dimension is 6.

(e) $\begin{bmatrix} 1 & 0 \\ 0 & 0 \end{bmatrix}, \begin{bmatrix} 0 & 0 \\ 0 & 1 \end{bmatrix}$, and $\begin{bmatrix} 0 & 1 \\ 1 & 0 \end{bmatrix}$ are a basis. The dimension is 3.

19. (b) If $f(x) = ag(x) + bh(x)$ then $3x^2 + 5x + 1 = (2a+b)x^2 + 3bx + 3a - b$, so that $3 = 2a + b$, $5 = 3b$, and $1 = 3a - b$. This system of equations has no solution, so $f(x)$ is not in the subspace spanned by $g(x)$ and $h(x)$.

(d) $f(x)$ and $g(x)$ are linearly independent and $h(x) = g(x) - f(x)$, so the set $\{f(x), g(x)\}$ is a basis.

20. (a) A basis for P_2 must consist of three linearly independent elements of P_2. $f(x) + g(x) - h(x) = 0$, so the three given functions are not linearly independent and therefore not a basis for P_2.

(c) $\begin{bmatrix} 1 & 2 \\ 0 & 1 \end{bmatrix} - \begin{bmatrix} 3 & 4 \\ 1 & 1 \end{bmatrix} + 2\begin{bmatrix} 1 & 2 \\ 1 & 1 \end{bmatrix} - \begin{bmatrix} 0 & 2 \\ 1 & 2 \end{bmatrix} = \begin{bmatrix} 0 & 0 \\ 0 & 0 \end{bmatrix}$, so these matrices are

linearly dependent and therefore not a basis for M_{22}.

(e) The set $\{(1,0), (0,1)\}$ is a basis for $\mathbf{C}^2$, so the dimension of $\mathbf{C}^2$ is 2, and any set of two linearly independent vectors in $\mathbf{C}^2$ is a basis. The given vectors are linearly independent since neither is a multiple of the other, so they are a basis.

21. If $af + bg + ch = 0$ then $af' + bg' + ch' = 0$ and $af'' + bg'' + ch'' = 0$. If the Wronskian is not the zero function this system of three homogeneous equations has the unique solution $a = b = c = 0$, so that f, g, and h are linearly independent.

22. (c) $$\begin{vmatrix} \sin x & \cos x & x\sin x \\ \cos x & -\sin x & \sin x + x\cos x \\ -\sin x & -\cos x & -x\sin x + 2\cos x \end{vmatrix} = \begin{vmatrix} \sin x & \cos x & x\sin x \\ \cos x & -\sin x & \sin x + x\cos x \\ 0 & 0 & 2\cos x \end{vmatrix}$$

$= -2\cos x \neq$ the zero function, so $\sin x$, $\cos x$, and $x \sin x$ are linearly independent.

23. If **u** and **v** are both in V but not in U, then it is possible for **u** + **v** to be in U. For example, let U be the subspace of V = $\mathbf{R}^2$ with basis (1,1). Then (1,0) and (0,1) are in $\mathbf{R}^2$ but not in U. However (1,0) + (0,1) = (1,1) is in U.

If **u** is in U then −**u** is in U and (**u** + **v**) + −**u** = **v**, so if **u** + **v** and **u** are in U then so is **v**. Therefore if **u** is in U and **v** is not then their sum is not in U.

If **cu** is in U then $\frac{1}{c}\mathbf{cu} = \mathbf{u}$ is in U, so if **u** is not in U then **cu** is also not in U for all $c \neq 0$.

25. (a) Let U be the subspace of $\mathbf{R}^2$ with basis (1,0) and let V be the subspace of $\mathbf{R}^2$ with basis (0,1). (1,0) and (0,1) are in the union of U and V but their sum (1,1) is not.

Exercise Set 6.2, page 248

3. $<\mathbf{u},\mathbf{u}> = 2x_1x_1 - x_2x_2 = 2x_1^2 - x_2^2 < 0$ for the vector **u** = (1,2), so condition 4 of the definition is not satisfied.

4. $\mathbf{u} = \begin{bmatrix} a & b \\ c & d \end{bmatrix}, \mathbf{v} = \begin{bmatrix} e & f \\ g & h \end{bmatrix}$, and $\mathbf{w} = \begin{bmatrix} j & k \\ l & m \end{bmatrix}$.

$<\mathbf{u}+\mathbf{v},\mathbf{w}> = (a+e)j + (b+f)k + (c+g)l + (d+h)m = aj + ej + bk + fk + cl + gl + dm + hm$

$= aj + bk + cl + dm + ej + fk + gl + hm = <\mathbf{u},\mathbf{w}> + <\mathbf{v},\mathbf{w}>$, so axiom 2 is satisfied.

5. (a) $<\mathbf{u},\mathbf{v}> = 1\text{x}4 + 2(2\text{x}1) + 3(0\text{x}{-3}) + 4(-3\text{x}2) = 4 + 4 + 0 - 24 = -16$.

8. (a) $<f,g> = \int_0^1 (2x+1)(3x-2)\,dx = \int_0^1 (6x^2-x-2)\,dx = \left[2x^3 - \frac{1}{2}x^2 - 2x\right]_0^1 = -\frac{1}{2}$.

(c) $<f,g> = \int_0^1 (x^2+3x-2)(x+1)\,dx = \int_0^1 (x^3+4x^2+x-2)\,dx = \left[\frac{1}{4}x^4 + \frac{4}{3}x^3 + \frac{1}{2}x^2 - 2x\right]_0^1$

$= \frac{1}{12}$.

9. (a) $\|f\|^2 = \int_0^1 (4x-2)^2\,dx = \int_0^1 (16x^2-16x+4)\,dx = \left[\frac{16}{3}x^3 - 8x^2 + 4x\right]_0^1 = \frac{4}{3}$, so $\|f\| = \frac{2}{\sqrt{3}}$.

(c) $\|f\|^2 = \int_0^1 (3x^2+2)^2\,dx = \int_0^1 (9x^4+12x^2+4)\,dx = \left[\frac{9}{5}x^5 + 4x^3 + 4x\right]_0^1 = \frac{49}{5}$, so $\|f\| = \frac{7}{\sqrt{5}}$.

10. $\langle f,g\rangle = \int_0^1 (x^2)(4x-3)\,dx = \int_0^1 (4x^3-3x^2)\,dx = [\,x^4-x^3\,]_0^1 = 0$, so the functions are orthogonal.

12. $\langle f,g\rangle = \int_0^1 (5x^2)(9x)\,dx = \int_0^1 (45x^3)\,dx = \left[\frac{45}{4}x^4\right]_0^1 = \frac{45}{4}$,

$\|f\|^2 = \int_0^1 (5x^2)^2\,dx = \int_0^1 (25x^4)\,dx = [5x^5\,]_0^1 = 5$, so $\|f\| = \sqrt{5}$, and

$\|g\|^2 = \int_0^1 (9x)^2\,dx = \int_0^1 (81x^2)\,dx = [27x^3\,]_0^1 = 27$, so $\|g\| = 3\sqrt{3}$. Thus

$\cos\theta = \frac{45}{4\ \sqrt{5}\ 3\sqrt{3}} = \frac{3x15}{4x3\sqrt{15}} = \frac{\sqrt{15}}{4}$.

13. $\langle f,g\rangle = \int_0^1 (6x+12)(ax+b)\,dx = \int_0^1 (6ax^2+12ax+6bx+12b)\,dx$

$= [2ax^3+6ax^2+3bx^2+12bx\,]_0^1 = 2a+6a+3b+12b$, so any choice of a and b that makes

$8a+15b = 0$ will do. Let $a = 15$ and $b = -8$. $g(x) = 15x - 8$ is orthogonal to $f(x)$.

14. $d(f,g) = \|f-g\| = \|2x+4\|$. $\|2x+4\|^2 = \int_0^1 (2x+4)^2\,dx = \int_0^1 (4x^2+16x+16)\,dx$

$= \left[\frac{4}{3}x^3 + 8x^2 + 16x\right]_0^1 = \frac{4}{3} + 8 + 16 = \frac{76}{3}$, so $d(f,g) = \sqrt{\frac{76}{3}}$.

15. $d(f,g)^2 = \| f-g \|^2 = \|-2x+3\|^2 = \int_0^1 (-2x+3)^2\,dx = \int_0^1 (4x^2-12x+9)\,dx$

$= \left[\frac{4}{3}x^3 - 6x^2 + 9x\right]_0^1 = \frac{4}{3} - 6 + 9 = \frac{13}{3}$, and $d(f,h)^2 = \| f-h \|^2 = \|3x-4\|^2 = \int_0^1 (3x-4)^2\,dx$

$= \int_0^1 (9x^2-24x+16)\,dx = \left[3x^3-12x^2+16x \right]_0^1 = 7 > \frac{13}{3}$, so g is closer to f.

16. (a) $\left\langle \begin{bmatrix} 1 & 2 \\ 3 & 4 \end{bmatrix}, \begin{bmatrix} -2 & 0 \\ -3 & 5 \end{bmatrix} \right\rangle = 1\text{x}-2 + 2\text{x}0 + 3\text{x}-3 + 4\text{x}5 = -2 + 0 - 9 + 20 = 9.$

17. (a) $\left\| \begin{bmatrix} 1 & 2 \\ 3 & 4 \end{bmatrix} \right\|^2 = 1^2 + 2^2 + 3^2 + 4^2 = 1 + 4 + 9 + 16 = 30$, so $\left\| \begin{bmatrix} 1 & 2 \\ 3 & 4 \end{bmatrix} \right\| = \sqrt{30}$.

(c) $\left\| \begin{bmatrix} 5 & -2 \\ -1 & 6 \end{bmatrix} \right\|^2 = 5^2 + (-2)^2 + (-1)^2 + 6^2 = 25 + 4 + 1 + 36 = 66$, so

$\left\| \begin{bmatrix} 5 & -2 \\ -1 & 6 \end{bmatrix} \right\| = \sqrt{66}$.

18. (a) $\left\langle \begin{bmatrix} 1 & 2 \\ -1 & 1 \end{bmatrix}, \begin{bmatrix} 2 & 4 \\ 3 & -7 \end{bmatrix} \right\rangle = 1\text{x}2 + 2\text{x}4 + -1\text{x}3 + 1\text{x}-7 = 2 + 8 - 3 - 7 = 0$, so the matrices are orthogonal.

19. $\left\langle \begin{bmatrix} 1 & 2 \\ 3 & 4 \end{bmatrix}, \begin{bmatrix} a & b \\ c & d \end{bmatrix} \right\rangle = a + 2b + 3c + 4d$, so any choice of a, b, c, and d satisfying the condition $a + 2b + 3c + 4d = 0$ will do. One such choice is $a = -3$, $b = 2$, $c = 1$, $d = -1$.

20. (a) $d\left(\begin{bmatrix} 4 & 0 \\ -1 & 3 \end{bmatrix}, \begin{bmatrix} 1 & 1 \\ 1 & 1 \end{bmatrix}\right) = \left\|\begin{bmatrix} 3 & -1 \\ -2 & 2 \end{bmatrix}\right\|.\ \left\|\begin{bmatrix} 3 & -1 \\ -2 & 2 \end{bmatrix}\right\|^2 = 18$, so

$d\left(\begin{bmatrix} 4 & 0 \\ -1 & 3 \end{bmatrix}, \begin{bmatrix} 1 & 1 \\ 1 & 1 \end{bmatrix}\right) = \sqrt{18} = 3\sqrt{2}$.

21. (a) $<\mathbf{u},\mathbf{v}> = (2-i)(3+2i) + (3+2i)(2-i) = 2(6 + 2 - 3i + 4i) = 16 + 2i$.

$||\mathbf{u}||^2 = (2-i)(2+i) + (3+2i)(3-2i) = 4 + 1 + 9 + 4 = 18$, so $||\mathbf{u}|| = \sqrt{18} = 3\sqrt{2}$.

$||\mathbf{v}||^2 = (3-2i)(3+2i) + (2+i)(2-i) = 18$, so $||\mathbf{v}|| = \sqrt{18} = 3\sqrt{2}$.

$d(\mathbf{u},\mathbf{v})^2 = ||\mathbf{u}-\mathbf{v}||^2 = ||(-1+i, 1+i)||^2 = (-1+i)(-1-i) + (1+i)(1-i) = 4$, so $d(\mathbf{u},\mathbf{v}) = 2$.

u and **v** are not orthogonal since $<\mathbf{u},\mathbf{v}> \neq 0$.

(d) $<\mathbf{u},\mathbf{v}> = (2-3i)(1) + (-2+3i)(1) = 2 - 3i - 2 + 3i = 0$.

$||\mathbf{u}||^2 = (2-3i)(2+3i) + (-2+3i)(-2-3i) = 4 + 9 + 4 + 9 = 26$, so $||\mathbf{u}|| = \sqrt{26}$.

$||\mathbf{v}||^2 = (1)(1) + (1)(1) = 2$, so $||\mathbf{v}|| = \sqrt{2}$.

$d(\mathbf{u},\mathbf{v})^2 = ||\mathbf{u}-\mathbf{v}||^2 = ||(1-3i,-3+3i)||^2 = (1-3i)(1+3i) + (-3+3i)(-3-3i) = 28$, so

$d(\mathbf{u},\mathbf{v}) = \sqrt{28} = 2\sqrt{7}$. **u** and **v** are orthogonal since $<\mathbf{u},\mathbf{v}> = 0$.

22. (a) $<\mathbf{u},\mathbf{v}> = (1+4i)(1-i) + (1+i)(-4-i) = 1 + 4 - i + 4i - 4 + 1 - i - 4i = 2-2i$.

$||\mathbf{u}||^2 = (1+4i)(1-4i) + (1+i)(1-i) = 1 + 16 + 1 + 1 = 19$, so $||\mathbf{u}|| = \sqrt{19}$.

$||\mathbf{v}||^2 = (1+i)(1-i) + (-4+i)(-4-i) = 1 + 1 + 16 + 1 = 19$, so $||\mathbf{v}|| = \sqrt{19}$.

$d(\mathbf{u},\mathbf{v})^2 = ||\mathbf{u}-\mathbf{v}||^2 = ||(3i,5)||^2 = (3i)(-3i) + (5)(5) = 34$, so

$d(\mathbf{u},\mathbf{v}) = \sqrt{34}$. **u** and **v** are not orthogonal since $<\mathbf{u},\mathbf{v}> \neq 0$.

(c) $<\mathbf{u},\mathbf{v}> = (1-3i)(2+i) + (1+i)(-5i) = 2 + 3 - 6i + i + 5 - 5i = 10-10i$.

$||\mathbf{u}||^2 = (1-3i)(1+3i) + (1+i)(1-i) = 1 + 9 + 1 + 1 = 12$, so $||\mathbf{u}|| = \sqrt{12} = 2\sqrt{3}$.

$||\mathbf{v}||^2 = (2-i)(2+i) + (5i)(-5i) = 4 + 1 + 25 = 30$, so $||\mathbf{v}|| = \sqrt{30}$.

$d(\mathbf{u},\mathbf{v})^2 = ||\mathbf{u}-\mathbf{v}||^2 = ||(-1-2i,1-4i)||^2 = (-1-2i)(-1+2i) + (1-4i)(1+4i) = 22$,

so $d(\mathbf{u},\mathbf{v}) = \sqrt{22}$. **u** and **v** are not orthogonal since $<\mathbf{u},\mathbf{v}> \neq 0$.

23. $\mathbf{u} = (x_1, \ldots, x_n)$, $\mathbf{v} = (y_1, \ldots, y_n)$, $\mathbf{w} = (z_1, \ldots, z_n)$, and c is a scalar.

$<\mathbf{u},\mathbf{v}> = x_1\overline{y}_1 + \ldots + x_n\overline{y}_n = \overline{y}_1 x_1 + \ldots + \overline{y}_n x_n = \overline{y}_1 \overline{\overline{x}}_1 + \ldots + \overline{y}_n \overline{\overline{x}}_n = <\overline{\mathbf{v}},\overline{\mathbf{u}}> = \overline{<\mathbf{v},\mathbf{u}>}$,

$<\mathbf{u}+\mathbf{v},\mathbf{w}> = (x_1 + y_1)\overline{z}_1 + \ldots + (x_n + y_n)\overline{z}_n = x_1\overline{z}_1 + y_1\overline{z}_1 + \ldots + x_n\overline{z}_n + y_n\overline{z}_n$

$= x_1\overline{z}_1 + \ldots + x_n\overline{z}_n + y_1\overline{z}_1 + \ldots + y_n\overline{z}_n = <\mathbf{u},\mathbf{w}> + <\mathbf{v},\mathbf{w}>$,

$<c\mathbf{u},\mathbf{v}> = cx_1\overline{y}_1 + \ldots + cx_n\overline{y}_n = c(x_1\overline{y}_1 + \ldots + x_n\overline{y}_n) = c<\mathbf{u},\mathbf{v}>$, and

$<\mathbf{u},\mathbf{u}> = x_1\overline{x}_1 + \ldots + x_n\overline{x}_n \geq 0$ since all $x_i\overline{x}_i \geq 0$, and equality holds if and only if all $x_i = 0$.

Thus the given function is an inner product on $\mathbf{C}^n$.

24. $<\mathbf{u},k\mathbf{v}> = x_1\overline{k}\,\overline{y}_1 + \ldots + x_n\overline{k}\,\overline{y}_n = \overline{k}x_1\overline{y}_1 + \ldots + \overline{k}x_n\overline{y}_n = \overline{k}(x_1\overline{y}_1 + \ldots + x_n\overline{y}_n)$

$= \overline{k}<\mathbf{u},\mathbf{v}>$.

26. (a) $<\mathbf{u},\mathbf{v}> = \mathbf{u}A\mathbf{v}^t = (\mathbf{u}A)\cdot\mathbf{v} = \mathbf{v}\cdot(\mathbf{u}A) = \mathbf{v}(\mathbf{u}A)^t = \mathbf{v}A\mathbf{u}^t = <\mathbf{v},\mathbf{u}>$,

$<\mathbf{u}+\mathbf{v},\mathbf{w}> = (\mathbf{u}+\mathbf{v})A\mathbf{w}^t = \mathbf{u}A\mathbf{w}^t + \mathbf{v}A\mathbf{w}^t = <\mathbf{u},\mathbf{w}> + <\mathbf{v},\mathbf{w}>$,

$<c\mathbf{u},\mathbf{v}> = (c\mathbf{u})A\mathbf{v}^t = c(\mathbf{u}A\mathbf{v}^t) = c<\mathbf{u},\mathbf{v}>$, and

$<\mathbf{u},\mathbf{u}> = \mathbf{u}A\mathbf{u}^t > 0$ if $\mathbf{u} \neq \mathbf{0}$ and $= 0$ if $\mathbf{u} = \mathbf{0}$, so this function is an inner product on $\mathbf{R}^n$.

(c) $<\mathbf{u},\mathbf{v}> = \mathbf{u}A\mathbf{v}^t = [1\ 0]\begin{bmatrix} 2 & 0 \\ 0 & 3 \end{bmatrix}\begin{bmatrix} 0 \\ 1 \end{bmatrix} = [2\ 0]\begin{bmatrix} 0 \\ 1 \end{bmatrix} = 0.$

$||\mathbf{u}||^2 = \mathbf{u}A\mathbf{u}^t = [1\ 0]\begin{bmatrix} 2 & 0 \\ 0 & 3 \end{bmatrix}\begin{bmatrix} 1 \\ 0 \end{bmatrix} = [2\ 0]\begin{bmatrix} 1 \\ 0 \end{bmatrix} = 2$, so $||\mathbf{u}|| = \sqrt{2}$.

$$\|\mathbf{v}\|^2 = \mathbf{v}A\mathbf{v}^t = [0\ 1]\begin{bmatrix} 2 & 0 \\ 0 & 3 \end{bmatrix}\begin{bmatrix} 0 \\ 1 \end{bmatrix} = [0\ 3]\begin{bmatrix} 0 \\ 1 \end{bmatrix} = 3, \text{ so } \|\mathbf{v}\| = \sqrt{3}.$$

$$d(\mathbf{u},\mathbf{v})^2 = \|\mathbf{u}-\mathbf{v}\|^2 = [1\ -1]\begin{bmatrix} 2 & 0 \\ 0 & 3 \end{bmatrix}\begin{bmatrix} 1 \\ -1 \end{bmatrix} = [2\ -3]\begin{bmatrix} 1 \\ -1 \end{bmatrix} = 5, \text{ so } d(\mathbf{u},\mathbf{v}) = \sqrt{5}.$$

$$<\mathbf{u},\mathbf{v}> = \mathbf{u}B\mathbf{v}^t = [1\ 0]\begin{bmatrix} 1 & 2 \\ 2 & 3 \end{bmatrix}\begin{bmatrix} 0 \\ 1 \end{bmatrix} = [1\ 2]\begin{bmatrix} 0 \\ 1 \end{bmatrix} = 2.$$

$$\|\mathbf{u}\|^2 = \mathbf{u}B\mathbf{u}^t = [1\ 0]\begin{bmatrix} 1 & 2 \\ 2 & 3 \end{bmatrix}\begin{bmatrix} 1 \\ 0 \end{bmatrix} = [1\ 2]\begin{bmatrix} 1 \\ 0 \end{bmatrix} = 1, \text{ so } \|\mathbf{u}\| = 1.$$

$$\|\mathbf{v}\|^2 = \mathbf{v}B\mathbf{v}^t = [0\ 1]\begin{bmatrix} 1 & 2 \\ 2 & 3 \end{bmatrix}\begin{bmatrix} 0 \\ 1 \end{bmatrix} = [2\ 3]\begin{bmatrix} 0 \\ 1 \end{bmatrix} = 3, \text{ so } \|\mathbf{v}\| = \sqrt{3}.$$

$$d(\mathbf{u},\mathbf{v})^2 = \|\mathbf{u}-\mathbf{v}\|^2 = [1\ -1]\begin{bmatrix} 1 & 2 \\ 2 & 3 \end{bmatrix}\begin{bmatrix} 1 \\ -1 \end{bmatrix} = [-1\ -1]\begin{bmatrix} 1 \\ -1 \end{bmatrix} = 0, \text{ so } d(\mathbf{u},\mathbf{v}) = 0.$$

$$<\mathbf{u},\mathbf{v}> = \mathbf{u}C\mathbf{v}^t = [1\ 0]\begin{bmatrix} 2 & 3 \\ 3 & 5 \end{bmatrix}\begin{bmatrix} 0 \\ 1 \end{bmatrix} = [2\ 3]\begin{bmatrix} 0 \\ 1 \end{bmatrix} = 3.$$

$$\|\mathbf{u}\|^2 = \mathbf{u}C\mathbf{u}^t = [1\ 0]\begin{bmatrix} 2 & 3 \\ 3 & 5 \end{bmatrix}\begin{bmatrix} 1 \\ 0 \end{bmatrix} = [2\ 3]\begin{bmatrix} 1 \\ 0 \end{bmatrix} = 2, \text{ so } \|\mathbf{u}\| = \sqrt{2}.$$

$$\|\mathbf{v}\|^2 = \mathbf{v}C\mathbf{v}^t = [0\ 1]\begin{bmatrix} 2 & 3 \\ 3 & 5 \end{bmatrix}\begin{bmatrix} 0 \\ 1 \end{bmatrix} = [3\ 5]\begin{bmatrix} 0 \\ 1 \end{bmatrix} = 5, \text{ so } \|\mathbf{v}\| = \sqrt{5}.$$

$$d(\mathbf{u},\mathbf{v})^2 = \|\mathbf{u}-\mathbf{v}\|^2 = [1\ -1]\begin{bmatrix} 2 & 3 \\ 3 & 5 \end{bmatrix}\begin{bmatrix} 1 \\ -1 \end{bmatrix} = [-1\ -2]\begin{bmatrix} 1 \\ -1 \end{bmatrix} = 1, \text{ so } d(\mathbf{u},\mathbf{v}) = 1.$$

(d) $\mathbf{u} = (x_1, x_2)$ and $\mathbf{v} = (y_1, y_2)$. $<\mathbf{u},\mathbf{v}> = x_1y_1 + 4x_2y_2 = [x_1\ \ x_2]\begin{bmatrix} a & b \\ c & d \end{bmatrix}\begin{bmatrix} y_1 \\ y_2 \end{bmatrix}$

$= [ax_1 + cx_2\ \ bx_1 + dx_2]\begin{bmatrix} y_1 \\ y_2 \end{bmatrix} = (ax_1 + cx_2)y_1 + (bx_1 + dx_2)y_2$

$= ax_1y_1 + cx_2y_1 + bx_1y_2 + dx_2y_2$ and equating coefficients $a = 1$, $b = 0$, $c = 0$, $d = 4$.

Exercise Set 6.3, page 256

1. $d((x_1,x_2), (0,0))^2 = ||(x_1,x_2)||^2 = x_1^{\ 2} + 4x_2^{\ 2}$, so the equation of the circle with radius 1 and center at the origin is $x_1^{\ 2} + 4x_2^{\ 2} = 1$.

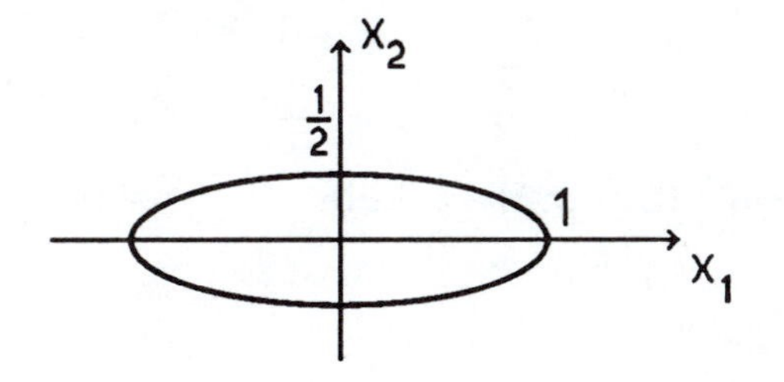

2. (a) $||(1,0)|| = \sqrt{4x1x1+9x0x0} = 2$, $||(0,1)|| = \sqrt{4x0x0+9x1x1} = 3$,

$||(1,1)|| = \sqrt{4x1x1+9x1x1} = \sqrt{13}$, and $||(2,3)|| = \sqrt{4x2x2+9x3x3} = \sqrt{97}$.

(c) $d((1,0),(0,1)) = ||(1,-1)|| = \sqrt{4x1x1+9x-1x-1} = \sqrt{13}$.

3. (c) $d((5,0),(0,4)) = ||(5,-4)|| = \sqrt{5x5+16x-4x-4} = \sqrt{281}$. In Euclidean space the distance is $\sqrt{5^2+4^2} = \sqrt{41}$.

(d) $d((x_1,x_2), (0,0))^2 = ||(x_1,x_2)||^2 = x_1^{\ 2} + 16x_2^{\ 2}$, so the equation of the circle with radius 1 and center at the origin is $x_1^{\ 2} + 16x_2^{\ 2} = 1$.

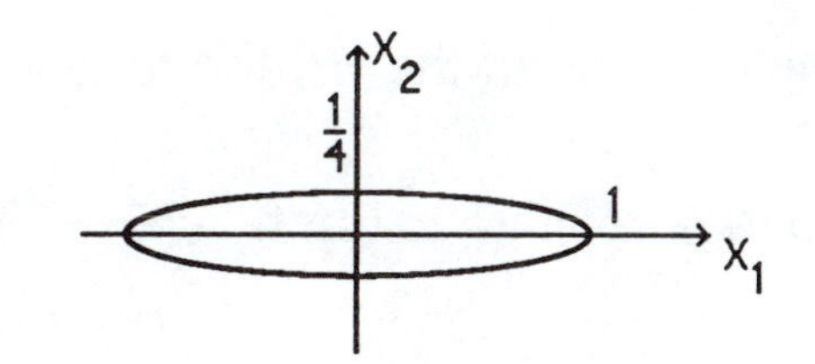

4. $<(x_1, x_2),(y_1, y_2)> = \frac{1}{4}x_1 y_1 + \frac{1}{25}x_2 y_2$.

6. $d(R,Q)^2 = ||(4,0,0,-5)||^2 = |-16 - 0 - 0 + 25| = 9$, so $d(R,Q) = 3$.

8. $<(2,0,0,1),(1,0,0,2)> -2 - 0 - 0 + 2 = 0$, so the vectors are orthogonal.

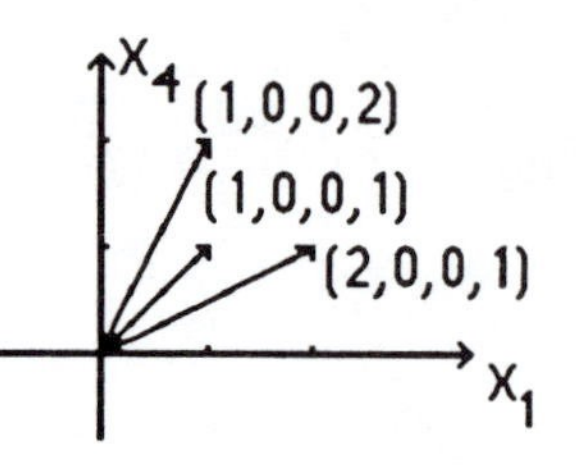

$<(a,0,0,b),(b,0,0,a)> = -ab - 0 - 0 + ba = 0$, so the vectors are orthogonal.

9. $d((x_1 ,0,0,x_4), (0,0,0,0))^2 = ||(x_1 ,0,0,x_4)||^2 = |-x_1{}^2 + x_4{}^2|$, so the equations of the circles with radii $a = 1,2,3$ and center at the origin are $|-x_1{}^2 + x_4{}^2| = a^2$.

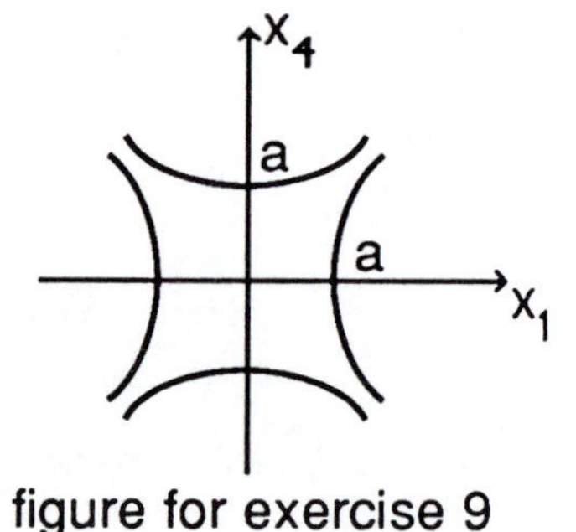

figure for exercise 9

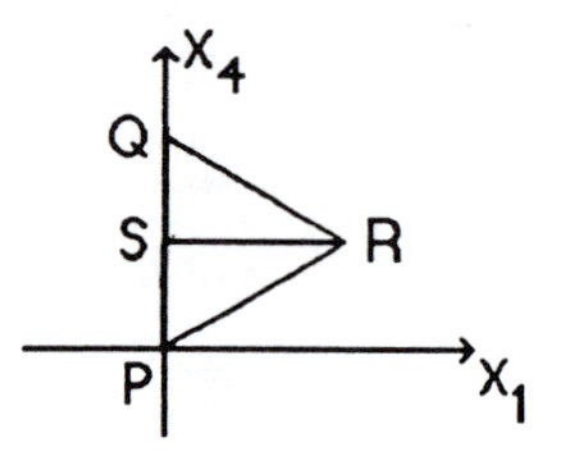

figure for exercises 10 and 12

10. PQ = (0,0,0,20), so PS = (0,0,0,10) and PR = (8,0,0,10). $||PR|| = \sqrt{-64-0-0+100} = 6$, so the duration of the voyage for a person on the spaceship is 2x6 = 12 years. Duration of voyage relative to Earth $= \dfrac{\text{distance in light years}}{\text{speed}}$, so speed $= \dfrac{16}{20}$ = .8 speed of light.

12. Let PS = (0,0,0,t). Then PR = (410,0,0,t) and $||PR||^2 = |-(410)^2-0-0+t^2| = 400$. We assume the speed of the traveler is less than the speed of light, so $t > 410$.

$-(410)^2 + t^2 = 400$, so $t^2 = 168{,}500$ and $t = 410.49$ years. Therefore the duration of the voyage from Earth's point of view is 820.98 years. More than 8 centuries will have passed on Earth.

Exercise Set 6.4, page 264

1. The set $\left\{\frac{1}{\sqrt{2}}, \frac{\sqrt{3}}{\sqrt{2}}x\right\}$ is an orthonormal basis for $P_1[-1,1]$ (see the text).

$$\left\langle x^2, \frac{1}{\sqrt{2}}\right\rangle = \frac{1}{\sqrt{2}}\int_{-1}^{1} x^2\,dx = \left[\frac{1}{3\sqrt{2}}x^3\right]_{-1}^{1} = \frac{2}{3\sqrt{2}} = \frac{\sqrt{2}}{3} \text{ and}$$

$$\left\langle x^2, \frac{\sqrt{3}}{\sqrt{2}}x\right\rangle = \frac{\sqrt{3}}{\sqrt{2}}\int_{-1}^{1} x^3\,dx = \left[\frac{\sqrt{3}}{\sqrt{2}}\frac{x^4}{4}\right]_{-1}^{1} = 0, \text{ so that } \operatorname{proj}_{P_1[-1,1]}x^2 = \frac{\sqrt{2}}{3}\frac{1}{\sqrt{2}} = \frac{1}{3}$$

is the least squares linear approximation to $f(x) = x^2$ over the interval $[-1,1]$.

2. The set $\left\{1, x-\frac{1}{2}\right\}$ is an orthogonal basis for $P_1[0,1]$, $||1|| = 1$, and

$$\left\|x-\frac{1}{2}\right\|^2 = \int_0^1 \left(x^2 - x + \frac{1}{4}\right)dx = \left[\frac{1}{3}x^3 - \frac{1}{2}x^2 + \frac{1}{4}x\right]_0^1 = \frac{1}{3} - \frac{1}{2} + \frac{1}{4} = \frac{1}{12}, \text{ so the set}$$

$\left\{1, 2\sqrt{3}\left(x-\frac{1}{2}\right)\right\}$ is an orthonormal basis for $P_1[0,1]$.

$$<e^x, 1> = \int_0^1 e^x\,dx = [e^x]_0^1 = e - 1, \text{ and } <e^x, 2\sqrt{3}x - \sqrt{3}> = \int_0^1 (2\sqrt{3}\,xe^x - \sqrt{3}\,e^x)\,dx$$

$$= [2\sqrt{3}(xe^x - e^x) - \sqrt{3}e^x]_0^1 = -\sqrt{3}e + 3\sqrt{3}, \text{ so}$$

$$\operatorname{proj}_{P_1[0,1]}e^x = e - 1 + (-\sqrt{3}e + 3\sqrt{3})2\sqrt{3}\left(x-\frac{1}{2}\right) = 4e - 10 + (18 - 6e)x \text{ is the least}$$

squares linear approximation to e^x over the interval $[0,1]$.

4. The set $\left\{1, x-\frac{\pi}{2}\right\}$ is an orthogonal basis for $P_1[0,\pi]$, $||1|| = \pi$, and

$$\left\|x-\frac{\pi}{2}\right\|^2 = \int_0^\pi \left(x^2 - \pi x + \frac{\pi^2}{4}\right)dx = \left[\frac{1}{3}x^3 - \frac{1}{2}\pi x^2 + \frac{\pi^2}{4}x\right]_0^\pi = \left(\frac{1}{3} - \frac{1}{2} + \frac{1}{4}\right)\pi^3 = \frac{\pi^3}{12}, \text{ so}$$

the set $\left\{\pi^{-\frac{1}{2}}, 2\sqrt{3}\,\pi^{-\frac{3}{2}}\left(x-\frac{\pi}{2}\right)\right\}$ is an orthonormal basis for $P_1\,[0,\pi]$.

$<\cos x, \pi^{-\frac{1}{2}}> = \pi^{-\frac{1}{2}}\int_0^\pi \cos x\,dx = \pi^{-\frac{1}{2}}\,[\sin x]_0^\pi = 0$, and

$$\left\langle \cos x,\ 2\sqrt{3}\,\pi^{-\frac{3}{2}}\left(x-\frac{\pi}{2}\right)\right\rangle = 2\sqrt{3}\,\pi^{-\frac{3}{2}}\int_0^\pi \left(x\cos x - \frac{\pi}{2}\cos x\right)dx$$

$= 2\sqrt{3}\,\pi^{-\frac{3}{2}}\left[x\sin x + \cos x - \frac{\pi}{2}\sin x\right]_0^\pi = 2\sqrt{3}\,\pi^{-\frac{3}{2}}\,[-1-1] = -4\sqrt{3}\,\pi^{-\frac{3}{2}}$, so that

$\text{proj}_{P_1[0,\pi]}\cos x = -4\sqrt{3}\,\pi^{-\frac{3}{2}}\,2\sqrt{3}\,\pi^{-\frac{3}{2}}\left(x-\frac{\pi}{2}\right) = -24\pi^{-3}x + 12\pi^{-2}$ is the least

squares linear approximation to cos x over the interval $[0,\pi]$.

5. (a) The set $\left\{\frac{1}{\sqrt{2}}, \frac{\sqrt{3}}{\sqrt{2}}x, \frac{3\sqrt{5}}{2\sqrt{2}}\left(x^2-\frac{1}{3}\right)\right\}$ is an orthonormal basis for $P_2\,[-1,1]$.

$\left\langle e^x, \frac{1}{\sqrt{2}}\right\rangle = \frac{1}{\sqrt{2}}\int_{-1}^1 e^x\,dx = \left[\frac{1}{\sqrt{2}}e^x\right]_{-1}^1 = \frac{1}{\sqrt{2}}(e-e^{-1})$, $\left\langle e^x, \frac{\sqrt{3}}{\sqrt{2}}x\right\rangle$

$= \frac{\sqrt{3}}{\sqrt{2}}\int_{-1}^1 xe^x\,dx = \frac{\sqrt{3}}{\sqrt{2}}[xe^x - e^x]_{-1}^1 = \frac{\sqrt{3}}{\sqrt{2}}(2e^{-1})$, and $\left\langle e^x, \frac{3\sqrt{5}}{2\sqrt{2}}\left(x^2-\frac{1}{3}\right)\right\rangle$

$= \frac{3\sqrt{5}}{2\sqrt{2}}\int_{-1}^1\left(x^2e^x - \frac{1}{3}e^x\right)dx = \frac{3\sqrt{5}}{2\sqrt{2}}\left[x^2e^x - 2xe^x + 2e^x - \frac{1}{3}e^x\right]_{-1}^1$

$= \frac{3\sqrt{5}}{2\sqrt{2}}\left(\frac{2}{3}e - \frac{14}{3}e^{-1}\right)$, so that $\text{proj}_{P_2[-1,1]}e^x$

$= \frac{1}{\sqrt{2}}(e-e^{-1})\frac{1}{\sqrt{2}} + \frac{\sqrt{3}}{\sqrt{2}}(2e^{-1})\frac{\sqrt{3}}{\sqrt{2}}x + \frac{3\sqrt{5}}{2\sqrt{2}}\left(\frac{2}{3}e - \frac{14}{3}e^{-1}\right)\frac{3\sqrt{5}}{2\sqrt{2}}\left(x^2-\frac{1}{3}\right)$

$= -\frac{3}{4}e + \frac{33}{4}e^{-1} + 3e^{-1}x + \frac{15}{4}(e-7e^{-1})x^2$ is the least squares quadratic

approximation to $f(x) = e^x$ over the interval $[-1,1]$.

6. The set $\left\{1,\ 2\sqrt{3}\left(x-\frac{1}{2}\right),\ 6\sqrt{5}\left(x^2-x+\frac{1}{6}\right)\right\}$ is an orthonormal basis for $P_2\,[0,1]$.

$\langle\sqrt{x}, 1\rangle = \frac{2}{3}$, $\langle\sqrt{x}, 2\sqrt{3}x-\sqrt{3}\rangle = \frac{2\sqrt{3}}{15}$, and $\left\langle\sqrt{x}, 6\sqrt{5}\left(x^2-x+\frac{1}{6}\right)\right\rangle$

$= 6\sqrt{5}\int_0^1\left(x^{\frac{5}{2}} - x^{\frac{3}{2}} + \frac{1}{6}x^{\frac{1}{2}}\right)dx = 6\sqrt{5}\left[\frac{2}{7}x^{\frac{7}{2}} - \frac{2}{5}x^{\frac{5}{2}} + \frac{1}{9}x^{\frac{3}{2}}\right]_0^1 = 6\sqrt{5}\left(\frac{2}{7}-\frac{2}{5}+\frac{1}{9}\right)$

$= -\frac{2\sqrt{5}}{105}$, so that $\text{proj}_{P_2[0,1]}\sqrt{x} = \frac{2}{3} + \frac{2\sqrt{3}}{15}\,2\sqrt{3}\left(x-\frac{1}{2}\right) - \frac{2\sqrt{5}}{105}\,6\sqrt{5}\left(x^2-x+\frac{1}{6}\right)$

$= -\frac{4}{7}x^2 + \frac{48}{35}x + \frac{6}{35}$ is the least squares quadratic approximation to $f(x) = \sqrt{x}$ over the interval [0,1].

8. The set $\left\{\pi^{-\frac{1}{2}},\ 2\sqrt{3}\,\pi^{-\frac{3}{2}}\left(x-\frac{\pi}{2}\right),\ 6\sqrt{5}\,\pi^{-\frac{5}{2}}\left(x^2-\pi x+\frac{\pi^2}{6}\right)\right\}$ is an orthonormal basis for $P_2\,[0,\pi]$. $\langle\sin x, \pi^{-\frac{1}{2}}\rangle = \pi^{-\frac{1}{2}}\int_0^\pi \sin x\,dx = \pi^{-\frac{1}{2}}\left[-\cos x\right]_0^\pi = 2\pi^{-\frac{1}{2}}$,

$\left\langle\sin x,\ 2\sqrt{3}\,\pi^{-\frac{3}{2}}\left(x-\frac{\pi}{2}\right)\right\rangle = 2\sqrt{3}\,\pi^{-\frac{3}{2}}\int_0^\pi\left(x\sin x - \frac{\pi}{2}\sin x\right)dx$

$= 2\sqrt{3}\,\pi^{-\frac{3}{2}}\left[-x\cos x + \sin x + \frac{\pi}{2}\cos x\right]_0^\pi = 0$, and

$\left\langle\sin x,\ 6\sqrt{5}\,\pi^{-\frac{5}{2}}\left(x^2-\pi x+\frac{\pi^2}{6}\right)\right\rangle$

$= 6\sqrt{5}\,\pi^{-\frac{5}{2}}\int_0^\pi\left(x^2\sin x - \pi x\sin x + \frac{\pi^2}{6}\sin x\right)dx$

$= 6\sqrt{5}\,\pi^{-\frac{5}{2}}\left[-x^2\cos x + 2x\sin x + 2\cos x + \pi x\cos x - \pi\sin x - \frac{\pi^2}{6}\cos x\right]_0^\pi$

$= 6\sqrt{5}\,\pi^{-\frac{5}{2}}\left(-4 + \frac{\pi^2}{3}\right)$, so

$$\text{proj}_{P_2[0,\pi]} \sin x = 2\pi^{-\frac{1}{2}}\pi^{-\frac{1}{2}} + 6\sqrt{5}\,\pi^{-\frac{5}{2}}\left(-4 + \frac{\pi^2}{3}\right)6\sqrt{5}\,\pi^{-\frac{5}{2}}\left(x^2 - \pi x + \frac{\pi^2}{6}\right)$$

$= 180\pi^{-5}\left(-4 + \frac{\pi^2}{3}\right)x^2 - 180\pi^{-4}\left(-4 + \frac{\pi^2}{3}\right)x - 120\pi^{-3} + 12\pi^{-1}$ is the least squares

quadratic approximation to cos x over the interval $[0,\pi]$.

9. The vectors which form an orthonormal basis over $[-\pi, \pi]$ also form an orthonormal basis over $[0,2\pi]$, so the Fourier approximations can be found in the same way, only with integration over $[0,2\pi]$.

$$a_0 = \frac{1}{2\pi}\int_0^{2\pi} x\,dx = \left[\frac{1}{4\pi}x^2\right]_0^{2\pi} = \pi,$$

$$a_k = \frac{1}{\pi}\int_0^{2\pi} x\cos kx\,dx = \frac{1}{\pi}\left[\frac{x}{k}\sin kx + \frac{1}{k^2}\cos kx\right]_0^{2\pi} = 0, \text{ and}$$

$$b_k = \frac{1}{\pi}\int_0^{2\pi} x\sin kx\,dx = \frac{1}{\pi}\left[-\frac{x}{k}\cos kx + \frac{1}{k^2}\sin kx\right]_0^{2\pi} = \frac{1}{\pi}\left(-\frac{2\pi}{k}\cos 2k\pi\right) = \frac{-2}{k},$$

so the fourth-order Fourier approximation to f(x) over $[0,2\pi]$ is

$$g(x) = \pi + \sum_{k=1}^{4}\frac{-2}{k}\sin kx = \pi - 2\left(\sin x + \frac{1}{2}\sin 2x + \frac{1}{3}\sin 3x + \frac{1}{4}\sin 4x\right).$$

11. $a_0 = \frac{1}{2\pi}\int_0^{2\pi} x^2\,dx = \left[\frac{1}{6\pi}x^3\right]_0^{2\pi} = \frac{4}{3}\pi^2,$

$$a_k = \frac{1}{\pi}\int_0^{2\pi} x^2\cos kx\,dx = \frac{1}{\pi}\left[\frac{x^2}{k}\sin kx - \frac{2}{k}\left(\frac{-x}{k}\cos kx + \frac{1}{k^2}\sin kx\right)\right]_0^{2\pi} = \frac{4}{k^2}, \text{ and}$$

$$b_k = \frac{1}{\pi}\int_0^{2\pi} x^2 \sin kx\,dx = \frac{1}{\pi}\left[\frac{-x^2}{k}\cos kx - \frac{2}{k}\left(\frac{x}{k}\sin kx + \frac{1}{k^2}\cos kx\right)\right]_0^{2\pi} = -\frac{4\pi}{k}, \text{ so}$$

$$g(x) = \frac{4}{3}\pi^2 + \sum_{k=1}^{4}\left(\frac{4}{k^2}\cos kx - \frac{4\pi}{k}\sin kx\right)$$

$$= \frac{4}{3}\pi^2 + 4\left(\cos x + \frac{1}{4}\cos 2x + \frac{1}{9}\cos 3x + \frac{1}{16}\cos 4x\right)$$

$$- 4\pi\left(\sin x + \frac{1}{2}\sin 2x + \frac{1}{3}\sin 3x + \frac{1}{4}\sin 4x\right).$$

13. (1,0,0,0,0,1,1), (0,1,0,0,1,0,1), (0,0,1,0,1,1,0), (0,0,0,1,1,1,1),
(0,1,1,1,1,0,0), (1,0,1,1,0,1,0), (1,1,0,1,0,0,1), (1,1,1,0,0,0,0),
(1,1,0,0,1,1,0), (1,0,1,0,1,0,1), (1,0,0,1,1,0,0), (1,1,1,1,1,1,1),
(0,0,1,1,0,0,1), (0,1,0,1,0,1,0), (0,1,1,0,0,1,1), (0,0,0,0,0,0,0)

14. (a) center = (0,0,1,0,1,1,0)
(1,0,1,0,1,1,0), (0,1,1,0,1,1,0), (0,0,0,0,1,1,0), (0,0,1,1,1,1,0), (0,0,1,0,0,1,0), (0,0,1,0,1,0,0), (0,0,1,0,1,1,1)

(b) center = (1,1,0,1,0,0,1)
(0,1,0,1,0,0,1), (1,0,0,1,0,0,1), (1,1,1,1,0,0,1), (1,1,0,0,0,0,1), (1,1,0,1,1,0,1), (1,1,0,1,0,1,1), (1,1,0,1,0,0,0)

15. (a) (0,0,0,0,0,0,0) (c) (0,0,1,0,1,1,0)

16. (a) Since each vector in V_{23} has 23 components and there are 2 possible values (zero and 1) for each component, there are 2^{23} vectors.

(c) There are 23 ways to change exactly one component of the center vector, 23x22/2 = 253 ways to change exactly two components, and 23x22x21/6 = 1771 ways to change exactly three components. 1 + 23 + 253 + 1771 = 2048 vectors.

Chapter 6 Review Exercises, page 264

1. $(f + g)(x) = 3x - 1 + 2x^2 + 3 = 2x^2 + 3x + 2$, $3f(x) = 3(3x - 1) = 9x - 3$, and $(2f - 3g)(x) = 2(3x - 1) - 3(2x^2 + 3) = -6x^2 + 6x - 11$.

2. $3x^2 + ax - b + 3x^2 + cx - d = 6x^2 + (a+c)x - (b+d)$ is not in S, so the set is not closed under addition and therefore is not a subspace of P_2.

3. $13x^2 + 8x - 21 = 2(2x^2 + x - 3) - 3(-3x^2 - 2x + 5)$.

4. $\begin{bmatrix} 1 & 0 & 0 \\ 0 & 0 & 0 \\ 0 & 0 & 0 \end{bmatrix}, \begin{bmatrix} 0 & 1 & 0 \\ 0 & 0 & 0 \\ 0 & 0 & 0 \end{bmatrix}, \begin{bmatrix} 0 & 0 & 1 \\ 0 & 0 & 0 \\ 0 & 0 & 0 \end{bmatrix}, \begin{bmatrix} 0 & 0 & 0 \\ 0 & 1 & 0 \\ 0 & 0 & 0 \end{bmatrix}, \begin{bmatrix} 0 & 0 & 0 \\ 0 & 0 & 1 \\ 0 & 0 & 0 \end{bmatrix}$, and $\begin{bmatrix} 0 & 0 & 0 \\ 0 & 0 & 0 \\ 0 & 0 & 1 \end{bmatrix}$ are a

basis for the vector space of upper triangular 3x3 matrices.

5. $a(x^2 + 2x - 3) + b(3x^2 + x - 1) + c(4x^2 + 3x - 3) = 0$ if and only if $a + 3b + 4c = 0$, $2a + b + 3c = 0$, and $-3a - b - 3c = 0$. This system of homogeneous equations has the unique solution $a = b = c = 0$. Thus the given functions are linearly independent. The dimension of P_2 is 3, so the three functions are a basis.

6. $\mathbf{u} = (x_1, x_2)$, $\mathbf{v} = (y_1, y_2)$ $\mathbf{w} = (z_1, z_2)$, and c is a scalar.

$<\mathbf{u},\mathbf{v}> = 2x_1 y_1 + 3x_2 y_2 = 2y_1 x_1 + 3y_2 x_2 = <\mathbf{v},\mathbf{u}>$,

$<\mathbf{u}+\mathbf{v},\mathbf{w}> = 2(x_1 + y_1)z_1 + 3(x_2 + y_2)z_2 = 2x_1 z_1 + 2y_1 z_1 + 3x_2 z_2 + 3y_2 z_2$

$= 2x_1 z_1 + 3x_2 z_2 + 2y_1 z_1 + 3y_2 z_2 = <\mathbf{u},\mathbf{w}> + <\mathbf{v},\mathbf{w}>$,

$<c\mathbf{u},\mathbf{v}> = 2cx_1 y_1 + 3cx_2 y_2 = c(2x_1 y_1 + 3x_2 y_2) = c<\mathbf{u},\mathbf{v}>$, and

$<\mathbf{u},\mathbf{u}> = 2x_1 x_1 + 3x_2 x_2 = 2x_1^2 + 3x_2^2 \geq 0$ and equality holds if and only if $2x_1^2 = 0$ and

$3x_2^2 = 0$, i.e. if and only if $x_1 = x_2 = 0$. Thus the given function is an inner product on $\mathbf{R}^2$.

7. $\langle f,g\rangle = \int_0^1 (3x-1)(5x+3)\,dx = \int_0^1 (15x^2+4x-3)\,dx = [\,5x^3+2x^2-3x]_0^1 = 4.$

$\|f\|^2 = \int_0^1 (3x-1)^2\,dx = \int_0^1 (9x^2-6x+1)\,dx = [\,3x^3-3x^2+x\,]_0^1 = 1$, so $\|f\| = 1$.

$\|g\|^2 = \int_0^1 (5x+3)^2\,dx = \int_0^1 (25x^2+30x+9)\,dx = \left[\frac{25}{3}x^3 + 15x^2 + 9x\right]_0^1 = \frac{97}{3}$, so $\|g\| = \frac{\sqrt{97}}{\sqrt{3}}$.

$d(f,g)^2 = \|f-g\|^2 = \|-2x-4\|^2 = \int_0^1 (-2x-4)^2\,dx = 4\int_0^1 (x^2+4x+4)\,dx$

$= 4\left[\frac{x^3}{3} + 2x^2 + 4x\right]_0^1 = 4\left(\frac{19}{3}\right)$, so $d(f,g) = 2\frac{\sqrt{19}}{\sqrt{3}}$.

8. $\|(1,-2)\| = \sqrt{2(1)+3(4)} = \sqrt{14}$, $\|(3,2)\| = \sqrt{2(9)+3(4)} = \sqrt{30}$, and

$d((1,-2),(3,2)) = \|(-2,-4)\| = \sqrt{2(4)+3(16)} = \sqrt{56} = 2\sqrt{14}$.

9. The condition is necessary because for every vector **u** in a subspace $0\mathbf{u} = \mathbf{0}$ must also be in the subspace. It is not sufficient. The subset of $\mathbf{R}^2$ consisting of the two vectors $\mathbf{0} = (0,0)$ and $\mathbf{u} = (0,5)$ contains the zero vector but is not a subspace of $\mathbf{R}^2$.

10. $d((x_1,x_2),(0,0))^2 = \|(x_1,x_2)\|^2 = 2x_1^2 + 3x_2^2$, so the equation of the circle with radius 1 and center at the origin is $2x_1^2 + 3x_2^2 = 1$. In the sketch below $a = 1/\sqrt{2}$ and $b = 1/\sqrt{3}$.

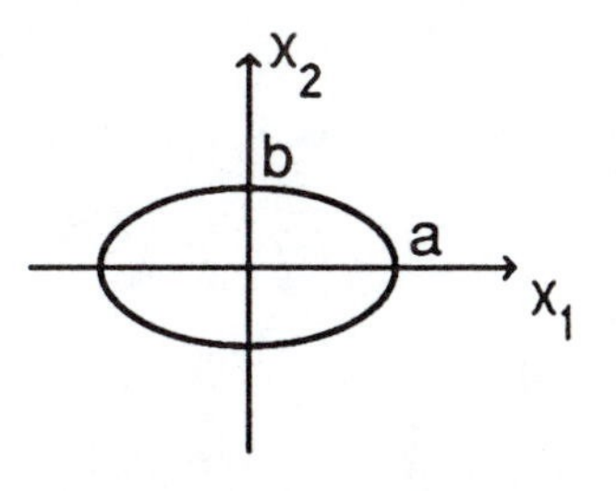

11. The set $\left\{\frac{1}{\sqrt{2}}, \frac{\sqrt{3}}{\sqrt{2}}x\right\}$ is an orthonormal basis for $P_1[-1,1]$ (see the text).

$$\left\langle x^2 + 2x - 1, \frac{1}{\sqrt{2}}\right\rangle = \frac{1}{\sqrt{2}}\int_{-1}^{1} (x^2 + 2x - 1)\,dx = \frac{1}{\sqrt{2}}\left[\frac{x^3}{3} + x^2 - x\right]_{-1}^{1} = -\frac{4}{3\sqrt{2}} \text{ and}$$

$$\left\langle x^2 + 2x - 1, \frac{\sqrt{3}}{\sqrt{2}}x\right\rangle = \frac{\sqrt{3}}{\sqrt{2}}\int_{-1}^{1} (x^3 + 2x^2 - x)\,dx = \frac{\sqrt{3}}{\sqrt{2}}\left[\frac{x^4}{4} + 2\frac{x^3}{3} - \frac{x^2}{2}\right]_{-1}^{1}$$

$$= \frac{\sqrt{3}}{\sqrt{2}}\frac{4}{3}, \text{ so that } \operatorname{proj}_{P_1[-1,1]} x^2 + 2x - 1 = -\frac{4}{3\sqrt{2}}\frac{1}{\sqrt{2}} + \frac{\sqrt{3}}{\sqrt{2}}\frac{4}{3}\frac{\sqrt{3}}{\sqrt{2}}x = 2x - \frac{2}{3}$$

is the least squares linear approximation to $f(x) = x^2 + 2x - 1$ over the interval $[-1,1]$.

12. $$a_0 = \frac{1}{2\pi}\int_0^{2\pi} (2x-1)\,dx = \frac{1}{2\pi}\left[x^2 - x\right]_0^{2\pi} = 2\pi - 1,$$

$$a_k = \frac{1}{\pi}\int_0^{2\pi} (2x-1)\cos kx\,dx = \frac{1}{\pi}\left[\frac{2x}{k}\sin kx + \frac{2}{k^2}\cos kx - \frac{1}{k}\sin kx\right]_0^{2\pi} = 0, \text{ and}$$

$$b_k = \frac{1}{\pi}\int_0^{2\pi} (2x-1)\sin kx\,dx = \frac{1}{\pi}\left[-\frac{2x}{k}\cos kx + \frac{2}{k^2}\sin kx + \frac{1}{k}\cos kx\right]_0^{2\pi} = \frac{-4}{k}, \text{ so the}$$

fourth-order Fourier approximation to $f(x) = 2x-1$ over $[0,2\pi]$ is

$$g(x) = 2\pi - 1 + \sum_{k=1}^{4} \frac{-4}{k}\sin kx = 2\pi - 1 - 4\left(\sin x + \frac{1}{2}\sin 2x + \frac{1}{3}\sin 3x + \frac{1}{4}\sin 4x\right).$$

13. center = (0,1,1,0,0,1,1)
(1,1,1,0,0,1,1), (0,0,1,0,0,1,1), (0,1,0,0,0,1,1), (0,1,1,1,0,1,1), (0,1,1,0,1,1,1),
(0,1,1,0,0,0,1), (0,1,1,0,0,1,0)

Chapter 7

Exercise Set 7.1, page 273

1. $T((x_1,y_1)+(x_2,y_2)) = T(x_1+x_2, y_1+y_2) = (2(x_1+x_2), x_1+x_2-y_1-y_2) = (2x_1+2x_2, x_1-y_1+x_2-y_2)$

   $= (2x_1,x_1-y_1) + (2x_2,x_2-y_2) = T(x_1,y_1) + T(x_2,y_2)$ and $T(c(x,y)) = T(cx,cy) = (2cx,cx-cy)$

   $= c(2x,x-y) = cT(x,y)$, so T is linear. $T(1,2) = (2,-1)$ and $T(-1,4) = (-2,-5)$.

3. $T((x_1,y_1,z_1)+(x_2,y_2,z_2)) = T(x_1+x_2, y_1+y_2, z_1+z_2) = (0,y_1+y_2,0) = (0,y_1,0) + (0,y_2,0)$

   $= T((x_1,y_1,z_1)+T(x_2,y_2,z_2)$ and $T(c(x,y,z) = T(cx,cy,cz) = (0,cy,0) = c(0,y,0) = cT(x,y,z)$, so

   T is linear. The image of (x,y,z) under T is the projection of (x,y,z) on the y axis.

4. (a) $T(c(x,y,z)) = T(cx,cy,cz) = (3cx,(cy)^2) = (3cx,c^2y^2) \neq c(3x,y^2) = cT(x,y,z)$,

   so T is not linear.

6. $T((x_1,y_1,z_1)+(x_2,y_2,z_2)) = T(x_1+x_2, y_1+y_2, z_1+z_2) = (2(x_1+x_2), y_1+y_2) = (2x_1+2x_2, y_1+y_2)$

   $= (2x_1,y_1) = (2x_2,y_2) = T((x_1,y_1,z_1)+T(x_2,y_2,z_2)$ and $T(c(x,y,z) = T(cx,cy,cz) = (2cx,cy)$

   $= c(2x,y) = cT(x,y,z)$, so T is linear.

8. (a) $T(x_1+x_2,y_1+y_2) = (x_1+x_2, y_1+y_2, 0) = (x_1,y_1,0) + (x_2,y_2,0) = T(x_1,y_1) + T(x_2,y_2)$, and
   $T(c(x,y)) = T(cx,cy) = (cx,cy,0) = c(x,y,0) = cT(x,y)$, so T is linear.

10. If $c \neq 0$ or 1, $T(c(x,y)) = T(cx,cy) = ((cx)^2,cy) = (c^2x^2,cy) = c(cx^2,y) \neq cT(x,y)$, so T is not linear.

11. $T((x_1,y_1,z_1)+(x_2,y_2,z_2)) = T(x_1+x_2, y_1+y_2, z_1+z_2)$

$= (x_1+x_2+2(y_1+y_2), x_1+x_2+y_1+y_2+z_1+z_2, 3(z_1+z_2))$

$= (x_1+x_2+2y_1+2y_2, x_1+x_2+y_1+y_2+z_1+z_2, 3z_1+3z_2)$

$= (x_1+2y_1, x_1+y_1+z_1, 3z_1) + (x_2+2y_2, x_2+y_2+z_2, 3z_2) = T((x_1,y_1,z_1)+T(x_2,y_2,z_2)$ and $T(c(x,y,z)$

$= T(cx,cy,cz) = (cx+2cy, cx+cy+cz, 3cz) = c(x+2y, x+y+z, 3z) = cT(x,y,z)$, so T is linear.

12. $A\mathbf{x} = \begin{bmatrix} 1 \\ 4 \\ 1 \end{bmatrix}$, $A\mathbf{y} = \begin{bmatrix} 8 \\ 7 \\ 8 \end{bmatrix}$, and $A\mathbf{z} = \begin{bmatrix} 9 \\ 1 \\ 9 \end{bmatrix}$.

14. $T(\mathbf{x}_1+\mathbf{x}_2) = A(\mathbf{x}_1+\mathbf{x}_2) + \mathbf{c} = A\mathbf{x}_1+A\mathbf{x}_2 + \mathbf{c} \neq T(\mathbf{x}_1)+T(\mathbf{x}_2)$ if $\mathbf{c} \neq 0$, so T is not linear.

17. $T(\mathbf{x}_1+\mathbf{x}_2) = (\mathbf{x}_1+\mathbf{x}_2)\cdot\mathbf{y} = \mathbf{x}_1\cdot\mathbf{y}+\mathbf{x}_2\cdot\mathbf{y} = T(\mathbf{x}_1)+T(\mathbf{x}_2)$ and $T(c\mathbf{x}) = c\mathbf{x}\cdot\mathbf{y} = c(\mathbf{x}\cdot\mathbf{y}) = cT(\mathbf{x})$, so T is a linear transformation from $\mathbf{R}^n$ to $\mathbf{R}$.

19. (a) $T(\mathbf{x}) = A_2A_1\mathbf{x} = \begin{bmatrix} -1 & 0 \\ 1 & 5 \end{bmatrix}\begin{bmatrix} 1 & 2 \\ 3 & 0 \end{bmatrix}\mathbf{x} = \begin{bmatrix} -1 & -2 \\ 16 & 2 \end{bmatrix}\mathbf{x}$, so

$T\left(\begin{bmatrix} 5 \\ 2 \end{bmatrix}\right) = \begin{bmatrix} -1 & -2 \\ 16 & 2 \end{bmatrix}\begin{bmatrix} 5 \\ 2 \end{bmatrix} = \begin{bmatrix} -9 \\ 84 \end{bmatrix}$.

(c) $T(\mathbf{x}) = A_2A_1\mathbf{x} = \begin{bmatrix} 2 & 2 \\ 1 & -1 \\ 0 & 4 \end{bmatrix}\begin{bmatrix} 3 & -2 \\ 0 & 1 \end{bmatrix}\mathbf{x} = \begin{bmatrix} 6 & -2 \\ 3 & -3 \\ 0 & 4 \end{bmatrix}\mathbf{x}$, so

$T\left(\begin{bmatrix} -3 \\ 2 \end{bmatrix}\right) = \begin{bmatrix} 6 & -2 \\ 3 & -3 \\ 0 & 4 \end{bmatrix}\begin{bmatrix} -3 \\ 2 \end{bmatrix} = \begin{bmatrix} -22 \\ -15 \\ 8 \end{bmatrix}$.

20. $T(x,y) = T_2 \circ T_1(x,y) = T_2(2x,-y) = (0, 2x-y)$, so $T(2,-3) = (0,7)$.

22. $T(x,y) = T_2 \circ T_1(x,y) = T_2(x+y, 2x, 3y) = (x+y, 3x+y, 2x-y)$, so $T(-2,5) = (3,-1,-9)$.

23. In general $T_2 \circ T_1 \neq T_1 \circ T_2$. In fact one of $T_2 \circ T_1$ and $T_1 \circ T_2$ may not exist, as in exercise 19(b), and if both do exist they will generally not be equal because composition of functions is not commutative.

25. $D(A+B) = |A+B|$ and $D(A) + D(B) = |A| + |B|$. In general $|A+B| \neq |A| + |B|$. For example, if A is the nxn diagonal matrix with 1 in all its diagonal positions and B is the nxn diagonal matrix with 2 in all its diagonal positions, then $|A| = 1$, $|B| = 2^n$, and A+B is the nxn diagonal matrix with 3 in all its diagonal positions, so $|A+B| = 3^n \neq 1 + 2^n$, for $n > 1$. Thus D is not a linear mapping.

Exercise Set 7.2, page 291

1. (a) $A = \begin{bmatrix} 0 & -1 \\ 1 & 0 \end{bmatrix}$ $\qquad A\begin{bmatrix} 2 \\ 1 \end{bmatrix} = \begin{bmatrix} -1 \\ 2 \end{bmatrix}$

(c) $A = \begin{bmatrix} \frac{1}{\sqrt{2}} & \frac{-1}{\sqrt{2}} \\ \frac{1}{\sqrt{2}} & \frac{1}{\sqrt{2}} \end{bmatrix}$ $\qquad A\begin{bmatrix} 2 \\ 1 \end{bmatrix} = \begin{bmatrix} \frac{1}{\sqrt{2}} \\ \frac{3}{\sqrt{2}} \end{bmatrix}$

(d) $A = \begin{bmatrix} -1 & 0 \\ 0 & -1 \end{bmatrix}$ $\qquad A\begin{bmatrix} 2 \\ 1 \end{bmatrix} = \begin{bmatrix} -2 \\ -1 \end{bmatrix}$

(f) $A = \begin{bmatrix} \frac{\sqrt{3}}{2} & \frac{-1}{2} \\ \frac{1}{2} & \frac{\sqrt{3}}{2} \end{bmatrix}$ $\qquad A\begin{bmatrix} 2 \\ 1 \end{bmatrix} = \begin{bmatrix} \sqrt{3} - \frac{1}{2} \\ 1 + \frac{\sqrt{3}}{2} \end{bmatrix}$

3. $\begin{bmatrix} 0 & -1 \\ 1 & 0 \end{bmatrix}\begin{bmatrix} x \\ y \end{bmatrix} = \begin{bmatrix} -y \\ x \end{bmatrix} = \begin{bmatrix} x' \\ y' \end{bmatrix}$. $\frac{x^2}{4} + \frac{y^2}{9} = 1$, so that $\frac{y'^2}{4} + \frac{x'^2}{9} = 1$. Thus the images of the points on the ellipse $\frac{x^2}{4} + \frac{y^2}{9} = 1$ are the points on the ellipse $\frac{x^2}{9} + \frac{y^2}{4} = 1$.

4. $\underset{\text{dilation}}{\begin{bmatrix} 2 & 0 \\ 0 & 2 \end{bmatrix}}\underset{\text{rotation}}{\begin{bmatrix} 0 & -1 \\ 1 & 0 \end{bmatrix}} = \begin{bmatrix} 0 & -2 \\ 2 & 0 \end{bmatrix}$

6. $\begin{bmatrix} 3 & 0 \\ 0 & 3 \end{bmatrix}\begin{bmatrix} x \\ y \end{bmatrix} = \begin{bmatrix} 3x \\ 3y \end{bmatrix} = \begin{bmatrix} x' \\ y' \end{bmatrix}$, so $\frac{x'^2}{9} + \frac{y'^2}{9} = 1$. Thus the images of the points on the circle $x^2 + y^2 = 1$ are the points on the circle $x^2 + y^2 = 9$.

7.

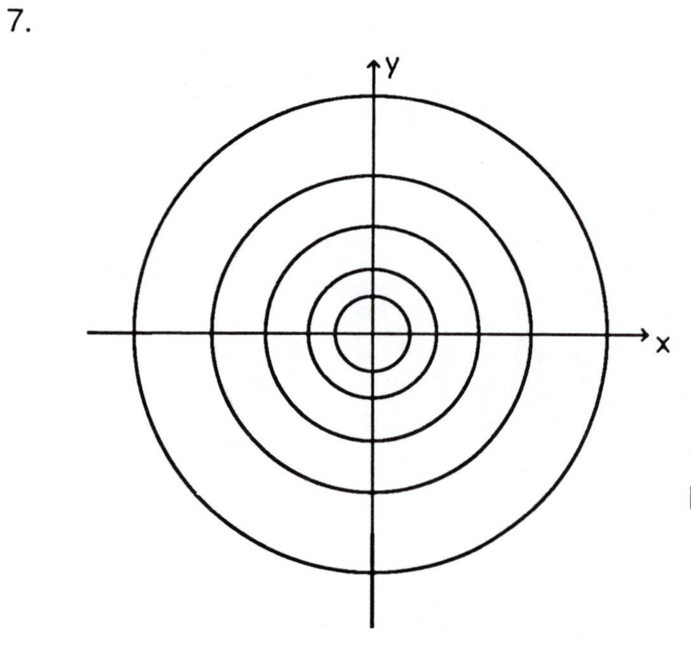

Radii are 1, 1.5, 2.25, 3.375 and 5.0625.

8. $\begin{bmatrix} 1 \\ 0 \end{bmatrix} \mapsto \begin{bmatrix} -1 \\ 0 \end{bmatrix}$ and $\begin{bmatrix} 0 \\ 1 \end{bmatrix} \mapsto \begin{bmatrix} 0 \\ 1 \end{bmatrix}$, so $A = \begin{bmatrix} -1 & 0 \\ 0 & 1 \end{bmatrix}$. $A\begin{bmatrix} 2 \\ 1 \end{bmatrix} = \begin{bmatrix} -2 \\ 1 \end{bmatrix}$.

10. Vertices of the image of the unit square ((1,0), (1,1), (0,1),(0,0)) are:

(a) (0,1), (−1,1), (−1,0), (0,0) (c) (3,1), (3,5), (0,4), (0,0)

(e) (−2,0), (−5,4), (−3,4), (0,0) (g) (0,2), (−2,2), (−2,0), (0,0)

11. (a) $\begin{bmatrix} 1 \\ 0 \end{bmatrix} \mapsto \begin{bmatrix} 2 \\ 1 \end{bmatrix}$ and $\begin{bmatrix} 0 \\ 1 \end{bmatrix} \mapsto \begin{bmatrix} 0 \\ -1 \end{bmatrix}$, so $A = \begin{bmatrix} 2 & 0 \\ 1 & -1 \end{bmatrix}$.

(c) $\begin{bmatrix} 1 \\ 0 \end{bmatrix} \mapsto \begin{bmatrix} 2 \\ 0 \end{bmatrix}$ and $\begin{bmatrix} 0 \\ 1 \end{bmatrix} \mapsto \begin{bmatrix} -5 \\ 3 \end{bmatrix}$, so $A = \begin{bmatrix} 2 & -5 \\ 0 & 3 \end{bmatrix}$.

13. $T\left(\begin{bmatrix} x \\ y \end{bmatrix}\right) = \begin{bmatrix} 0 \\ y \end{bmatrix}$, so $\begin{bmatrix} 1 \\ 0 \end{bmatrix} \mapsto \begin{bmatrix} 0 \\ 0 \end{bmatrix}$ and $\begin{bmatrix} 0 \\ 1 \end{bmatrix} \mapsto \begin{bmatrix} 0 \\ 1 \end{bmatrix}$, and $A = \begin{bmatrix} 0 & 0 \\ 0 & 1 \end{bmatrix}$.

14. $T\left(\begin{bmatrix} x \\ y \end{bmatrix}\right) = \begin{bmatrix} \frac{x+y}{2} \\ \frac{x+y}{2} \end{bmatrix}$, so $\begin{bmatrix} 1 \\ 0 \end{bmatrix} \mapsto \begin{bmatrix} \frac{1}{2} \\ \frac{1}{2} \end{bmatrix}$ and $\begin{bmatrix} 0 \\ 1 \end{bmatrix} \mapsto \begin{bmatrix} \frac{1}{2} \\ \frac{1}{2} \end{bmatrix}$, and $A = \begin{bmatrix} \frac{1}{2} & \frac{1}{2} \\ \frac{1}{2} & \frac{1}{2} \end{bmatrix}$.

$\begin{bmatrix} 4 \\ 2 \end{bmatrix} \mapsto \begin{bmatrix} 3 \\ 3 \end{bmatrix}$.

15. $\begin{bmatrix} 1 \\ 0 \end{bmatrix} \mapsto \begin{bmatrix} a \\ 0 \end{bmatrix}$ and $\begin{bmatrix} 0 \\ 1 \end{bmatrix} \mapsto \begin{bmatrix} 0 \\ b \end{bmatrix}$, so $A = \begin{bmatrix} a & 0 \\ 0 & b \end{bmatrix}$.

If a = 3 and b = 2 the unit square becomes a 3x2 rectangle.

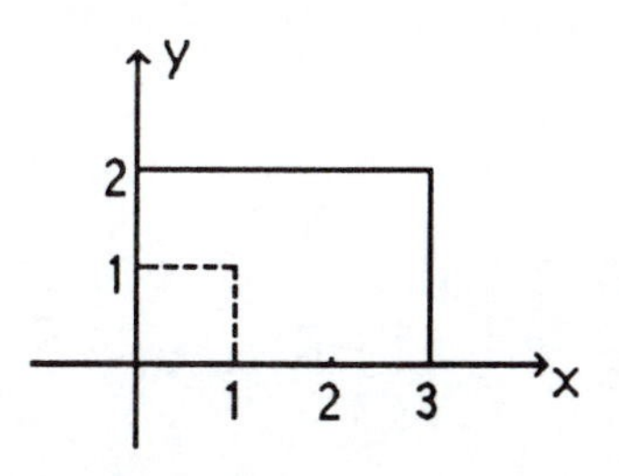

16. $\begin{bmatrix} 2 & 0 \\ 0 & 3 \end{bmatrix}\begin{bmatrix} x \\ y \end{bmatrix} = \begin{bmatrix} 2x \\ 3y \end{bmatrix} = \begin{bmatrix} x' \\ y' \end{bmatrix}$. $y = 2x$, so $\frac{y'}{3} = 2\frac{x'}{2}$, and $y' = 3x'$. Thus the images of the points on the line $y = 2x$ are the points on the line $y = 3x$.

18. $\begin{bmatrix} 1 \\ 0 \end{bmatrix} \mapsto \begin{bmatrix} 1 \\ 0 \end{bmatrix}$ and $\begin{bmatrix} 0 \\ 1 \end{bmatrix} \mapsto \begin{bmatrix} c \\ 1 \end{bmatrix}$, so $A = \begin{bmatrix} 1 & c \\ 0 & 1 \end{bmatrix}$. If $c = 2$, then

$\begin{bmatrix} 1 \\ 0 \end{bmatrix} \mapsto \begin{bmatrix} 1 \\ 0 \end{bmatrix}$, $\begin{bmatrix} 1 \\ 1 \end{bmatrix} \mapsto \begin{bmatrix} 3 \\ 1 \end{bmatrix}$, and $\begin{bmatrix} 0 \\ 1 \end{bmatrix} \mapsto \begin{bmatrix} 2 \\ 1 \end{bmatrix}$.

20. $\begin{bmatrix} 1 & 5 \\ 0 & 1 \end{bmatrix}\begin{bmatrix} x \\ y \end{bmatrix} = \begin{bmatrix} x+5y \\ y \end{bmatrix} = \begin{bmatrix} x' \\ y' \end{bmatrix}$. $y = 3x$, so $y' = 3(x' - 5y')$ and $16y' = 3x'$. Thus the images of the points on the line $y = 3x$ are the points on the line $16y = 3x$.

23. $\sin^2 \theta + \cos^2 \theta = 1$, so the norm of each column vector is 1, and the columns are orthogonal, so the matrix is orthogonal.

26. (a) $T\left(\begin{bmatrix} x \\ y \end{bmatrix}\right) = \begin{bmatrix} x \\ y \end{bmatrix} + \begin{bmatrix} 2 \\ 5 \end{bmatrix} = \begin{bmatrix} x+2 \\ y+5 \end{bmatrix} = \begin{bmatrix} x' \\ y' \end{bmatrix}$. $y = 3x + 1$, so $y' - 5 = 3(x' - 2) + 1$ and thus $y' = 3x'$, so the image of the line $y = 3x + 1$ is the line $y = 3x$.

27. (a) $\begin{bmatrix} 1 \\ 0 \end{bmatrix} \mapsto \begin{bmatrix} 6 \\ 4 \end{bmatrix}$, $\begin{bmatrix} 1 \\ 1 \end{bmatrix} \mapsto \begin{bmatrix} 6 \\ 6 \end{bmatrix}$, $\begin{bmatrix} 0 \\ 1 \end{bmatrix} \mapsto \begin{bmatrix} 4 \\ 6 \end{bmatrix}$, and $\begin{bmatrix} 0 \\ 0 \end{bmatrix} \mapsto \begin{bmatrix} 4 \\ 4 \end{bmatrix}$.

$\begin{bmatrix} x \\ y \end{bmatrix} \mapsto \begin{bmatrix} 2x+4 \\ 2y+4 \end{bmatrix} = \begin{bmatrix} x' \\ y' \end{bmatrix}$, so the image of $x^2 + y^2 = 1$ is $(x-4)^2 + (y-4)^2 = 4$.

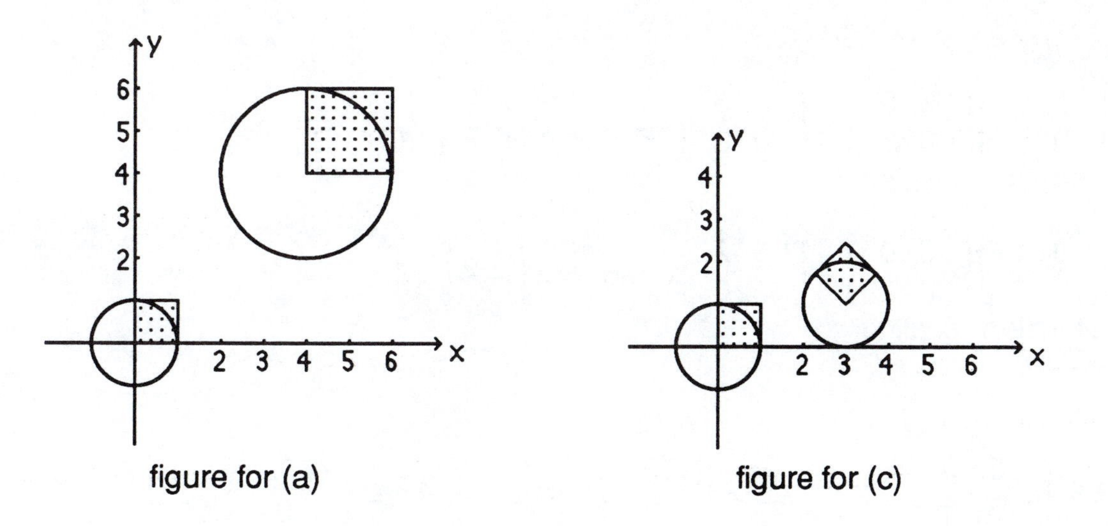

figure for (a)

figure for (c)

(c) $\begin{bmatrix} 1 \\ 0 \end{bmatrix} \mapsto \begin{bmatrix} \frac{1}{\sqrt{2}} + 3 \\ \frac{1}{\sqrt{2}} + 1 \end{bmatrix}$, $\begin{bmatrix} 1 \\ 1 \end{bmatrix} \mapsto \begin{bmatrix} 3 \\ \frac{2}{\sqrt{2}} + 1 \end{bmatrix}$, $\begin{bmatrix} 0 \\ 1 \end{bmatrix} \mapsto \begin{bmatrix} \frac{-1}{\sqrt{2}} + 3 \\ \frac{1}{\sqrt{2}} + 1 \end{bmatrix}$, and

$\begin{bmatrix} 0 \\ 0 \end{bmatrix} \mapsto \begin{bmatrix} 3 \\ 1 \end{bmatrix}$. $\begin{bmatrix} x \\ y \end{bmatrix} \mapsto \begin{bmatrix} \frac{x}{\sqrt{2}} - \frac{y}{\sqrt{2}} + 3 \\ \frac{x}{\sqrt{2}} + \frac{y}{\sqrt{2}} + 1 \end{bmatrix} = \begin{bmatrix} x' \\ y' \end{bmatrix}$, so the image of the circle

$x^2 + y^2 = 1$ is $\frac{(x+y-4)^2}{2} + \frac{(y-x+2)^2}{2} = 1$, which is the circle $(x-3)^2 + (y-1)^2 = 1$.

29. To reverse the operations on $\mathbf{u}$, one must first subtract $\mathbf{v}$ then multiply by A^{-1}, so $T^{-1}(\mathbf{u}) = A^{-1}(\mathbf{u}-\mathbf{v})$. Checking, $T^{-1} \circ T(\mathbf{u}) = T^{-1}(A\mathbf{u} + \mathbf{v}) = A^{-1}(A\mathbf{u} + \mathbf{v} - \mathbf{v}) = A^{-1}A\mathbf{u} = \mathbf{u}$ and $T \circ T^{-1}(\mathbf{u}) = T(A^{-1}(\mathbf{u}-\mathbf{v})) = AA^{-1}(\mathbf{u}-\mathbf{v}) + \mathbf{v} = \mathbf{u} - \mathbf{v} + \mathbf{v} = \mathbf{u}$. T^{-1} is also an affine transformation since$T^{-1}(\mathbf{u}) = A^{-1}\mathbf{u} + (-A^{-1}\mathbf{v}) = A^{-1}\mathbf{u} + \mathbf{w}$, where $\mathbf{w} = -A^{-1}\mathbf{v}$.

30. (a) $\underbrace{\begin{bmatrix} 2 & 0 \\ 0 & 2 \end{bmatrix}}_{\text{dilation}} \underbrace{\begin{bmatrix} 0 & -1 \\ 1 & 0 \end{bmatrix}}_{\text{rotation}} = \begin{bmatrix} 0 & -2 \\ 2 & 0 \end{bmatrix}$ and $\begin{bmatrix} 2 \\ 1 \end{bmatrix} \mapsto \begin{bmatrix} -2 \\ 4 \end{bmatrix}$.

(c) $\underbrace{\begin{bmatrix} -1 & 0 \\ 0 & -1 \end{bmatrix}}_{\text{rotation}} \underbrace{\begin{bmatrix} 0 & 1 \\ 1 & 0 \end{bmatrix}}_{\text{reflection}} = \begin{bmatrix} 0 & -1 \\ -1 & 0 \end{bmatrix}$ and $\begin{bmatrix} 2 \\ 1 \end{bmatrix} \mapsto \begin{bmatrix} -1 \\ -2 \end{bmatrix}$.

31. (a) $\begin{bmatrix} 1 & 2 \\ 0 & 1 \end{bmatrix}\begin{bmatrix} 3 & 0 \\ 0 & 3 \end{bmatrix} = \begin{bmatrix} 3 & 6 \\ 0 & 3 \end{bmatrix}$ and $\begin{bmatrix} 3 \\ 2 \end{bmatrix} \mapsto \begin{bmatrix} 21 \\ 6 \end{bmatrix}$.

shear dilation

(b) $\begin{bmatrix} 0 & 1 \\ 1 & 0 \end{bmatrix}\begin{bmatrix} 3 & 0 \\ 0 & 2 \end{bmatrix} = \begin{bmatrix} 0 & 2 \\ 3 & 0 \end{bmatrix}$ and $\begin{bmatrix} 3 \\ 2 \end{bmatrix} \mapsto \begin{bmatrix} 4 \\ 9 \end{bmatrix}$.

reflection scaling

36. (a) $A = \begin{bmatrix} .5 & 0 \\ 0 & .5 \end{bmatrix}$ $\quad A^2 = \begin{bmatrix} .25 & 0 \\ 0 & .25 \end{bmatrix}$ $\quad A^3 = \begin{bmatrix} .0625 & 0 \\ 0 & .0625 \end{bmatrix}$

$\begin{bmatrix} 2.5 \\ 2 \end{bmatrix} = A\begin{bmatrix} 1 \\ 0 \end{bmatrix} + \begin{bmatrix} 2 \\ 2 \end{bmatrix}$, $\begin{bmatrix} 2.5 \\ 2.5 \end{bmatrix} = A\begin{bmatrix} 1 \\ 1 \end{bmatrix} + \begin{bmatrix} 2 \\ 2 \end{bmatrix}$, $\begin{bmatrix} 2 \\ 2.5 \end{bmatrix} = A\begin{bmatrix} 0 \\ 1 \end{bmatrix} + \begin{bmatrix} 2 \\ 2 \end{bmatrix}$,

$\begin{bmatrix} 2 \\ 2 \end{bmatrix} = A\begin{bmatrix} 0 \\ 0 \end{bmatrix} + \begin{bmatrix} 2 \\ 2 \end{bmatrix}$. Thus the image of the unit square is the square with vertices (2.5,2), (2.5,2.5), (2,2.5), (2,2), when $T = A\mathbf{u} + \mathbf{v}$. Likewise the images of the unit square are the squares with vertices (2.25,2), (2.25,2.25), (2,2.25), (2,2), when $T = A^2\mathbf{u} + \mathbf{v}$ and (2.0625,2), (2.0625,2.0625), (2,2.0625), (2,2), when $T = A^3\mathbf{u} + \mathbf{v}$.

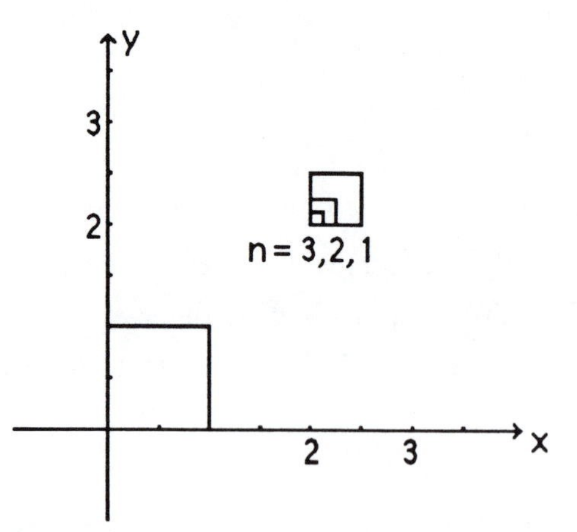

38. The pairs which commute are D and R, D and F, D and S, and D and H.

39. If $\begin{bmatrix} x \\ y \\ z \end{bmatrix} \mapsto \begin{bmatrix} -y \\ x \\ z \end{bmatrix}$ then $\begin{bmatrix} 1 \\ 0 \\ 0 \end{bmatrix} \mapsto \begin{bmatrix} 0 \\ 1 \\ 0 \end{bmatrix}$, $\begin{bmatrix} 0 \\ 1 \\ 0 \end{bmatrix} \mapsto \begin{bmatrix} -1 \\ 0 \\ 0 \end{bmatrix}$, and $\begin{bmatrix} 0 \\ 0 \\ 1 \end{bmatrix} \mapsto \begin{bmatrix} 0 \\ 0 \\ 1 \end{bmatrix}$

so that $A = \begin{bmatrix} 0 & -1 & 0 \\ 1 & 0 & 0 \\ 0 & 0 & 1 \end{bmatrix}$. If $\begin{bmatrix} x \\ y \\ z \end{bmatrix} \mapsto \begin{bmatrix} y \\ -x \\ z \end{bmatrix}$ then $A = \begin{bmatrix} 0 & 1 & 0 \\ -1 & 0 & 0 \\ 0 & 0 & 1 \end{bmatrix}$.

41. Let $\theta = \frac{\pi}{2}$, $h = 5$ and $k = 1$ in the matrix $\begin{bmatrix} \cos\theta & -\sin\theta & -h\cos\theta+k\sin\theta+h \\ \sin\theta & \cos\theta & -h\sin\theta-k\cos\theta+k \\ 0 & 0 & 1 \end{bmatrix}$.

The resulting matrix is $\begin{bmatrix} 0 & -1 & 6 \\ 1 & 0 & -4 \\ 0 & 0 & 1 \end{bmatrix}$, and $\begin{bmatrix} 1 \\ 0 \\ 1 \end{bmatrix} \mapsto \begin{bmatrix} 6 \\ -3 \\ 1 \end{bmatrix}$, $\begin{bmatrix} 1 \\ 1 \\ 1 \end{bmatrix} \mapsto \begin{bmatrix} 5 \\ -3 \\ 1 \end{bmatrix}$,

$\begin{bmatrix} 0 \\ 1 \\ 1 \end{bmatrix} \mapsto \begin{bmatrix} 5 \\ -4 \\ 1 \end{bmatrix}$, and $\begin{bmatrix} 0 \\ 0 \\ 1 \end{bmatrix} \mapsto \begin{bmatrix} 6 \\ -4 \\ 1 \end{bmatrix}$, so that the image of the unit square is the square

with vertices (6,−3), (5,−3), (5,−4), and (6,−4).

42. $T^{-1} = \begin{bmatrix} 1 & 0 & -h \\ 0 & 1 & -k \\ 0 & 0 & 1 \end{bmatrix}$.

44. $SRT = \begin{bmatrix} 3 & 0 & 0 \\ 0 & 5 & 0 \\ 0 & 0 & 1 \end{bmatrix}\begin{bmatrix} 0 & 1 & 0 \\ -1 & 0 & 0 \\ 0 & 0 & 1 \end{bmatrix}\begin{bmatrix} 1 & 0 & 4 \\ 0 & 1 & -3 \\ 0 & 0 & 1 \end{bmatrix} = \begin{bmatrix} 0 & 3 & -9 \\ -5 & 0 & -20 \\ 0 & 0 & 1 \end{bmatrix}$

$\begin{bmatrix} 1 \\ 6 \\ 1 \end{bmatrix} \mapsto \begin{bmatrix} 9 \\ -25 \\ 1 \end{bmatrix}$, $\begin{bmatrix} 3 \\ 0 \\ 1 \end{bmatrix} \mapsto \begin{bmatrix} -9 \\ -35 \\ 1 \end{bmatrix}$, and $\begin{bmatrix} 4 \\ 6 \\ 1 \end{bmatrix} \mapsto \begin{bmatrix} 9 \\ -40 \\ 1 \end{bmatrix}$, so that the image of the given triangle is the triangle with vertices (9,–25), (–9,–35), and (9,–40).

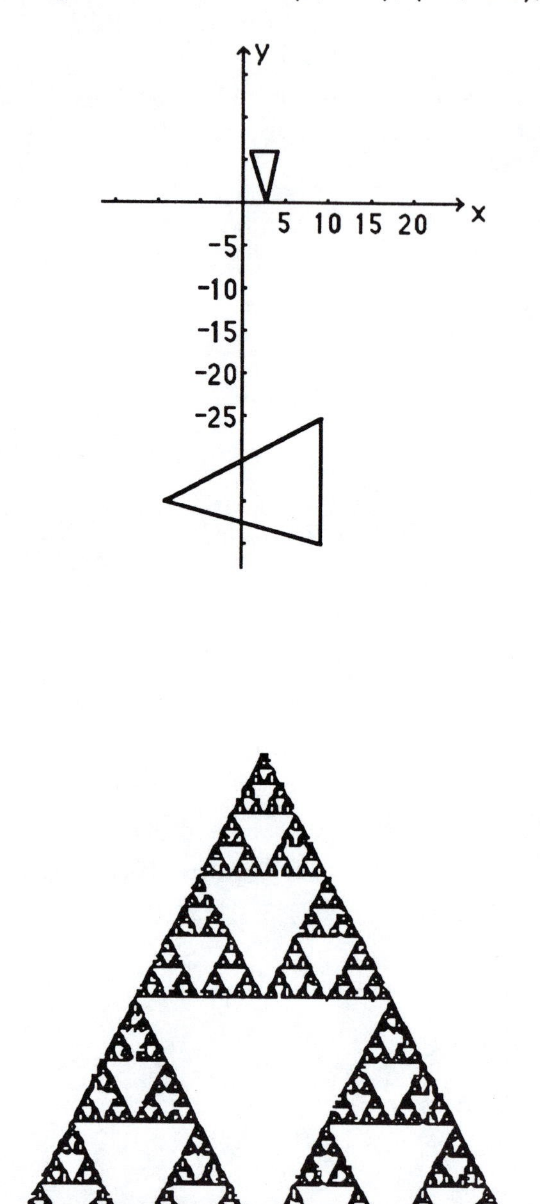

45.

Sierpinski triangle:
triangles in six sizes

46.

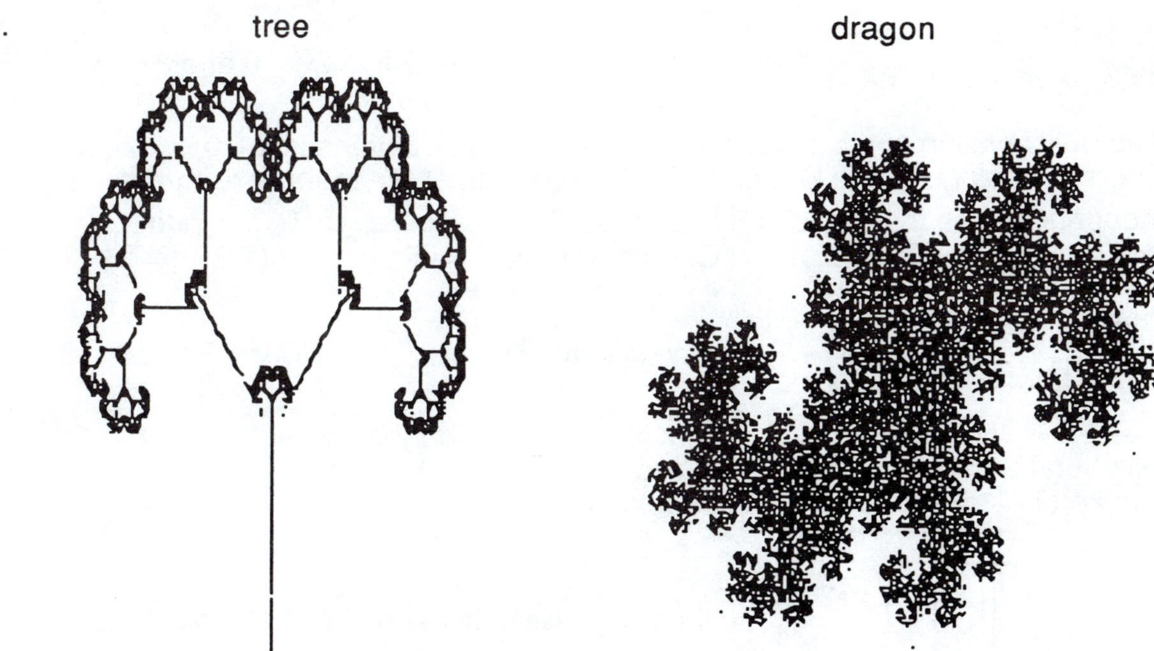

Exercise Set 7.3, page 303

1. (a) $\begin{bmatrix} 1 & 2 \\ 3 & 0 \end{bmatrix}\begin{bmatrix} x \\ y \end{bmatrix} = \begin{bmatrix} x+2y \\ 3x \end{bmatrix}$; thus (x,y) is in the kernel of T if x+2y = 0 and 3x = 0.

The only solution of this system of homogeneous equations is x = 0, y = 0, so only the zero vector is in the kernel of T. The columns of the matrix are linearly independent, so the range is $\mathbf{R}^2$. Dim ker(T) = 0, dim range(T) = 2, and dim domain(T) = 2, so dim ker(T) + dim range(T) = dim domain(T).

(c) $\begin{bmatrix} 2 & 4 \\ 4 & 8 \end{bmatrix}\begin{bmatrix} x \\ y \end{bmatrix} = \begin{bmatrix} 2x+4y \\ 4x+8y \end{bmatrix}$; thus (x,y) is in the kernel of T if 2x+4y = 0 and 4x+8y = 0,

i.e., if x = −2y. The kernel of T is the set {(−2r,r)} and the range is the set {(r,2r)}. Dim ker(T) = 1, dim range(T) = 1, and dim domain(T) = 2, so dim ker(T) + dim range(T) = dim domain(T).

(e) $\begin{bmatrix} 1 & 2 & 3 \\ 0 & 1 & 2 \end{bmatrix}\begin{bmatrix} x \\ y \\ z \end{bmatrix} = \begin{bmatrix} x+2y+3z \\ y+2z \end{bmatrix}$; thus (x,y,z) is in the kernel of T if x+2y+3z = 0 and

y+2z = 0. The general solution to this system of equations is x = r, y = −2r, and z = r, so the kernel of T is the set {(r,−2r,r)}. The range is $\mathbf{R}^2$. Dim ker(T) = 1, dim range(T) = 2, and dim domain(T) = 3, so dim ker(T) + dim range(T) = dim domain(T).

(g) $\begin{bmatrix} 0 & 1 & 0 \\ 0 & 2 & 0 \\ 0 & 0 & 4 \end{bmatrix}\begin{bmatrix} x \\ y \\ z \end{bmatrix} = \begin{bmatrix} y \\ 2y \\ 4z \end{bmatrix}$; thus (x,y,z) is in the kernel of T if y = 0, 2y = 0, and 4z = 0,

i.e., if y = z = 0. The kernel of T is the set {(r,0,0)}. The range is the set {(a,2a,b)}. Dim ker(T) = 1, dim range(T) = 2, and dim domain(T) = 3, so dim ker(T) + dim range(T) = dim domain(T).

(h) $\begin{bmatrix} 1 & 2 & 1 \\ -1 & -2 & 0 \\ 2 & 4 & 1 \end{bmatrix}\begin{bmatrix} x \\ y \\ z \end{bmatrix} = \begin{bmatrix} x+2y+z \\ -x-2y \\ 2x+4y+z \end{bmatrix}$; thus (x,y,z) is in the kernel of T if x+2y+z = 0,

−x−2y = 0 and 2x+4y+z = 0, i.e., if x = −2y and z = 0. The kernel of T is the set {(−2r,r,0)} and the range is the set {(a+b,−a,2a+b)}, using the first and third columns of the matrix as a basis. Dim ker(T) = 1, dim range(T) = 2, and dim domain(T) = 3, so dim ker(T) + dim range(T) = dim domain(T).

2. (a) The kernel is the set {(0,r,s)} and the range is the set {(a,0,0)}. Dim ker(T) = 2, dim range(T) = 1, and dim domain(T) = 3, so dim ker(T) + dim range(T) = dim domain(T).

   (b) The kernel is the set {(r,−r,0)} and the range is $\mathbf{R}^2$. Dim ker(T) = 1, dim range(T) = 2, and dim domain(T) = 3, so dim ker(T) + dim range(T) = dim domain(T).

   (e) The kernel is the zero vector and the range is the set {(3a,a−b,b)}. Dim ker(T) = 0, dim range(T) = 2, and dim domain(T) = 2, so dim ker(T) + dim range(T) = dim domain(T).

3. (a) T(**u**+**w**) = 5(**u**+**w**) = 5**u** + 5**w** = T(**u**) + T(**w**) and T(c**u**) = 5(c**u**) = 5c**u** = c(5**u**) = cT(**u**), so T is linear. ker(T) is the zero vector and range(T) is U (since for any vector **u** in U, **u** = T(.2**u**)).

   (b) T(c**u**) = 2(c**u**) + 3**v** ≠ c(2**u** + 3**v**) = cT(**u**), so T is not linear.

4. (a) The dimension of the range of the transformation is the rank of the matrix = 2. The domain is $\mathbf{R}^3$, which has dimension 3, so the dimension of the kernel is 1. This transformation is not one-to-one.

   (c) The dimension of the range of the transformation is the rank of the matrix = 3. The domain is $\mathbf{R}^4$, which has dimension 4, so the dimension of the kernel is 1. This transformation is not one-to-one.

   (f) The dimension of the range of the transformation is the rank of the matrix = 3. The domain is $\mathbf{R}^3$, which has dimension 3, so the dimension of the kernel is 0. This transformation is one-to-one.

5. (a) $|A| = 12 \neq 0$, so the transformation is nonsingular and therefore one-to-one.

   (c) $|C| = 0$, so the transformation is not one-to-one.

   (f) $|F| = 212$, so the transformation is one-to-one.

6. (a) The set of fixed points is the set of all points for which (x,y) = (x,3y). This is the set {(r,0)}.

   (c) This transformation has no fixed points since there are no solutions to the equation y = y + 1.

(f) The set of fixed points is the set of all points for which $(x,y) = (x+y, x-y)$. This set contains only the zero vector.

9. $\dim \ker(T) + \dim \text{range}(T) = \dim \text{domain}(T)$. $\dim \ker(T) \geq 0$, so $\dim \text{range}(T) \leq \dim \text{domain}(T)$.

11. $T(x,y,z) = (x,y,0) = (1,2,0)$ if $x = 1$, $y = 2$, $z = r$. Thus the set of vectors mapped by T into $(1,2,0)$ is the set $\{(1,2,r)\}$.

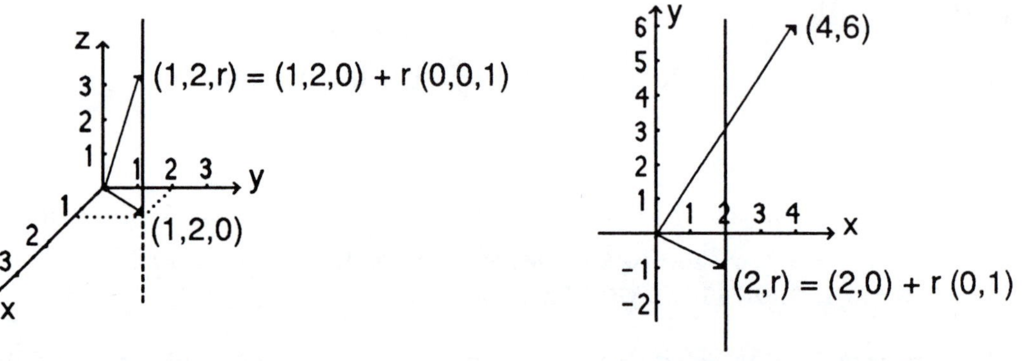

figure for exercise 11

figure for exercise 13

13. $T(x,y) = (2x,3x) = (4,6)$ if $2x = 4$ and $3x = 6$, i.e., if $x = 2$. Thus the set of vectors mapped by T into $(4,6)$ is the set $\{(2, r)\}$. This set is not a subspace of $\mathbf{R}^2$. It does not contain the zero vector.

15. This set is not a subspace because it does not contain the zero vector.

16. $T(a_2x^2 + a_1x + a_0) + T(b_2x^2 + b_1x + b_0) = (a_2 + a_1)x^2 + a_1x + 2a_0 + (b_2 + b_1)x^2 + b_1x + 2b_0$
$= (a_2 + a_1 + b_2 + b_1)x^2 + (a_1 + b_1)x + 2(a_0 + b_0) = T((a_2 + b_2)x^2 + (a_1 + b_1)x + a_0 + b_0)$
$= T(a_2x^2 + a_1x + a_0 + b_2x^2 + b_1x + b_0)$ and $T(c(a_2x^2 + a_1x + a_0)) = T(ca_2x^2 + ca_1x + ca_0)$
$= (ca_2 + ca_1)x^2 + ca_1x + 2ca_0 = c((a_2 + a_1)x^2 + a_1x + 2a_0) = cT(a_2x^2 + a_1x + a_0)$, so T is linear.

$T(a_2x^2 + a_1x + a_0) = 0$ if $a_2 = a_1 = a_0 = 0$, so $\ker(T)$ is the zero polynomial and range(T) is P_2. A basis for P_2 is the set $\{1, x, x^2\}$.

18. $g(a_2 x^2 + a_1 x + a_0) + g(b_2 x^2 + b_1 x + b_0) = 2a_2 x^3 + a_1 x + 3a_0 + 2b_2 x^3 + b_1 x + 3b_0$
$= 2(a_2 + b_2)x^3 + (a_1 + b_1)x + 3(a_0 + b_0) = g((a_2 + b_2)x^2 + (a_1 + b_1)x + a_0 + b_0)$
$= g(a_2 x^2 + a_1 x + a_0 + b_2 x^2 + b_1 x + b_0)$ and $g(c(a_2 x^2 + a_1 x + a_0))$
$= g(ca_2 x^2 + ca_1 x + ca_0) = 2ca_2 x^3 + ca_1 x + 3ca_0 = c(2a_2 x^3 + a_1 x + 3a_0)$
$= cg(a_2 x^2 + a_1 x + a_0)$, so g is linear.

$g(a_2 x^2 + a_1 x + a_0) = 0$ if $a_2 = a_1 = a_0 = 0$, so ker(g) is the zero polynomial and range(g) $= \{a_3 x^3 + a_1 x + a_0\}$. A basis for range(g) is the set $\{1, x, x^3\}$.

19. If T is a linear transformation T(0) = 0. For the transformation T given in this exercise T(0) = 4 ≠ 0, so T is not linear.

21. $D(x^3 - 3x^2 + 2x + 1) = 3x^2 - 6x + 2$.

$D(a_n x^n + \ldots + a_1 x + a_0) = 0$ if $a_n = \ldots = a_1 = 0$; i.e., ker(D) = the set of constant polynomials. Range(D) = P_{n-1}, because every polynomial of degree less than or equal to n−1 is the derivative of a polynomial of degree one larger than its own degree, and no polynomial of degree n is the derivative of a polynomial of degree n or less.

22. (a) $D^2(2x^3 + 3x^2 - 5x + 4) = D(6x^2 + 6x - 5) = 12x + 6$.

$D^2(a_n x^n + \ldots + a_1 x + a_0) = D(na_n x^{n-1} + \ldots + 2a_2 x + a_1)$
$= n(n-1)a_n x^{n-2} + \ldots + 6a_3 x + 2a_2$.

$D^2(a_n x^n + \ldots + a_1 x + a_0) + D^2(b_n x^n + \ldots + b_1 x + b_0)$
$= n(n-1)a_n x^{n-2} + \ldots + 6a_3 x + 2a_2 + n(n-1)b_n x^{n-2} + \ldots + 6b_3 x + 2b_2$
$= n(n-1)(a_n + b_n)x^{n-2} + \ldots + 6(a_3 + b_3)x + 2(a_2 + b_2)$
$= D^2((a_n + b_n)x^n + \ldots + (a_1 + b_1)x + a_0 + b_0)$ and
$D^2(c(a_n x^n + \ldots + a_1 x + a_0)) = D^2(ca_n x^n + \ldots + ca_1 x + ca_0)$
$= n(n-1)ca_n x^{n-2} + \ldots + 6ca_3 x + 2ca_2 = c(n(n-1)a_n x^{n-2} + \ldots + 6a_3 x + 2a_2)$
$= cD^2(a_n x^n + \ldots + a_1 x + a_0)$, so D^2 is linear.

$D^2(a_n x^n + \ldots + a_1 x + a_0) = n(n-1)a_n x^{n-2} + \ldots + 6a_3 x + 2a_2 = 0$ if

$a_n = \ldots = a_3 = a_2 = 0$. Thus $\ker(D^2) = \{a_1 x + a_0\} = P_1$. Range$(D^2) = P_{n-2}$ because every polynomial of degree less than or equal to n−2 is the second derivative of a polynomial of degree two larger than its own degree, and no polynomial of degree n−1 or larger is the second derivative of a polynomial of degree n or less.

23. $D(a_n x^n + \ldots + a_1 x + a_0) = na_n x^{n-1} + \ldots + 3a_3 x^2 + 2a_2 x + a_1 = 3x^2 - 4x + 7$ if $a_n = a_{n-1} = \ldots = a_4 = 0$, $a_3 = 1$, $a_2 = -2$, $a_1 = 7$, and $a_0 = r$, where r is any real number. Thus the set of polynomials mapped into $3x^2 - 4x + 7$ is the set $\{x^3 - 2x^2 + 7x + r\}$.

25. $\int_0^1 (8x^3 + 6x^2 + 4x + 1)\, dx = [\, 2x^4 + 2x^3 + 2x^2 + x\,]_0^1 = 7.$

$$\int_0^1 (a_n x^n + a_{n-1} x^{n-1} + \ldots + a_1 x + a_0)\, dx = \frac{a_n}{n+1} + \frac{a_{n-1}}{n} + \ldots + \frac{a_1}{2} + a_0,$$

so ker(T) is all polynomials $a_n x^n + a_{n-1} x^{n-1} + \ldots + a_1 x + a_0$ for which

$\frac{a_n}{n+1} + \frac{a_{n-1}}{n} + \ldots + \frac{a_1}{2} = -a_0$. For any real number c, let p(x) = c. Then

$\int_0^1 p(x)\, dx = \int_0^1 c\, dx = [\, cx\,]_0^1 = c$, so range(T) = **R**. The set of elements mapped into

2 is the set of all polynomials $a_n x^n + a_{n-1} x^{n-1} + \ldots + a_1 x + a_0$ for which

$$\frac{a_n}{n+1} + \frac{a_{n-1}}{n} + \ldots + \frac{a_1}{2} + a_0 = 2.$$

26. (b) $T(A+C) = |A+C|$ and $T(A) + T(C) = |A| + |C| \neq |A+C|$, so T is not linear.

(d) $T(A+C) = (A+C)^2 \neq A^2 + C^2 = T(A) + T(C)$, so T is not linear.

(f) $T(A+C) = a_{11} + c_{11} = T(A) + T(C)$ and $T(cA) = ca_{11} = cT(A)$, so T is linear.

$\text{Ker}(T) = \left\{ \begin{bmatrix} 0 & a \\ b & c \end{bmatrix} \right\}$ and range(T) = **R**.

(i) $T(A+C) = A+C + (A+C)^t = A+C + A^t + C^t = A + A^t + C + C^t = T(A) + T(C)$ and $T(cA) = cA + (cA)^t = cA + cA^t = c(A + A^t) = cT(A)$, so T is linear.

$\begin{bmatrix} a & b \\ c & d \end{bmatrix} + \begin{bmatrix} a & c \\ b & d \end{bmatrix} = \begin{bmatrix} 2a & b+c \\ b+c & 2d \end{bmatrix}$, so $\ker(T) = \left\{ \begin{bmatrix} 0 & r \\ -r & 0 \end{bmatrix} \right\}$ and range(T) is the set of all symmetric matrices.

(j) If $A = \begin{bmatrix} 1 & 0 \\ 0 & 1 \end{bmatrix}$ and $C = \begin{bmatrix} 1 & 0 \\ 0 & 1 \end{bmatrix}$ then $A+C = \begin{bmatrix} 2 & 0 \\ 0 & 2 \end{bmatrix}$ and

$T(A+C) = I_2 \neq I_2 + I_2 = T(A) + T(C)$, so T is not linear.

27. (a) $T(\mathbf{u}+\mathbf{w}) = ||\mathbf{u}+\mathbf{w}|| \neq ||\mathbf{u}|| + ||\mathbf{w}||$, so T is not linear.

(b) $T(\mathbf{u}+\mathbf{w}) = \langle(\mathbf{u}+\mathbf{w}),\mathbf{v}\rangle = \langle\mathbf{u},\mathbf{v}\rangle + \langle\mathbf{w},\mathbf{v}\rangle = T(\mathbf{u}) + T(\mathbf{w})$ and $T(c\mathbf{u}) = \langle c\mathbf{u},\mathbf{v}\rangle = c\langle\mathbf{u},\mathbf{v}\rangle = cT(\mathbf{u})$, so T is linear. $\langle\mathbf{u},\mathbf{v}\rangle = \mathbf{0}$ if and only if **u** is orthogonal to **v**, so ker(T) is the set of all vectors in U which are orthogonal to **v**. Range(T) = **R**.

28. T is one-to-one if and only if ker(T) is the zero vector if and only if dim ker(T) = 0 if and only if dim range(T) = dim domain(T).

Exercise Set 7.4, page 312

1. (r, r, 1) = r(1,1,0) + (0,0,1).

3. (r+1, 2r, r) = r(1,2,1) + (1,0,0).

5. (-4r+3s-2, -5r+5s-6, r, s) = r(-4,-5,1,0) + s(3,5,0,1) + (-2,-6,0,0).

7. The system $A\mathbf{x} = \mathbf{y}$ has solutions if the vector **y** is in range(T), i.e. if the coordinates of $\mathbf{y} = (y_1, y_2, y_3)$ satisfy the condition $y_3 = y_1 + y_2$.

(a) 2 = 1 + 1, so solutions exist.

(d) $5 \neq 2 + 4$, so there are no solutions.

8. Any particular solution of the system will do in place of $\mathbf{x}_1$. (2,3,1) is another particular solution (obtained from the general solution by letting r = 1).

10. The coefficient matrix is A. Since $\mathbf{x}_1 = (1,-1,4)$ is a solution substitute $x_1 = 1$, $x_2 = -1$, $x_3 = 4$ in the left side of each equation to find the number on the right.

$$\begin{aligned} x_1 + 2x_2 + x_3 &= 3 \\ x_2 + 2x_3 &= 7 \\ x_1 + x_2 - x_3 &= -4 \end{aligned}$$

13. (a) $(D^2+D-2)e^{mx} = 0$, so $(m^2+m-2)e^{mx} = 0$, which gives $(m+2)(m-1) = 0$, so that $m = -2$ or $m = 1$. Thus a basis for the kernel is the set $\{e^{-2x}, e^x\}$.
$\ker(D^2+D-2) = \{ae^{-2x} + be^x\}$.

(c) $(D^2+2D-8)e^{mx} = 0$, so $(m^2+2m-8)e^{mx} = 0$, which gives $(m+4)(m-2) = 0$, so that $m = -4$ or $m = 2$. Thus a basis for the kernel is the set $\{e^{-4x}, e^{2x}\}$.
$\ker(D^2+2D-8) = \{ae^{-4x} + be^{2x}\}$.

14. (b) $(D^2-3D)e^{mx} = 0$ gives $(m^2-3m)e^{mx} = 0$, which gives $(m-3)m = 0$, so that $m = 3$ or $m = 0$. Thus a basis for the kernel is the set $\{e^{3x}, 1\}$.

A particular solution is given by $-3dy/dx = 8$ or $y = -8x/3$, so the general solution is $y = re^{3x} + s - 8x/3$.

(d) $(D^2+8D)e^{mx} = 0$ gives $(m^2+8m)e^{mx} = 0$, which gives $(m+8)m = 0$, so that $m = -8$ or $m = 0$. Thus a basis for the kernel is the set $\{e^{-8x}, 1\}$.

A particular solution is of the form $y = ke^{2x}$. We determine k. $dy/dx = 2ke^{2x}$ and $d^2y/dx^2 = 4ke^{2x}$, so that $d^2y/dx^2 + 8dy/dx = 20ke^{2x} = 3e^{2x}$. Thus $k = 3/20$, and the general solution is $y = re^{-8x} + s + 3e^{2x}/20$.

<u>Exercise Set 7.5</u>, page 323

1. $$\text{Pinv}\begin{bmatrix} 1 & 3 \\ 0 & 1 \\ -1 & 2 \end{bmatrix} = \left(\begin{bmatrix} 1 & 0 & -1 \\ 3 & 1 & 2 \end{bmatrix}\begin{bmatrix} 1 & 3 \\ 0 & 1 \\ -1 & 2 \end{bmatrix}\right)^{-1}\begin{bmatrix} 1 & 0 & -1 \\ 3 & 1 & 2 \end{bmatrix} = \begin{bmatrix} 2 & 1 \\ 1 & 14 \end{bmatrix}^{-1}\begin{bmatrix} 1 & 0 & -1 \\ 3 & 1 & 2 \end{bmatrix}$$

$$= \frac{1}{27}\begin{bmatrix} 14 & -1 \\ -1 & 2 \end{bmatrix}\begin{bmatrix} 1 & 0 & -1 \\ 3 & 1 & 2 \end{bmatrix} = \frac{1}{27}\begin{bmatrix} 11 & -1 & -16 \\ 5 & 2 & 5 \end{bmatrix}.$$

3. The pseudoinverse does not exist because $\left(\begin{bmatrix} 1 & 2 & 3 \\ 2 & 4 & 6 \end{bmatrix}\begin{bmatrix} 1 & 2 \\ 2 & 4 \\ 3 & 6 \end{bmatrix}\right)^{-1}$ does not exist.

6. There is no pseudoinverse. This is not the coefficient matrix of an overdetermined system of equations.

8. $\text{Pinv}\begin{bmatrix} 1 & 2 \\ -3 & 5 \end{bmatrix} = \left(\begin{bmatrix} 1 & -3 \\ 2 & 5 \end{bmatrix}\begin{bmatrix} 1 & 2 \\ -3 & 5 \end{bmatrix}\right)^{-1}\begin{bmatrix} 1 & -3 \\ 2 & 5 \end{bmatrix} = \frac{1}{11}\begin{bmatrix} 5 & -2 \\ 3 & 1 \end{bmatrix}.$

9. The equation of the line will be $y = a + bx$. The data points give $0 = a + b$, $4 = a + 2b$, and $7 = a + 3b$. Thus the system of equations is given by $A\begin{bmatrix} a \\ b \end{bmatrix} = \begin{bmatrix} 0 \\ 4 \\ 7 \end{bmatrix}$, where $A = \begin{bmatrix} 1 & 1 \\ 1 & 2 \\ 1 & 3 \end{bmatrix}$.

$$\text{Pinv } A = \left(\begin{bmatrix} 1 & 1 & 1 \\ 1 & 2 & 3 \end{bmatrix}\begin{bmatrix} 1 & 1 \\ 1 & 2 \\ 1 & 3 \end{bmatrix}\right)^{-1}\begin{bmatrix} 1 & 1 & 1 \\ 1 & 2 & 3 \end{bmatrix} = \frac{1}{6}\begin{bmatrix} 8 & 2 & -4 \\ -3 & 0 & 3 \end{bmatrix}.$$

$\text{Pinv } A\begin{bmatrix} 0 \\ 4 \\ 7 \end{bmatrix} = \frac{1}{6}\begin{bmatrix} -20 \\ 21 \end{bmatrix}$, so $a = \frac{-10}{3}$, $b = \frac{7}{2}$, and the least squares line is $y = \frac{-10}{3} + \frac{7}{2}x$.

11. The system of equations is given by $A\begin{bmatrix} a \\ b \end{bmatrix} = \begin{bmatrix} 1 \\ 5 \\ 9 \end{bmatrix}$, with A given in exercise 9.

$\text{Pinv } A\begin{bmatrix} 1 \\ 5 \\ 9 \end{bmatrix} = \frac{1}{6}\begin{bmatrix} 8 & 2 & -4 \\ -3 & 0 & 3 \end{bmatrix}\begin{bmatrix} 1 \\ 5 \\ 9 \end{bmatrix} = \frac{1}{6}\begin{bmatrix} -18 \\ 24 \end{bmatrix}$, so $a = -3$, $b = 4$, and the least squares line is $y = -3 + 4x$.

13. The system of equations is given by $A\begin{bmatrix} a \\ b \end{bmatrix} = \begin{bmatrix} 7 \\ 2 \\ 0 \end{bmatrix}$, with A given in exercise 9.

$$\text{Pinv } A\begin{bmatrix} 7 \\ 2 \\ 0 \end{bmatrix} = \frac{1}{6}\begin{bmatrix} 8 & 2 & -4 \\ -3 & 0 & 3 \end{bmatrix}\begin{bmatrix} 7 \\ 2 \\ 0 \end{bmatrix} = \frac{1}{6}\begin{bmatrix} 60 \\ -21 \end{bmatrix}, \text{ so}$$

a = 10, b = −7/2, and the least squares line is y = 10 − 7x/2.

15. The equation of the line will be y = a + bx. The data points give 0 = a + b, 3 = a + 2b,

5 = a + 3b, and 6 = a + 4b. Thus the system of equations is given by $A\begin{bmatrix} a \\ b \end{bmatrix} = \begin{bmatrix} 0 \\ 3 \\ 5 \\ 6 \end{bmatrix}$,

$$\text{where } A = \begin{bmatrix} 1 & 1 \\ 1 & 2 \\ 1 & 3 \\ 1 & 4 \end{bmatrix}. \text{ Pinv } A = \left(\begin{bmatrix} 1 & 1 & 1 & 1 \\ 1 & 2 & 3 & 4 \end{bmatrix}\begin{bmatrix} 1 & 1 \\ 1 & 2 \\ 1 & 3 \\ 1 & 4 \end{bmatrix}\right)^{-1}\begin{bmatrix} 1 & 1 & 1 & 1 \\ 1 & 2 & 3 & 4 \end{bmatrix}$$

$$= \frac{1}{10}\begin{bmatrix} 10 & 5 & 0 & -5 \\ -3 & -1 & 1 & 3 \end{bmatrix}. \text{ Pinv } A\begin{bmatrix} 0 \\ 3 \\ 5 \\ 6 \end{bmatrix} = \frac{1}{10}\begin{bmatrix} -15 \\ 20 \end{bmatrix}, \text{ so}$$

a = −3/2, b = 2, and the least squares line is y = −3/2 + 2x.

17. The system of equations is given by $A\begin{bmatrix} a \\ b \end{bmatrix} = \begin{bmatrix} 9 \\ 7 \\ 3 \\ 2 \end{bmatrix}$, with A given in exercise 15.

$$\text{Pinv } A\begin{bmatrix} 9 \\ 7 \\ 3 \\ 2 \end{bmatrix} = \frac{1}{10}\begin{bmatrix} 10 & 5 & 0 & -5 \\ -3 & -1 & 1 & 3 \end{bmatrix}\begin{bmatrix} 9 \\ 7 \\ 3 \\ 2 \end{bmatrix} = \frac{1}{2}\begin{bmatrix} 23 \\ -5 \end{bmatrix}, \text{ so}$$

a = 23/2, b = −5/2, and the least squares line is y = 23/2 − 5x/2.

19. The equation of the line will be $y = a + bx$. The data points give $0 = a + b$, $1 = a + 2b$, $4 = a + 3b$, $7 = a + 4b$, and $9 = a + 5b$. Thus the system of equations is given by

$$A\begin{bmatrix} a \\ b \end{bmatrix} = \begin{bmatrix} 0 \\ 1 \\ 4 \\ 7 \\ 9 \end{bmatrix}, \text{ where } A = \begin{bmatrix} 1 & 1 \\ 1 & 2 \\ 1 & 3 \\ 1 & 4 \\ 1 & 5 \end{bmatrix}.$$

$$\text{Pinv } A = \left(\begin{bmatrix} 1 & 1 & 1 & 1 & 1 \\ 1 & 2 & 3 & 4 & 5 \end{bmatrix} \begin{bmatrix} 1 & 1 \\ 1 & 2 \\ 1 & 3 \\ 1 & 4 \\ 1 & 5 \end{bmatrix} \right)^{-1} \begin{bmatrix} 1 & 1 & 1 & 1 & 1 \\ 1 & 2 & 3 & 4 & 5 \end{bmatrix}$$

$$= \frac{1}{10} \begin{bmatrix} 8 & 5 & 2 & -1 & -4 \\ -2 & -1 & 0 & 1 & 2 \end{bmatrix}. \quad \text{Pinv } A \begin{bmatrix} 0 \\ 1 \\ 4 \\ 7 \\ 9 \end{bmatrix} = \frac{1}{10} \begin{bmatrix} -30 \\ 24 \end{bmatrix}, \text{ so}$$

$a = -3$, $b = 12/5$, and the least squares line is $y = -3 + 12x/5$.

21. The equation of the parabola will be $y = a + bx + cx^2$. The data points give $5 = a + b + c$, $2 = a + 2b + 4c$, $3 = a + 3b + 9c$, and $8 = a + 4b + 16c$. Thus the system of equations is given by

$$A\begin{bmatrix} a \\ b \\ c \end{bmatrix} = \begin{bmatrix} 5 \\ 2 \\ 3 \\ 8 \end{bmatrix}, \text{ where } A = \begin{bmatrix} 1 & 1 & 1 \\ 1 & 2 & 4 \\ 1 & 3 & 9 \\ 1 & 4 & 16 \end{bmatrix}.$$

$$\text{Pinv } A = \left(\begin{bmatrix} 1 & 1 & 1 & 1 \\ 1 & 2 & 3 & 4 \\ 1 & 4 & 9 & 16 \end{bmatrix} \begin{bmatrix} 1 & 1 & 1 \\ 1 & 2 & 4 \\ 1 & 3 & 9 \\ 1 & 4 & 16 \end{bmatrix} \right)^{-1} \begin{bmatrix} 1 & 1 & 1 & 1 \\ 1 & 2 & 3 & 4 \\ 1 & 4 & 9 & 16 \end{bmatrix}$$

$$= \begin{bmatrix} 4 & 10 & 30 \\ 10 & 30 & 100 \\ 30 & 100 & 354 \end{bmatrix}^{-1} \begin{bmatrix} 1 & 1 & 1 & 1 \\ 1 & 2 & 3 & 4 \\ 1 & 4 & 9 & 16 \end{bmatrix} = \frac{1}{80} \begin{bmatrix} 620 & -540 & 100 \\ -540 & 516 & -100 \\ 100 & -100 & 20 \end{bmatrix} \begin{bmatrix} 1 & 1 & 1 & 1 \\ 1 & 2 & 3 & 4 \\ 1 & 4 & 9 & 16 \end{bmatrix}$$

$$= \frac{1}{20}\begin{bmatrix} 45 & -15 & -25 & 15 \\ -31 & 23 & 27 & -19 \\ 5 & -5 & -5 & 5 \end{bmatrix}. \text{ Pinv A}\begin{bmatrix} 5 \\ 2 \\ 3 \\ 8 \end{bmatrix} = \begin{bmatrix} 12 \\ -9 \\ 2 \end{bmatrix}, \text{ so}$$

$a = 12$, $b = -9$, $c = 2$, and the least squares parabola is $y = 12 - 9x + 2x^2$.

23. The system of equations is given by $A\begin{bmatrix} a \\ b \\ c \end{bmatrix} = \begin{bmatrix} 2 \\ 5 \\ 7 \\ 1 \end{bmatrix}$ with A given in exercise 21.

$$\text{Pinv A}\begin{bmatrix} 2 \\ 5 \\ 7 \\ 1 \end{bmatrix} = \frac{1}{20}\begin{bmatrix} 45 & -15 & -25 & 15 \\ -31 & 23 & 27 & -19 \\ 5 & -5 & -5 & 5 \end{bmatrix}\begin{bmatrix} 2 \\ 5 \\ 7 \\ 1 \end{bmatrix} = \frac{1}{20}\begin{bmatrix} -145 \\ 223 \\ -45 \end{bmatrix}, \text{ so } a = \frac{-29}{4},\ b = \frac{223}{20},$$

$c = \frac{-9}{4}$, and the least squares parabola is $y = \frac{-29}{4} + \frac{223}{20}x - \frac{9}{4}x^2$.

25. (a) The equation of the line will be $L = a + bF$. The data points give $5.9 = a + 2b$, $8 = a + 4b$, $10.3 = a + 6b$, and $12.2 = a + 8b$. Thus the system of equations is given by $A\begin{bmatrix} a \\ b \end{bmatrix} = \begin{bmatrix} 5.9 \\ 8 \\ 10.3 \\ 12.2 \end{bmatrix}$, where $A = \begin{bmatrix} 1 & 2 \\ 1 & 4 \\ 1 & 6 \\ 1 & 8 \end{bmatrix}$. Pinv $A = \begin{bmatrix} 1 & .5 & 0 & -.5 \\ -.15 & -.05 & .05 & .15 \end{bmatrix}$, as given in the text. Pinv $A\begin{bmatrix} 5.9 \\ 8 \\ 10.3 \\ 12.2 \end{bmatrix} = \begin{bmatrix} 3.8 \\ 1.06 \end{bmatrix}$, so $a = 3.8$, $b = 1.06$, and the least squares equation is $L = 3.8 + 1.06F$. If the force is 15 ounces, then the predicted length of the spring is $L = 3.8 + 1.06 \times 15 = 19.7$ inches.

26. The equation of the line will be $G = a + bS$. The data points give $18.25 = a + 30b$, $20 = a + 40b$, $16.32 = a + 50b$, $15.77 = a + 60b$, and $13.61 = a + 70b$. Thus the system of equations is given by $A\begin{bmatrix} a \\ b \end{bmatrix} = \begin{bmatrix} 18.25 \\ 20.00 \\ 16.32 \\ 15.77 \\ 13.61 \end{bmatrix}$, where $A = \begin{bmatrix} 1 & 30 \\ 1 & 40 \\ 1 & 50 \\ 1 & 60 \\ 1 & 70 \end{bmatrix}$. Pinv $A = (A^tA)^{-1} A^t$

$$= \frac{1}{5000}\begin{bmatrix} 13500 & -250 \\ -250 & 5 \end{bmatrix}\begin{bmatrix} 1 & 1 & 1 & 1 & 1 \\ 30 & 40 & 50 & 60 & 70 \end{bmatrix} = \frac{1}{50}\begin{bmatrix} 60 & 35 & 10 & -15 & -40 \\ 1 & .5 & 0 & .5 & 1 \end{bmatrix}.$$

Pinv $A\begin{bmatrix} 18.25 \\ 20.00 \\ 16.32 \\ 15.77 \\ 13.61 \end{bmatrix} = \frac{1}{50}\begin{bmatrix} 1177.25 \\ -6.755 \end{bmatrix}$, so $a = 23.545$, $b = -.1351$, and the least squares line is $G = 23.545 - .1351S$. The predicted mileage per gallon of this car at 55 mph is $G = 23.545 - .1351 \times 55 = 16.1145$ miles per gallon.

28. The equation of the line will be $U = a + bD$. The data points give $520 = a + 1000b$, $540 = a + 1500b$, $582 = a + 2000b$, $600 = a + 2500b$, $610 = a + 3000b$, and $615 = a + 3500b$. Thus the system of equations is given by $A\begin{bmatrix} a \\ b \end{bmatrix} = \begin{bmatrix} 520 \\ 540 \\ 582 \\ 600 \\ 610 \\ 615 \end{bmatrix}$,

where $A = \begin{bmatrix} 1 & 1000 \\ 1 & 1500 \\ 1 & 2000 \\ 1 & 2500 \\ 1 & 3000 \\ 1 & 3500 \end{bmatrix}$. Pinv $A = (A^tA)^{-1} A^t$

$$= \frac{1}{26250000}\begin{bmatrix} 34750000 & -13500 \\ -13500 & 6 \end{bmatrix}\begin{bmatrix} 1 & 1 & 1 & 1 & 1 & 1 \\ 1000 & 1500 & 2000 & 2500 & 3000 & 3500 \end{bmatrix}$$

$$= \frac{1}{262500}\begin{bmatrix} 212500 & 145000 & 77500 & 10000 & -57500 & -125000 \\ -75 & -45 & -15 & 15 & 45 & 75 \end{bmatrix}.$$

$$\text{Pinv } A\begin{bmatrix} 520 \\ 540 \\ 582 \\ 600 \\ 610 \\ 615 \end{bmatrix} = \frac{1}{262500}\begin{bmatrix} 127955000 \\ 10545 \end{bmatrix}, \text{ so } a = \frac{127955000}{262500},\ b = \frac{10545}{262500},$$

and the least squares line is $U = \frac{127955000 + 10545D}{262500}$. Predicted sales when

\$5000 is spent on advertising is $U = \frac{127955000 + 10545x5000}{262500} = 688$ units. This is

probably an unrealistic prediction. The data do not appear to be linear.

30. The equation of the least squares line will be % = a + bY. The system of equations is

given by $A\begin{bmatrix} a \\ b \end{bmatrix} = C$, where $C = \begin{bmatrix} .18 \\ .19 \\ .21 \\ .25 \\ .35 \\ .54 \\ .83 \end{bmatrix}$ and $A = \begin{bmatrix} 1 & 20 \\ 1 & 25 \\ 1 & 30 \\ 1 & 35 \\ 1 & 40 \\ 1 & 45 \\ 1 & 50 \end{bmatrix}$. Pinv $A = (A^tA)^{-1}A^t$

$$= \frac{1}{4900}\begin{bmatrix} 4375 & 3150 & 1925 & 700 & -525 & -1750 & -2975 \\ -105 & -70 & -35 & 0 & 35 & 70 & 105 \end{bmatrix}, \text{ as in exercise 27.}$$

$$(\text{Pinv } A)\ C = \frac{1}{4900}\begin{bmatrix} -1632.75 \\ 97.65 \end{bmatrix}, \text{ so the least squares line is given by}$$

$\% = \frac{-1632.75 + 97.65Y}{4900}$. The predicted percentage of deaths at 23 years of

age is $\frac{-1632.75 + 97.65x23}{4900} = .125$.

It is clear from the drawing below that the least squares line does not fit the data well.

The parabola fits better. The equation of the least squares parabola will be

$\% = a + bY + cY^2$. The system of equations is given by $A\begin{bmatrix} a \\ b \\ c \end{bmatrix} = C$, where

$$A = \begin{bmatrix} 1 & 20 & 400 \\ 1 & 25 & 625 \\ 1 & 30 & 900 \\ 1 & 35 & 1225 \\ 1 & 40 & 1600 \\ 1 & 45 & 2025 \\ 1 & 50 & 2500 \end{bmatrix} \text{ and C is given above. } (\text{Pinv } A)C = \begin{bmatrix} .93643 \\ -.05907 \\ .00113 \end{bmatrix}, \text{ so}$$

the least squares parabola is $\% = .93643 - .05907Y + .00113Y^2$. The predicted percentage of deaths at 23 years of age is $.93643 - .05907 \times 23 + .00113 \times 529 = .17559$.

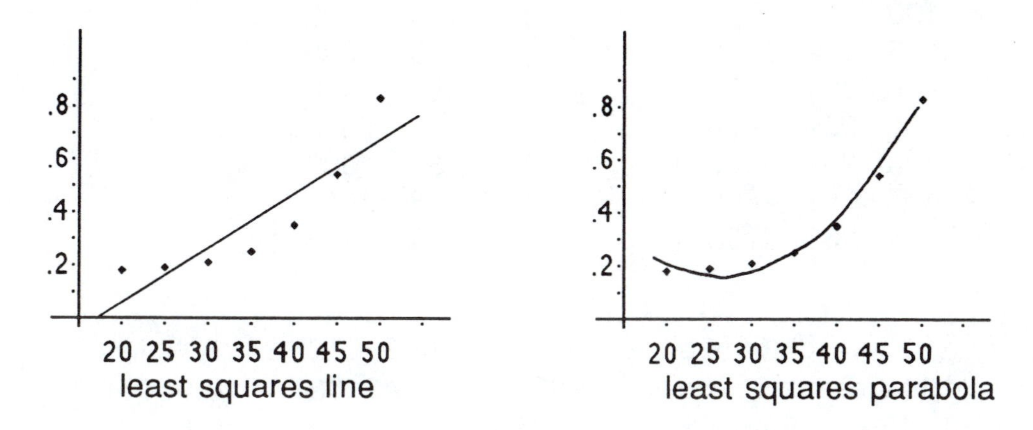

least squares line

least squares parabola

32. The equations will be of the form $N = a + bY$. Let 1975 be year 0. Then 1980 is year 5, 1983 is year 8 and 1984 is year 9. The system of equations for a and b is therefore

$$A\begin{bmatrix} a \\ b \end{bmatrix} = C, \text{ where C is the appropriate column of } D = \begin{bmatrix} 1308 & 5179 & 2017 \\ 1478 & 5313 & 2065 \\ 1784 & 5322 & 2026 \\ 1795 & 5185 & 2006 \end{bmatrix}$$

$$\text{and } A = \begin{bmatrix} 1 & 0 \\ 1 & 5 \\ 1 & 8 \\ 1 & 9 \end{bmatrix}. \quad (\text{Pinv } A)D = (A^tA)^{-1}A^tD = \frac{1}{196}\begin{bmatrix} 170 & -22 \\ -22 & 4 \end{bmatrix} A^tD$$

$$= \frac{1}{196}\begin{bmatrix} 250076 & 1022098 & 398466 \\ 11238 & 1246 & -160 \end{bmatrix}, \text{ so the least squares lines are}$$

$$\text{U.K.} \quad N = \frac{125038 + 5619Y}{98} \qquad \text{U.S.A.} \quad N = \frac{511049 + 623Y}{98} \qquad \text{Japan} \quad N = \frac{199233 - 80Y}{98}$$

The year 2000 is year 25. The least squares predictions are N = 2709 for U.K., 5374 for U.S.A., and 2013 for Japan.

35. The equations will be of the form $N = a + bY + cY^2$. Let 1950 be year 0. The system of equations for a, b, and c is $A \begin{bmatrix} a \\ b \\ c \end{bmatrix} = C$, where C is the appropriate column of

$$D = \begin{bmatrix} 554 & 84 & 122 & 166 & 2504 \\ 801 & 94 & 135 & 199 & 3014 \\ 873 & 99 & 143 & 214 & 3324 \\ 984 & 104 & 148 & 227 & 3683 \\ 1102 & 112 & 152 & 239 & 4076 \\ 1183 & 117 & 154 & 252 & 4453 \\ 1225 & 119 & 154 & 259 & 4685 \\ 1239 & 119 & 154 & 261 & 4763 \end{bmatrix}, \text{ and } A = \begin{bmatrix} 1 & 0 & 0 \\ 1 & 10 & 100 \\ 1 & 15 & 225 \\ 1 & 20 & 400 \\ 1 & 25 & 625 \\ 1 & 30 & 900 \\ 1 & 33 & 1089 \\ 1 & 34 & 1156 \end{bmatrix}.$$

$$(\text{Pinv } A)D = \begin{bmatrix} 555.61302 & 83.65595 & 121.31114 & 166.28642 & 2493.40194 \\ 24.11237 & 1.05255 & 1.78878 & 3.42106 & 48.06250 \\ -0.11360 & 0.00066 & -0.02385 & -0.01882 & 0.56248 \end{bmatrix},$$

so the least squares parabolas are

China $N = 555.61302 + 24.11237Y - 0.11360Y^2$
Japan $N = 83.65595 + 1.05255Y + 0.00066Y^2$
W. Europe $N = 121.31114 + 1.78878Y - 0.02385Y^2$
N. America $N = 166.28642 + 3.42106Y - 0.01882Y^2$
World $N = 2493.40194 + 48.06250Y + 0.56248Y^2$

The year 2000 is year 50. The least squares predictions are N = 1477 for China, 138 for Japan, 151 for W. Europe, 290 for N. America, and 6303 for the world.

36. (a) Let W be the plane given by the equation $x - y - z = 0$. For all points in this plane $x = y + z$, so $W = \{(y + z, y, z)\} = \{y(1,1,0) + z(1,0,1)\}$. Thus a basis for W is the set

$\{(1,1,0), (1,0,1)\}$ and $A = \begin{bmatrix} 1 & 1 \\ 1 & 0 \\ 0 & 1 \end{bmatrix}$. $(A^tA)^{-1} = \frac{1}{3}\begin{bmatrix} 2 & -1 \\ -1 & 2 \end{bmatrix}$, and the projection

matrix $P = A(A^tA)^{-1}A^t = \frac{1}{3}\begin{bmatrix} 2 & 1 & 1 \\ 1 & 2 & -1 \\ 1 & -1 & 2 \end{bmatrix}$. $P\begin{bmatrix} 1 \\ 2 \\ 0 \end{bmatrix} = \frac{1}{3}\begin{bmatrix} 4 \\ 5 \\ -1 \end{bmatrix}$, so the projection of

the vector $(1,2,0)$ onto the plane $x - y - z = 0$ is the vector $(4/3, 5/3, -1/3)$.

(c) Let W be the plane given by the equation $x - 2y + z = 0$. For all points in this plane $x = 2y - z$, so $W = \{(2y - z, y, z)\} = \{y(2,1,0) + z(-1,0,1)\}$. Thus a basis for W is the

set $\{(2,1,0), (-1,0,1)\}$ and $A = \begin{bmatrix} 2 & -1 \\ 1 & 0 \\ 0 & 1 \end{bmatrix}$. $(A^tA)^{-1} = \frac{1}{6}\begin{bmatrix} 2 & 2 \\ 2 & 5 \end{bmatrix}$, and the projection

matrix $P = A(A^tA)^{-1}A^t = \frac{1}{6}\begin{bmatrix} 5 & 2 & -1 \\ 2 & 2 & 2 \\ -1 & 2 & 5 \end{bmatrix}$. $P\begin{bmatrix} 0 \\ 3 \\ 0 \end{bmatrix} = \frac{1}{6}\begin{bmatrix} 6 \\ 6 \\ 6 \end{bmatrix}$, so the projection of

the vector $(0,3,0)$ onto the plane $x - 2y + z = 0$ is the vector $(1,1,1)$.

37. (a) $A = \begin{bmatrix} 1 & 1 \\ -1 & 1 \\ 1 & -1 \end{bmatrix}$ and $(A^tA)^{-1} = \frac{1}{8}\begin{bmatrix} 3 & 1 \\ 1 & 3 \end{bmatrix}$. Thus the projection matrix

$P = A(A^tA)^{-1}A^t = \frac{1}{8}\begin{bmatrix} 8 & 0 & 0 \\ 0 & 4 & -4 \\ 0 & -4 & 4 \end{bmatrix}$. $P\begin{bmatrix} -2 \\ 1 \\ 3 \end{bmatrix} = \frac{1}{8}\begin{bmatrix} -16 \\ -8 \\ 8 \end{bmatrix}$, so the projection of

the vector $(-2,1,3)$ onto W is the vector $(-2,-1,1)$.

39. $A\mathbf{x} = \mathbf{y}$, so $A^tA\mathbf{x} = A^t\mathbf{y}$, and if $(A^tA)^{-1}$ exists then $\mathbf{x} = (A^tA)^{-1}A^tA\mathbf{x} = (A^tA)^{-1}A^t\mathbf{y}$.

$\mathbf{x} = (A^tA)^{-1}A^t\mathbf{y}$ solves the system $A^tA\mathbf{x} = A^t\mathbf{y}$ but does not solve the original system unless A^t (and therefore A) is an invertible matrix. The first step of multiplying by A^t is not reversible unless A^t is invertible.

41. $pinv(AB) = ((AB)^t(AB))^{-1}(AB)^t$. If A is mxn and B is nxk then pinv(A) is nxm and pinv(B) is kxn, AB is mxk, and pinv (AB) is kxm. In general pinv(A)pinv(B) does not exist and pinv(B)pinv(A) is kxm, so (a) is not possible, but (b) is. In fact (b) obviously holds if A and B are invertible. (b) also holds for some fairly simple examples where A and B are not invertible, such as $A = \begin{bmatrix} 1 & 0 & 0 \\ 0 & 1 & 0 \\ 0 & 0 & 1 \\ 0 & 0 & 0 \end{bmatrix}$ and $B = \begin{bmatrix} 1 & 0 \\ 0 & 1 \\ 0 & 0 \end{bmatrix}$. However it is easy to find examples where (b) does not hold. If $A = \begin{bmatrix} 1 & 0 & 1 \\ 1 & 0 & 0 \\ 1 & 1 & 0 \\ 0 & 1 & 0 \end{bmatrix}$ and $B = \begin{bmatrix} 1 & 1 \\ 0 & 1 \\ 1 & 0 \end{bmatrix}$ then

$$pinv(B)pinv(A) = \frac{1}{3}\begin{bmatrix} 1 & -1 & 2 \\ 1 & 2 & -1 \end{bmatrix}\frac{1}{3}\begin{bmatrix} 0 & 2 & 1 & -1 \\ 0 & -1 & 1 & 2 \\ 3 & -2 & -1 & 1 \end{bmatrix} = \frac{1}{9}\begin{bmatrix} 6 & -1 & -2 & -1 \\ -3 & 2 & 4 & 2 \end{bmatrix}$$ and

$$pinv(AB) = \frac{1}{17}\begin{bmatrix} 9 & 2 & -3 & -5 \\ -4 & 1 & 7 & 6 \end{bmatrix}.$$

42. (b) Pinv(A) exists if and only if $(A^tA)^{-1}$ exists, and $(A^t A)^{-1}$ exists if and only if $|A^t A| \neq 0$. Since A (and therefore A^t) is square, $|A^tA| = |A^t||A|$, so $|A^tA| \neq 0$ if and only if $|A| = |A^t| \neq 0$. Thus pinv(A) exists if and only if A is invertible, and therefore $pinv(A) = A^{-1}$ (see exercise 40). Thus $pinv(pinv(A)) = pinv(A^{-1}) = (A^{-1})^{-1} = A$.

(c) As in (b), $pinv(A) = A^{-1}$ and $pinv(A^t) = (A^t)^{-1} = (A^{-1})^t = (pinv(A))^t$.

44. $P = A(A^tA)^{-1}A^t$, where the columns of A are a basis for the vector space.

(a) $P^t = (A(A^tA)^{-1}A^t)^t = (A^t)^t((A^tA)^{-1})^t A^t = A((A^tA)^t)^{-1}A^t = A(A^tA)^{-1}A^t = P$.

Exercise Set 7.6, page 334

1. $(2,-3) = 2(1,0) + -3(0,1)$, so $\mathbf{u}_B = \begin{bmatrix} 2 \\ -3 \end{bmatrix}$.

2. $(8,-1) = a(3,-1) + b(2,1) = (3a+2b,-a+b)$, so $8 = 3a+2b$ and $-1 = -a+b$. This system of equations has the unique solution $a = 2$, $b = 1$, so $\mathbf{u}_B = \begin{bmatrix} 2 \\ 1 \end{bmatrix}$.

6. $(6,-3,1) = a(1,-1,0) + b(2,1,-1) + c(2,0,0)$, so $6 = a+2b+2c$, $-3 = -a+b$, and $1 = -b$. This system of equations has the unique solution $a = 2$, $b = -1$, $c = 3$, so $\mathbf{u}_B = \begin{bmatrix} 2 \\ -1 \\ 3 \end{bmatrix}$.

7. $(3,-1,7) = a(2,0,-1) + b(0,1,3) + c(1,1,1)$, so $3 = 2a+c$, $-1 = b+c$, and $7 = -a+3b+c$. This system of equations has the unique solution $a = 16/3$, $b = 20/3$, and $c = -23/3$, so

$$\mathbf{u}_B = \begin{bmatrix} 16/3 \\ 20/3 \\ -23/3 \end{bmatrix}.$$

10. $2x - 6 = a(3x - 5) + b(x - 1)$, so $2 = 3a+b$ and $-6 = -5a-b$. This system of equations has the unique solution $a = 2$, $b = -4$, so $\mathbf{u}_B = \begin{bmatrix} 2 \\ -4 \end{bmatrix}$.

12. $3x^2 - 6x - 2 = a(x^2) + b(x - 1) + c(2x)$, so $3 = a$, $-6 = b+2c$, and $-2 = -b$. This system of equations has the unique solution $a = 3$, $b = 2$, $c = -4$, so $\mathbf{u}_B = \begin{bmatrix} 3 \\ 2 \\ -4 \end{bmatrix}$.

13. $\mathbf{u}\cdot(0,-1,0) = 0$, $\mathbf{u}\cdot\left(\frac{3}{5}, 0, -\frac{4}{5}\right) = 11$, $\mathbf{u}\cdot\left(\frac{4}{5}, 0, \frac{3}{5}\right) = -2$, so $\mathbf{u}_B = \begin{bmatrix} 0 \\ 11 \\ -2 \end{bmatrix}$.

15. $\mathbf{u}\cdot(1,0,0) = 2$, $\mathbf{u}\cdot\left(0, \frac{1}{\sqrt{2}}, \frac{1}{\sqrt{2}}\right) = \frac{5}{\sqrt{2}}$, $\mathbf{u}\cdot\left(0, \frac{1}{\sqrt{2}}, -\frac{1}{\sqrt{2}}\right) = -\frac{3}{\sqrt{2}}$, so $\mathbf{u}_B = \begin{bmatrix} 2 \\ 5/\sqrt{2} \\ -3/\sqrt{2} \end{bmatrix}$.

16. $P = \begin{bmatrix} 2 & 1 \\ 3 & 2 \end{bmatrix}$, and $\mathbf{u}_{B'} = P\mathbf{u}_B = \begin{bmatrix} 4 \\ 7 \end{bmatrix}$, $\mathbf{v}_{B'} = P\mathbf{v}_B = \begin{bmatrix} 5 \\ 7 \end{bmatrix}$, and $\mathbf{w}_{B'} = P\mathbf{w}_B = \begin{bmatrix} 8 \\ 14 \end{bmatrix}$.

18. $P = \begin{bmatrix} 1 & 2 \\ 1 & -3 \end{bmatrix}$, and $\mathbf{u}_{B'} = P\mathbf{u}_B = \begin{bmatrix} 2 \\ -3 \end{bmatrix}$, $\mathbf{v}_{B'} = P\mathbf{v}_B = \begin{bmatrix} -3 \\ 7 \end{bmatrix}$, and $\mathbf{w}_{B'} = P\mathbf{w}_B = \begin{bmatrix} 5 \\ 0 \end{bmatrix}$.

20. $P = \begin{bmatrix} 2 & -3 \\ -3 & 4 \end{bmatrix}$, and $\mathbf{u}_{B'} = P\mathbf{u}_B = \begin{bmatrix} -1 \\ 1 \end{bmatrix}$, $\mathbf{v}_{B'} = P\mathbf{v}_B = \begin{bmatrix} 6 \\ -9 \end{bmatrix}$, and $\mathbf{w}_{B'} = P\mathbf{w}_B = \begin{bmatrix} 2 \\ -4 \end{bmatrix}$.

21. The transition matrix from B' to B is $P = \begin{bmatrix} 5 & 3 \\ 3 & 2 \end{bmatrix}$, so the transition matrix from B to B' is $P^{-1} = \begin{bmatrix} 2 & -3 \\ -3 & 5 \end{bmatrix}$, and $\mathbf{u}_{B'} = P^{-1}\mathbf{u}_B = \begin{bmatrix} -19 \\ 31 \end{bmatrix}$.

23. The transition matrix from B to the standard basis is $R = \begin{bmatrix} 1 & 3 \\ 2 & 0 \end{bmatrix}$ and the transition matrix from B' to the standard basis is $Q = \begin{bmatrix} 2 & 3 \\ 1 & 2 \end{bmatrix}$. $P = Q^{-1}R = \begin{bmatrix} 2 & -3 \\ -1 & 2 \end{bmatrix}\begin{bmatrix} 1 & 3 \\ 2 & 0 \end{bmatrix}$ $= \begin{bmatrix} -4 & 6 \\ 3 & -3 \end{bmatrix}$, and $\mathbf{u}_{B'} = P\mathbf{u}_B = \begin{bmatrix} 24 \\ -15 \end{bmatrix}$.

25. $3x^2 = 3(x^2) + 0(x) + 0(1)$, $x - 1 = 0(x^2) + 1(x) + -1(1)$, and $4 = 0(x^2) + 0(x) + 4(1)$, so the transition matrix from B' to B is $P = \begin{bmatrix} 3 & 0 & 0 \\ 0 & 1 & 0 \\ 0 & -1 & 4 \end{bmatrix}$. The transition matrix from B to B' is $P^{-1} = \begin{bmatrix} 1/3 & 0 & 0 \\ 0 & 1 & 0 \\ 0 & 1/4 & 1/4 \end{bmatrix}$. The coordinate vectors relative to B' of $3x^2 + 4x + 8$, $6x^2 + 4$, $8x + 12$, and $3x^2 + 4x + 4$ are respectively $P^{-1}\begin{bmatrix} 3 \\ 4 \\ 8 \end{bmatrix} = \begin{bmatrix} 1 \\ 4 \\ 3 \end{bmatrix}$, $P^{-1}\begin{bmatrix} 6 \\ 0 \\ 4 \end{bmatrix} = \begin{bmatrix} 2 \\ 0 \\ 1 \end{bmatrix}$, $P^{-1}\begin{bmatrix} 0 \\ 8 \\ 12 \end{bmatrix} = \begin{bmatrix} 0 \\ 8 \\ 5 \end{bmatrix}$, and $P^{-1}\begin{bmatrix} 3 \\ 4 \\ 4 \end{bmatrix} = \begin{bmatrix} 1 \\ 4 \\ 2 \end{bmatrix}$.

26. $x + 2 = 1(x) + 2(1)$ and $3 = 0(x) + 3(1)$, so the transition matrix from B' to B is $P = \begin{bmatrix} 1 & 0 \\ 2 & 3 \end{bmatrix}$ and the transition matrix from B to B' is $P^{-1} = \frac{1}{3}\begin{bmatrix} 3 & 0 \\ -2 & 1 \end{bmatrix}$. The coordinate vectors relative to B' of $3x + 3$, $6x$, $6x + 9$, and $12x - 3$ are respectively $P^{-1}\begin{bmatrix} 3 \\ 3 \end{bmatrix} = \begin{bmatrix} 3 \\ -1 \end{bmatrix}$, $P^{-1}\begin{bmatrix} 6 \\ 0 \end{bmatrix} = \begin{bmatrix} 6 \\ -4 \end{bmatrix}$, $P^{-1}\begin{bmatrix} 6 \\ 9 \end{bmatrix} = \begin{bmatrix} 6 \\ -1 \end{bmatrix}$, and $P^{-1}\begin{bmatrix} 12 \\ -3 \end{bmatrix} = \begin{bmatrix} 12 \\ -9 \end{bmatrix}$.

27. Let $T\left(\begin{bmatrix} a & 0 \\ 0 & b \end{bmatrix}\right) = (a,b)$. $T\left(\begin{bmatrix} a & 0 \\ 0 & b \end{bmatrix} + \begin{bmatrix} d & 0 \\ 0 & e \end{bmatrix}\right) = T\left(\begin{bmatrix} a+d & 0 \\ 0 & b+e \end{bmatrix}\right) = (a+d,b+e)$

$= (a,b) + (d,e) = T\left(\begin{bmatrix} a & 0 \\ 0 & b \end{bmatrix}\right) + T\left(\begin{bmatrix} d & 0 \\ 0 & e \end{bmatrix}\right)$ and $T\left(c\begin{bmatrix} a & 0 \\ 0 & b \end{bmatrix}\right) = T\left(\begin{bmatrix} ca & 0 \\ 0 & cb \end{bmatrix}\right)$

$= (ca,cb) = c(a,b) = c\,T\left(\begin{bmatrix} a & 0 \\ 0 & b \end{bmatrix}\right)$, so T is linear. T is one-to-one because if

$T\left(\begin{bmatrix} x & 0 \\ 0 & y \end{bmatrix}\right) = (a,b)$ then x = a and y = b, and T is onto because if (a,b) is an element of

$\mathbf{R}^2$ then $T\left(\begin{bmatrix} a & 0 \\ 0 & b \end{bmatrix}\right) = (a,b)$. Thus T is an isomorphism.

Exercise Set 7.7, page 345

1. $T(0,1,-1) = T(0(1,0,0) + 1(0,1,0) - 1(0,0,1)) = T(0,1,0) - T(0,0,1) = (0,-2) - (-1,1)$

$= (1,-3)$. Alternatively, the matrix of T with respect to the standard basis is

$A = \begin{bmatrix} 2 & 0 & -1 \\ 1 & -2 & 1 \end{bmatrix}$. $A\begin{bmatrix} 0 \\ 1 \\ -1 \end{bmatrix} = \begin{bmatrix} 1 \\ -3 \end{bmatrix}$, so $T(0,1,-1) = (1,-3)$.

2. $T(3,-2) = T(3(1,0) - 2(0,1)) = 3T(1,0) - 2T(0,1) = 3(4) - 2(-3) = 18$. Alternatively, the matrix of T with respect to the standard basis is $[\,4\ \ -3\,]$. $[\,4\ \ -3\,]\begin{bmatrix} 3 \\ -2 \end{bmatrix} = 18$, so $T(3,-2) = 18$.

4. $T(3x^2 - 2x + 1) = T(3(x^2) - 2(x) + 1(1)) = 3T(x^2) - 2T(x) + T(1) = 3(3x+1) - 2(2) + 2x-5$

$= 9x + 3 - 4 + 2x - 5 = 11x - 6$. Alternatively, the matrix of T with respect to the standard basis is $A = \begin{bmatrix} 3 & 0 & 2 \\ 1 & 2 & -5 \end{bmatrix}$. $A\begin{bmatrix} 3 \\ -2 \\ 1 \end{bmatrix} = \begin{bmatrix} 11 \\ -6 \end{bmatrix}$, so $T(3x^2 - 2x + 1) = 11x - 6$.

6. The coordinate vectors of $T(\mathbf{u}_1)$ and $T(\mathbf{u}_2)$ relative to the basis $\{\mathbf{v}_1, \mathbf{v}_2\}$ are $\begin{bmatrix} 2 \\ 3 \end{bmatrix}$ and $\begin{bmatrix} 4 \\ -1 \end{bmatrix}$, so $A = \begin{bmatrix} 2 & 4 \\ 3 & -1 \end{bmatrix}$. The coordinate vector of $\mathbf{u}$ relative to the basis $\{\mathbf{u}_1, \mathbf{u}_2\}$ is $\begin{bmatrix} 2 \\ 5 \end{bmatrix}$, so the coordinate vector of $T(\mathbf{u})$ relative to $\{\mathbf{v}_1, \mathbf{v}_2\}$ is $A\begin{bmatrix} 2 \\ 5 \end{bmatrix} = \begin{bmatrix} 24 \\ 1 \end{bmatrix}$. Thus $T(\mathbf{u}) = 24\mathbf{v}_1 + \mathbf{v}_2$.

8. The coordinate vectors of $T(\mathbf{u}_1)$, $T(\mathbf{u}_2)$, and $T(\mathbf{u}_3)$ relative to the basis $\{\mathbf{v}_1, \mathbf{v}_2, \mathbf{v}_3\}$ are $\begin{bmatrix} 1 \\ 1 \\ 1 \end{bmatrix}$, $\begin{bmatrix} 3 \\ -2 \\ 0 \end{bmatrix}$, and $\begin{bmatrix} 1 \\ 2 \\ -1 \end{bmatrix}$, so $A = \begin{bmatrix} 1 & 3 & 1 \\ 1 & -2 & 2 \\ 1 & 0 & -1 \end{bmatrix}$. The coordinate vector of $\mathbf{u}$ relative to the basis $\{\mathbf{u}_1, \mathbf{u}_2, \mathbf{u}_3\}$ is $\begin{bmatrix} 3 \\ 2 \\ -5 \end{bmatrix}$, so the coordinate vector of $T(\mathbf{u})$ relative to $\{\mathbf{v}_1, \mathbf{v}_2, \mathbf{v}_3\}$ is

$A\begin{bmatrix} 3 \\ 2 \\ -5 \end{bmatrix} = \begin{bmatrix} 4 \\ -11 \\ 8 \end{bmatrix}$. Thus $T(\mathbf{u}) = 4\mathbf{v}_1 - 11\mathbf{v}_2 + 8\mathbf{v}_3$.

9. (a) $T(1,0,0) = (1,0) = 1(1,0) + 0(0,1)$, $T(0,1,0) = (0,0) = 0(1,0) + 0(0,1)$, and $T(0,0,1) = (0,1) = 0(1,0) + 1(0,1)$, so $A = \begin{bmatrix} 1 & 0 & 0 \\ 0 & 0 & 1 \end{bmatrix}$. $(1,2,3) = 1(1,0,0) + 2(0,1,0) + 3(0,0,1)$, so the coordinate vector of $T(1,2,3)$ relative to the standard basis of $\mathbf{R}^2$ is $A\begin{bmatrix} 1 \\ 2 \\ 3 \end{bmatrix} = \begin{bmatrix} 1 \\ 3 \end{bmatrix}$. Thus $T(1,2,3) = 1(1,0) + 3(0,1) = (1,3)$.

(c) $T(1,0,0) = (1,2) = 1(1,0) + 2(0,1)$, $T(0,1,0) = (1,-1) = 1(1,0) - 1(0,1)$, and $T(0,0,1) = (0,0) = 0(1,0) + 0(0,1)$, so $A = \begin{bmatrix} 1 & 1 & 0 \\ 2 & -1 & 0 \end{bmatrix}$. $(1,2,3) = 1(1,0,0) + 2(0,1,0) + 3(0,0,1)$, so the coordinate vector of $T(1,2,3)$ relative to the standard basis of $\mathbf{R}^2$ is $A\begin{bmatrix}1\\2\\3\end{bmatrix}$ $=\begin{bmatrix}3\\0\end{bmatrix}$. Thus $T(1,2,3) = 3(1,0) + 0(0,1) = (3,0)$.

10. (a) $T(1,0,0) = (1,0,0) = 1(1,0,0) + 0(0,1,0) + 0(0,0,1)$, $T(0,1,0) = (0,2,0) = 0(1,0,0)$ $+ 2(0,1,0)$, and $T(0,0,1) = (0,0,3) = 0(1,0,0) + 0(0,1,0) + 3(0,0,1)$, so $A = \begin{bmatrix}1&0&0\\0&2&0\\0&0&3\end{bmatrix}$.

$(-1,5,2) = -1(1,0,0) + 5(0,1,0) + 2(0,0,1)$, so the coordinate vector of $T(-1,5,2)$ relative to the standard basis of $\mathbf{R}^3$ is $A\begin{bmatrix}-1\\5\\2\end{bmatrix} = \begin{bmatrix}-1\\10\\6\end{bmatrix}$. Thus $T(-1,5,2) = -1(1,0,0)$ $+ 10(0,1,0) + 6(0,0,1) = (-1,10,6)$.

(c) $T(1,0,0) = (1,0,0)$, $T(0,1,0) = (0,0,0)$, and $T(0,0,1) = (0,0,0)$, so $A = \begin{bmatrix}1&0&0\\0&0&0\\0&0&0\end{bmatrix}$.

The coordinate vector of $T(-1,5,2)$ relative to the standard basis of $\mathbf{R}^3$ is

$A\begin{bmatrix}-1\\5\\2\end{bmatrix} = \begin{bmatrix}-1\\0\\0\end{bmatrix}$, so $T(-1,5,2) = (-1,0,0)$.

11. $T(\mathbf{u}_1) = (2,1) = -2\mathbf{u}_1' + 1\mathbf{u}_2'$, $T(\mathbf{u}_2) = (2,3) = -2\mathbf{u}_1' + 3\mathbf{u}_2'$, and $T(\mathbf{u}_3) = (-1,2) = 1\mathbf{u}_1' + 2\mathbf{u}_2'$,

so $A = \begin{bmatrix}-2&-2&1\\1&3&2\end{bmatrix}$. $\mathbf{u} = (3,-4,0) = 2\mathbf{u}_1 + \mathbf{u}_2 - \mathbf{u}_3$, so the coordinate vector of $\mathbf{u}$

relative to the basis $\{\mathbf{u}_1, \mathbf{u}_2, \mathbf{u}_3\}$ is $\begin{bmatrix}2\\1\\-1\end{bmatrix}$ and the coordinate vector of $T(\mathbf{u})$ relative to the basis $\{\mathbf{u}_1', \mathbf{u}_2'\}$ is $A\begin{bmatrix} 2 \\ 1 \\ -1 \end{bmatrix} = \begin{bmatrix} -7 \\ 3 \end{bmatrix}$. Thus $T(\mathbf{u}) = -7\mathbf{u}_1' + 3\mathbf{u}_2' = (7,3)$.

13. $T(\mathbf{u}_1) = T(1,2) = (2,3) = 2\mathbf{u}_1 + 1\mathbf{u}_2$ and $T(\mathbf{u}_2) = T(0,-1) = (0,-1) = 0\mathbf{u}_1 + 1\mathbf{u}_2$, so

$A = \begin{bmatrix} 2 & 0 \\ 1 & 1 \end{bmatrix}$. $\mathbf{u} = (-1,3) = -1\mathbf{u}_1 - 5\mathbf{u}_2$, so the coordinate vector of $\mathbf{u}$ relative to the

basis $\{\mathbf{u}_1, \mathbf{u}_2\}$ is $\begin{bmatrix} -1 \\ -5 \end{bmatrix}$ and the coordinate vector of $T(\mathbf{u})$ relative to the basis $\{\mathbf{u}_1, \mathbf{u}_2\}$ is

$A\begin{bmatrix} -1 \\ -5 \end{bmatrix} = \begin{bmatrix} -2 \\ -6 \end{bmatrix}$. Thus $T(\mathbf{u}) = -2\mathbf{u}_1 - 6\mathbf{u}_2 = (-2,2)$.

14. $D(2x^2) = 4x = 0(2x^2) + 4(x) + 0\,(-1)$, $D(x) = 1 = 0(2x^2) + 0(x) - (-1)$, and $D(-1) = 0$

$= 0(2x^2) + 0(x) + 0\,(-1)$, so $A = \begin{bmatrix} 0 & 0 & 0 \\ 4 & 0 & 0 \\ 0 & -1 & 0 \end{bmatrix}$. $3x^2 - 2x + 4 = \frac{3}{2}(2x^2) - 2(x) - 4(-1)$, so

the coordinate vector of $3x^2 - 2x + 4$ relative to the given basis is $\begin{bmatrix} 3/2 \\ -2 \\ -4 \end{bmatrix}$, and the

coordinate vector of $T(3x^2 - 2x + 4)$ is $A\begin{bmatrix} 3/2 \\ -2 \\ -4 \end{bmatrix} = \begin{bmatrix} 0 \\ 6 \\ 2 \end{bmatrix}$. Thus $T(3x^2 - 2x + 4)$

$= 0(2x^2) + 6(x) + 2(-1) = 6x - 2$.

16. (a) $T(x^2) = 0 = 0(x^2) + 0(x) + 0(1)$, $T(x) = x^2 + x = 1(x^2) + 1(x) + 0(1)$, and $T(1) = x^2 - x$

$= 1(x^2) - 1(x) + 0(1)$, so $A = \begin{bmatrix} 0 & 1 & 1 \\ 0 & 1 & -1 \\ 0 & 0 & 0 \end{bmatrix}$.

(b) $T(x) = x = 0(x^2) + 1(x) + 0(1)$ and $T(1) = x^2 + 1 = 1(x^2) + 0(x) + 1(1)$, so $A = \begin{bmatrix} 0 & 1 \\ 1 & 0 \\ 0 & 1 \end{bmatrix}$.

17. $T(x^2 + x) = x = 1(x) + 0(1)$, $T(x) = 0 = 0(x) + 0(1)$, and $T(1) = 1 = 0(x) + 1(1)$, so

$A = \begin{bmatrix} 1 & 0 & 0 \\ 0 & 0 & 1 \end{bmatrix}$. $3x^2 + 2x - 1 = 3(x^2 + x) - 1(x) - 1(1)$, so the coordinate vector for

$3x^2 + 2x - 1$ relative to the basis $\{x^2 + x, x, 1\}$ is $\begin{bmatrix} 3 \\ -1 \\ -1 \end{bmatrix}$, and the coordinate vector for

$T(3x^2 + 2x - 1)$ relative to the basis $\{x, 1\}$ is $A\begin{bmatrix} 3 \\ -1 \\ -1 \end{bmatrix} = \begin{bmatrix} 3 \\ -1 \end{bmatrix}$. Thus $T(3x^2 + 2x - 1)$

$= 3(x) - 1(1) = 3x - 1$.

19. $T(x) = x = 1(x) + 0\ (1)$ and $T(1) = x - 1 = 1(x) - 1(1)$, so $A = \begin{bmatrix} 1 & 1 \\ 0 & -1 \end{bmatrix}$.

$x + 1 = 1(x) + 1(1)$ and $x - 1 = 1(x) - 1(1)$, so $P = \begin{bmatrix} 1 & 1 \\ 1 & -1 \end{bmatrix}$ and $P^{-1} = \frac{1}{2}\begin{bmatrix} 1 & 1 \\ 1 & -1 \end{bmatrix}$.

$P^{-1}AP = \frac{1}{2}\begin{bmatrix} 1 & 1 \\ 3 & -1 \end{bmatrix}$, which is the matrix of T with respect to the basis $\{x+1, x-1\}$.

20. $T(1,0) = (2,1) = 2(1,0) + 1(0,1)$ and $T(0,1) = (0,1) = 0(1,0) + 1(0,1)$, so $A = \begin{bmatrix} 2 & 0 \\ 1 & 1 \end{bmatrix}$.

$(1,1) = 1(1,0) + 1(0,1)$ and $(2,1) = 2(1,0) + 1(0,1)$, so $P = \begin{bmatrix} 1 & 2 \\ 1 & 1 \end{bmatrix}$ and $P^{-1} = \begin{bmatrix} -1 & 2 \\ 1 & -1 \end{bmatrix}$.

$P^{-1}AP = \begin{bmatrix} 2 & 2 \\ 0 & 1 \end{bmatrix}$, which is the matrix of T with respect to the basis $\{(1,1), (2,1)\}$.

22. (a) Let $\{\mathbf{u}_1, \mathbf{u}_2, \ldots, \mathbf{u}_m\}$ be a basis for U, and let $\mathbf{v}_{m+1}, \ldots, \mathbf{v}_n$ be additional linearly independent vectors in V so that $\{\mathbf{u}_1, \mathbf{u}_2, \ldots, \mathbf{u}_m, \mathbf{v}_{m+1}, \ldots, \mathbf{v}_n\}$ is a basis for V. Let T be a linear transformation which maps each of $\mathbf{u}_1, \mathbf{u}_2, \ldots, \mathbf{u}_m$ into the zero vector of W. Then each element of U will be in the kernel of T. If $\dim(W) \geq n-m$, then $\mathbf{v}_{m+1}, \ldots, \mathbf{v}_n$ can be mapped by T into linearly independent vectors in W, and $U = \ker(T)$. If $\dim(W) < n-m$ then $T(\mathbf{v}_{m+1}), \ldots, T(\mathbf{v}_n)$ will be linearly dependent and there will be constants $a_{m+1}, \ldots, a_n$, not all zero, such that $T(a_{m+1}\mathbf{v}_{m+1}+ \ldots +a_n\mathbf{v}_n)$ $= a_{m+1}T(\mathbf{v}_{m+1})+ \ldots +a_nT(\mathbf{v}_n) = \mathbf{0}$. The vector $a_{m+1}\mathbf{v}_{m+1}+ \ldots +a_n\mathbf{v}_n$ is not in U, for if it were there would be constants $b_1, b_2, \ldots, b_m$ such that $b_1\mathbf{u}_1+b_2\mathbf{u}_2+ \ldots +b_m\mathbf{u}_m$ $= a_{m+1}\mathbf{v}_{m+1}+ \ldots +a_n\mathbf{v}_n$; i.e., $b_1\mathbf{u}_1+b_2\mathbf{u}_2+ \ldots +b_m\mathbf{u}_m-a_{m+1}\mathbf{v}_{m+1}- \ldots -a_n\mathbf{v}_n = \mathbf{0}$. This cannot be because the vectors $\mathbf{u}_1, \mathbf{u}_2, \ldots, \mathbf{u}_m, \mathbf{v}_{m+1}, \ldots, \mathbf{v}_n$ are linearly independent. Thus in this case U is contained in but not equal to ker(T).

(b) The set $\{(1,3,-1), (1,0,0), (0,1,0)\}$ is a basis for $\mathbf{R}^3$. Let T be the linear transformation given by $T(1,3,-1) = (0,0)$, $T(1,0,0) = (1,0)$ and $T(0,1,0) = (0,1)$. Let $\mathbf{u} = a_1(1,3,-1) + a_2(1,0,0) + a_3(0,1,0)$ and suppose $T(\mathbf{u}) = (0,0)$. Thus $a_1T(1,3,-1) + a_2T(1,0,0) + a_3T(0,1,0) = a_1(0,0) + a_2(1,0) + a_3(0,1) = (0,0)$, so that $a_2(1,0) + a_3(0,1) = (0,0)$. Since $(1,0)$ and $(0,1)$ are linearly independent, a_2 and a_3 must be zero and thus $\mathbf{u}$ is a multiple of $(1,3,-1)$.

23. The set $\{(2,-1), (1,0)\}$ is a basis for $\mathbf{R}^2$. Let T be the linear transformation given by $T(2,-1) = (0,0)$ and $T(1,0) = (1,0)$. Let $\mathbf{u} = a_1(2,-1) + a_2(1,0)$ and suppose $T(\mathbf{u}) = (0,0)$. Thus $a_1T(2,-1) + a_2T(1,0) = a_1(0,0) + a_2(1,0) = (0,0)$, so that $a_2(1,0) = (0,0)$, and therefore $a_2 = 0$. Thus $\mathbf{u}$ is a multiple of $(2,-1)$.

25. $T(\mathbf{u} + \mathbf{v}) = \mathbf{u} + \mathbf{v} = T(\mathbf{u}) + T(\mathbf{v})$ and $T(c\mathbf{u}) = c\mathbf{u} = cT(\mathbf{u})$, so T is linear. Let $B = \{\mathbf{u}_1, \mathbf{u}_2, \ldots, \mathbf{u}_n\}$ be a basis for U. $T(\mathbf{u}_i) = \mathbf{u}_i$, so if A is the matrix of T with respect to B, its ith column will have a 1 in the ith position and zeros everywhere else. Thus $A = I_n$.

28. If $\mathbf{u}$ is an element of U and $\mathbf{u}_B$ is the coordinate vector of $\mathbf{u}$ relative to basis B, then the coordinate vector of $T(\mathbf{u})$ relative to the basis B' is given by $A\mathbf{u}_B$, where A is the matrix of T with respect to the bases B and B'. If A is also the matrix of L with respect to the same two bases, then the coordinate vector of $L(\mathbf{u})$ relative to the basis B' will also be given by $A\mathbf{u}_B$. But this means that the coordinate vector of $L(\mathbf{u})$ relative to the basis B' is the same as the coordinate vector of $T(\mathbf{u})$ relative to the basis B'. Thus $L(\mathbf{u}) = T(\mathbf{u})$ for every vector $\mathbf{u}$ in U and therefore $L = T$.

30. $\begin{bmatrix} 1 & 0 & 0 \\ 0 & 0 & 1 \\ 0 & 1 & 0 \end{bmatrix}\begin{bmatrix} a & b & c \\ d & e & f \\ g & h & i \end{bmatrix} = \begin{bmatrix} a & b & c \\ g & h & i \\ d & e & f \end{bmatrix}$. $\begin{bmatrix} 1 & 0 & 0 \\ 2 & 1 & 0 \\ 0 & 0 & 1 \end{bmatrix}\begin{bmatrix} a & b & c \\ d & e & f \\ g & h & i \end{bmatrix} = \begin{bmatrix} a & b & c \\ d+2a & e+2b & f+2c \\ g & h & i \end{bmatrix}$.

31. The elementary matrix of T_1 is $E_1 = \begin{bmatrix} -3 & 0 & 0 \\ 0 & 1 & 0 \\ 0 & 0 & 1 \end{bmatrix}$.

$$\begin{bmatrix} -3 & 0 & 0 \\ 0 & 1 & 0 \\ 0 & 0 & 1 \end{bmatrix}\begin{bmatrix} a & b & c \\ d & e & f \\ g & h & i \end{bmatrix} = \begin{bmatrix} -3a & -3b & -3c \\ d & e & f \\ g & h & i \end{bmatrix}.$$

32. (a) $\begin{bmatrix} 0 & 0 & 1 \\ 0 & 1 & 0 \\ 1 & 0 & 0 \end{bmatrix}$ interchanges rows 1 and 3. $\begin{bmatrix} 0 & 0 & 1 \\ 0 & 1 & 0 \\ 1 & 0 & 0 \end{bmatrix}^{-1} = \begin{bmatrix} 0 & 0 & 1 \\ 0 & 1 & 0 \\ 1 & 0 & 0 \end{bmatrix}$.

(c) $\begin{bmatrix} 1 & 0 & 0 \\ 0 & 1 & 0 \\ 0 & 0 & 4 \end{bmatrix}$ multiplies row 3 by 4. $\begin{bmatrix} 1 & 0 & 0 \\ 0 & 1 & 0 \\ 0 & 0 & 4 \end{bmatrix}^{-1} = \begin{bmatrix} 0 & 1 & 0 \\ 1 & 0 & 0 \\ 0 & 0 & 1/4 \end{bmatrix}$.

(e) $\begin{bmatrix} 1 & 0 & 0 \\ 0 & 1 & 0 \\ -4 & 0 & 1 \end{bmatrix}$ adds -4 times row 1 to row 3. $\begin{bmatrix} 1 & 0 & 0 \\ 0 & 1 & 0 \\ -4 & 0 & 1 \end{bmatrix}^{-1} = \begin{bmatrix} 1 & 0 & 0 \\ 0 & 1 & 0 \\ 4 & 0 & 1 \end{bmatrix}$.

The inverse of an elementary matrix will have ones and zeros in exactly the same positions as the given matrix. Any other number on the diagonal of the original will be replaced by its reciprocal, and any other number not on the diagonal will be replaced by its negative.

Chapter 7 Review Exercises, page 348

1. (a) $T((x_1,y_1)+(x_2,y_2)) = T(x_1+x_2,y_1+y_2) = (2(x_1+x_2), y_1+y_2, y_1+y_2-x_1-x_2)$

$= (2x_1+2x_2, y_1+y_2, y_1-x_1+y_2-x_2) = (2x_1,y_1,y_1-x_1) + (2x_2,y_2,y_2-x_2) = T(x_1,y_1) + T(x_2,y_2)$

and $T(c(x,y)) = T(cx,cy) = (2cx,cy,cy-cx) = c(2x,y,y-x) = cT(x,y)$, so T is linear.

(b) $T(c(x,y)) = T(cx,cy) = (cx+cy, 2cy+3)$ and $cT(x,y) = c(x+y, 2y+3) = (cx+cy, 2cy+3c)$, so T is not linear.

2. $\begin{bmatrix} a & b \\ c & d \end{bmatrix}\begin{bmatrix} 1 \\ 2 \end{bmatrix} = \begin{bmatrix} 5 \\ 1 \end{bmatrix}$, so $a + 2b = 5$ and $c + 2d = 1$, and $\begin{bmatrix} a & b \\ c & d \end{bmatrix}\begin{bmatrix} 3 \\ -2 \end{bmatrix} = \begin{bmatrix} -1 \\ 11 \end{bmatrix}$, so $3a - 2b = -1$ and $3c - 2d = 11$. Thus $a = 1$, $b = 2$, $c = 3$, and $d = -1$, and the matrix is $\begin{bmatrix} a & b \\ c & d \end{bmatrix} = \begin{bmatrix} 1 & 2 \\ 3 & -1 \end{bmatrix}$.

3. If T is linear, then $T(a\mathbf{u} + b\mathbf{v}) = T(a\mathbf{u}) + T(b\mathbf{v}) = aT(\mathbf{u}) + bT(\mathbf{v})$.
If $T(a\mathbf{u} + b\mathbf{v}) = aT(\mathbf{u}) + bT(\mathbf{v})$, then $T(\mathbf{u} + \mathbf{v}) = T(\mathbf{u}) + T(\mathbf{v})$ and $T(c\mathbf{u}) = cT(\mathbf{u})$, so T is linear.

4. $\begin{bmatrix} 3 & 0 \\ 0 & 3 \end{bmatrix}\begin{bmatrix} \frac{\sqrt{3}}{2} & \frac{-1}{2} \\ \frac{1}{2} & \frac{\sqrt{3}}{2} \end{bmatrix} = \begin{bmatrix} \frac{3\sqrt{3}}{2} & \frac{-3}{2} \\ \frac{3}{2} & \frac{3\sqrt{3}}{2} \end{bmatrix}$.

5. $\begin{bmatrix} 1 \\ 0 \end{bmatrix} \mapsto \begin{bmatrix} 0 \\ -1 \end{bmatrix}$ and $\begin{bmatrix} 0 \\ 1 \end{bmatrix} \mapsto \begin{bmatrix} -1 \\ 0 \end{bmatrix}$, so $A = \begin{bmatrix} 0 & -1 \\ -1 & 0 \end{bmatrix}$.

6. $\begin{bmatrix} 1 \\ 0 \end{bmatrix} \mapsto \begin{bmatrix} \frac{1}{2} \\ \frac{-1}{2} \end{bmatrix}$ and $\begin{bmatrix} 0 \\ 1 \end{bmatrix} \mapsto \begin{bmatrix} \frac{-1}{2} \\ \frac{1}{2} \end{bmatrix}$, so $A = \begin{bmatrix} \frac{1}{2} & \frac{-1}{2} \\ \frac{-1}{2} & \frac{1}{2} \end{bmatrix}$.

7. $\begin{bmatrix} 5 & 0 \\ 0 & 2 \end{bmatrix}\begin{bmatrix} x \\ y \end{bmatrix} = \begin{bmatrix} 5x \\ 2y \end{bmatrix} = \begin{bmatrix} x' \\ y' \end{bmatrix}$. $y = -5x + 1$, so $\frac{y'}{2} = -5\frac{x'}{5} + 1$. Thus the images of the points on the line $y = -5x + 1$ are the points on the line $\frac{y}{2} = -x + 1$.

8. $\begin{bmatrix} 1 & 0 \\ 3 & 1 \end{bmatrix}\begin{bmatrix} x \\ y \end{bmatrix} = \begin{bmatrix} x \\ 3x+y \end{bmatrix} = \begin{bmatrix} x' \\ y' \end{bmatrix}$. $y = 2x + 3$, so $y' - 3x' = 2x' + 3$ and $y' = 5x' + 3$. Thus the images of the points on the line $y = 2x + 3$ are the points on the line $y = 5x + 3$.

9. $\begin{bmatrix} -1 & 0 \\ 0 & 1 \end{bmatrix}\begin{bmatrix} 2 & 0 \\ 0 & 1 \end{bmatrix}\begin{bmatrix} 1 & 0 \\ 3 & 1 \end{bmatrix} = \begin{bmatrix} -2 & 0 \\ 3 & 1 \end{bmatrix}$.

10. $\begin{bmatrix} 6 & 4 \\ 3 & 2 \end{bmatrix}\begin{bmatrix} x \\ y \end{bmatrix} = \begin{bmatrix} 6x+4y \\ 3x+2y \end{bmatrix}$; thus (x,y) is in the kernel of T if $3x+2y = 0$, i.e. if $y = -3x/2$.

The kernel of T is the set $\{(r,-3r/2)\}$ and the range is the set $\{(2r,r)\}$. Dim ker(T) = 1, dim range(T) = 1, and dim domain(T) = 2, so dim ker(T) + dim range(T) = dim domain(T).

11. $\begin{bmatrix} 1 & 1 & 1 \\ 0 & 1 & -1 \\ 2 & 3 & 1 \end{bmatrix}\begin{bmatrix} x \\ y \\ z \end{bmatrix} = \begin{bmatrix} x+y+z \\ y-z \\ 2x+3y+z \end{bmatrix}$; thus (x,y,z) is in the kernel of T if $x+y+z = 0$, $y-z = 0$, and $2x+3y+z = 0$, i.e., if $x = -2y$ and $z = y$. The kernel of T is the set $\{(-2r,r,r)\}$ with basis $\{(-2,1,1)\}$, and a basis for the range is the set $\{(1,0,2), (1,1,3)\}$.

12. (a) The dimension of the range of the transformation is the rank of the matrix = 2. The domain is $\mathbf{R}^4$, which has dimension 4, so the dimension of the kernel is 2. This transformation is not one-to-one.

(b) $|B| \neq 0$, so the transformation is one-to-one.

13. The kernel is the set $\{(0,r,r)\}$ with basis $\{(0,1,1)\}$ and the range is the set $\{(a,2a,b)\}$ with basis $\{(1,2,0), (0,0,1)\}$.

14. $g(a_2x^2 + a_1x + a_0) + g(b_2x^2 + b_1x + b_0) = (a_2 - a_1)x^3 - a_1x + 2a_0 + (b_2 - b_1)x^3 - b_1x + 2b_0$
$= (a_2 + b_2 - a_1 - b_1)x^3 - (a_1 + b_1)x + 2(a_0 + b_0) = g((a_2 + b_2)x^2 + (a_1 + b_1)x + a_0 + b_0)$
$= g(a_2x^2 + a_1x + a_0 + b_2x^2 + b_1x + b_0)$ and $g(c(a_2x^2 + a_1x + a_0)) = g(ca_2x^2 + ca_1x + ca_0)$
$= (ca_2 - ca_1)x^3 - ca_1x + 2ca_0 = c((a_2 - a_1)x^3 - a_1x + 2a_0) = cg(a_2x^2 + a_1x + a_0)$, so g is linear.

Ker(g) is the set of all polynomials $a_2x^2 + a_1x + a_0$ with $a_2 - a_1 = 0$, $a_1 = 0$, and $a_0 = 0$, i.e., $a_2 = a_1 = a_0 = 0$, so ker(g) is the zero polynomial. Range(g) is the set of all polynomials $ax^3 + bx + c$. The set $\{x^3, x, 1\}$ is a basis for range(g).

15. $(D^2 - 2D + 1)(a_nx^n + \ldots + a_1x + a_0) = n(n-1)a_nx^{n-2} + \ldots + 6a_3x + 2a_2$
$- 2(na_nx^{n-1} + \ldots + 2a_2x + a_1) + (a_nx^n + \ldots + a_1x + a_0)$.

$(D^2 - 2D + 1)(a_nx^n + \ldots + a_1x + a_0) + (D^2 - 2D + 1)(b_nx^n + \ldots + b_1x + b_0)$
$= n(n-1)a_nx^{n-2} + \ldots + 6a_3x + 2a_2 - 2(na_nx^{n-1} + \ldots + 2a_2x + a_1)$
$+ (a_nx^n + \ldots + a_1x + a_0) + n(n-1)b_nx^{n-2} + \ldots + 6b_3x + 2b_2$
$- 2(nb_nx^{n-1} + \ldots + 2b_2x + b_1) + (b_nx^n + \ldots + b_1x + b_0)$
$= n(n-1)(a_n + b_n)x^{n-2} + \ldots + 6(a_3 + b_3)x + 2(a_2 + b_2)$
$- 2(n(a_n + b_n)x^{n-1} + \ldots + 2(a_2 + b_2)x + a_1 + b_1) + ((a_n + b_n)x^n + \ldots + (a_1 + b_1)x + a_0 + b_0)$
$= (D^2 - 2D + 1)((a_n + b_n)x^n + \ldots + (a_1 + b_1)x + a_0 + b_0)$ and

$(D^2 - 2D + 1)(c(a_nx^n + \ldots + a_1x + a_0)) = (D^2 - 2D + 1)(ca_nx^n + \ldots + ca_1x + ca_0)$
$= n(n-1)ca_nx^{n-2} + \ldots + 6ca_3x + 2ca_2 - 2(nca_nx^{n-1} + \ldots + 2ca_2x + ca_1)$
$+ (ca_nx^n + \ldots + ca_1x + ca_0) = c(n(n-1)a_nx^{n-2} + \ldots + 6a_3x + 2a_2$
$- 2(na_nx^{n-1} + \ldots + 2a_2x + a_1) + (a_nx^n + \ldots + a_1x + a_0))$
$= c(D^2 - 2D + 1)(a_nx^n + \ldots + a_1x + a_0)$, so $(D^2 - 2D + 1)$ is linear.

$(D^2 - 2D + 1)(a_nx^n + \ldots + a_1x + a_0) = a_nx^n + (a_{n-1} - 2na_n)x^{n-1}$
$+(a_{n-2} - 2(n-1)a_{n-1} + n(n-1)a_n)x^{n-2} + \ldots + (a_1 - 4a_2 + 6a_3)x + (a_0 - 2a_1 + 2a_2)$. Thus if $(D^2 - 2D + 1)(a_nx^n + \ldots + a_1x + a_0) = 12x - 4$, then $a_n = a_{n-1} = \ldots = a_2 = 0$, $a_1 = 12$, and $a_0 = 20$. Thus the only polynomial mapped into $12x - 4$ is the polynomial $12x + 20$.

16. $\text{Pinv}\begin{bmatrix} 1 & 2 \\ 3 & 1 \\ 4 & 2 \end{bmatrix} = \left(\begin{bmatrix} 1 & 3 & 4 \\ 2 & 1 & 2 \end{bmatrix}\begin{bmatrix} 1 & 2 \\ 3 & 1 \\ 4 & 2 \end{bmatrix}\right)^{-1}\begin{bmatrix} 1 & 3 & 4 \\ 2 & 1 & 2 \end{bmatrix} = \frac{1}{65}\begin{bmatrix} -17 & 14 & 10 \\ 39 & -13 & 0 \end{bmatrix}$.

17. The equation of the parabola will be $y = a + bx + cx^2$. The data points give $6 = a + b + c$, $2 = a + 2b + 4c$, $5 = a + 3b + 9c$, and $9 = a + 4b + 16c$. Thus the system of equations is given by $A\begin{bmatrix} a \\ b \\ c \end{bmatrix} = \begin{bmatrix} 6 \\ 2 \\ 5 \\ 9 \end{bmatrix}$, where $A = \begin{bmatrix} 1 & 1 & 1 \\ 1 & 2 & 4 \\ 1 & 3 & 9 \\ 1 & 4 & 16 \end{bmatrix}$.

$$\text{Pinv } A = \left(\begin{bmatrix} 1 & 1 & 1 & 1 \\ 1 & 2 & 3 & 4 \\ 1 & 4 & 9 & 16 \end{bmatrix}\begin{bmatrix} 1 & 1 & 1 \\ 1 & 2 & 4 \\ 1 & 3 & 9 \\ 1 & 4 & 16 \end{bmatrix}\right)^{-1}\begin{bmatrix} 1 & 1 & 1 & 1 \\ 1 & 2 & 3 & 4 \\ 1 & 4 & 9 & 16 \end{bmatrix}$$

$$= \begin{bmatrix} 4 & 10 & 30 \\ 10 & 30 & 100 \\ 30 & 100 & 354 \end{bmatrix}^{-1}\begin{bmatrix} 1 & 1 & 1 & 1 \\ 1 & 2 & 3 & 4 \\ 1 & 4 & 9 & 16 \end{bmatrix} = \frac{1}{80}\begin{bmatrix} 620 & -540 & 100 \\ -540 & 516 & -100 \\ 100 & -100 & 20 \end{bmatrix}\begin{bmatrix} 1 & 1 & 1 & 1 \\ 1 & 2 & 3 & 4 \\ 1 & 4 & 9 & 16 \end{bmatrix}$$

$= \frac{1}{20}\begin{bmatrix} 45 & -15 & -25 & 15 \\ -31 & 23 & 27 & -19 \\ 5 & -5 & -5 & 5 \end{bmatrix}$. $\text{Pinv } A\begin{bmatrix} 6 \\ 2 \\ 5 \\ 9 \end{bmatrix} = \begin{bmatrix} 12.5 \\ -8.8 \\ 2 \end{bmatrix}$, so $a = 12.5$, $b = -8.8$, $c = 2$,

and the least squares parabola is $y = 12.5 - 8.8x + 2x^2$.

18. $B = \text{Pinv } A = (A^tA)^{-1}A^t$. If A is nxm then B is mxn.

(a) $ABA = A(A^tA)^{-1}A^tA = AI_m = A$. (b) $BAB = (A^tA)^{-1}A^tAB = I_mB = B$.

(c) AB is the projection matrix for the subspace spanned by the columns of A. This matrix was shown to be symmetric in exercise 43(a) of section 7.5.

(d) $BA = (A^tA)^{-1}A^tA = I_m$ and is therefore symmetric.

19. $(-1,18) = a(1,3) + b(-1,4)$, so $-1 = a - b$ and $18 = 3a + 4b$. This system of equations has the unique solution $a = 2$, $b = 3$, so the coordinate vector is $\begin{bmatrix} a \\ b \end{bmatrix} = \begin{bmatrix} 2 \\ 3 \end{bmatrix}$.

20. $3x^2 + 2x - 13 = a(x^2 + 1) + b(x + 2) + c(x - 3)$, so $3 = a$, $2 = b + c$, and $-13 = a + 2b - 3c$. This system of equations has the unique solution $a = 3$, $b = -2$, $c = 4$, so the coordinate vector is $\begin{bmatrix} a \\ b \\ c \end{bmatrix} = \begin{bmatrix} 3 \\ -2 \\ 4 \end{bmatrix}$.

21. $(0,5,-15)\cdot(0,1,0) = 5$, $(0,5,-15)\cdot(-3/5,0,4/5) = -12$, and $(0,5,-15)\cdot(4/5,0,3/5) = -9$, so the coordinate vector is $\begin{bmatrix} 5 \\ -12 \\ -9 \end{bmatrix}$.

22. $P = \begin{bmatrix} 1 & 5 \\ 3 & 2 \end{bmatrix}$, and $\mathbf{u}_{B'} = P\mathbf{u}_B = \begin{bmatrix} 8 \\ 11 \end{bmatrix}$, $\mathbf{v}_{B'} = P\mathbf{v}_B = \begin{bmatrix} -5 \\ 11 \end{bmatrix}$, and $\mathbf{w}_{B'} = P\mathbf{w}_B = \begin{bmatrix} 9 \\ 14 \end{bmatrix}$.

23. The transition matrix from B to the standard basis is $R = \begin{bmatrix} -1 & 2 \\ 2 & 1 \end{bmatrix}$ and the transition matrix from B' to the standard basis is $Q = \begin{bmatrix} 4 & -3 \\ 3 & 2 \end{bmatrix}$. $P = Q^{-1}R = \frac{1}{17}\begin{bmatrix} 2 & 3 \\ -3 & 4 \end{bmatrix}\begin{bmatrix} -1 & 2 \\ 2 & 1 \end{bmatrix}$ $= \frac{1}{17}\begin{bmatrix} 4 & 7 \\ 11 & -2 \end{bmatrix}$, and $\mathbf{u}_{B'} = P\mathbf{u}_B = \frac{1}{17}\begin{bmatrix} 23 \\ 42 \end{bmatrix}$.

24. The matrix with respect to the standard basis is $A = \begin{bmatrix} 3 & -1 \\ 2 & 4 \end{bmatrix}$. $A\begin{bmatrix} 2 \\ 7 \end{bmatrix} = \begin{bmatrix} -1 \\ 32 \end{bmatrix}$, so $T(2,7) = (-1,32)$.

25. The matrix of T with respect to the given bases is $A = \begin{bmatrix} 1 & 3 \\ 5 & -1 \\ -2 & 2 \end{bmatrix}$. $A\begin{bmatrix} 2 \\ -3 \end{bmatrix} = \begin{bmatrix} -7 \\ 13 \\ -10 \end{bmatrix}$, so $T(\mathbf{u}) = -7\mathbf{v}_1 + 13\mathbf{v}_2 - 10\mathbf{v}_3$.

26. $T(1,0,0) = (2,0)$, $T(0,1,0) = (0,-3)$, and $T(0,0,1) = (0,0)$, so the matrix of T with respect to the standard bases is $A = \begin{bmatrix} 2 & 0 & 0 \\ 0 & -3 & 0 \end{bmatrix}$. $A\begin{bmatrix} 1 \\ 2 \\ 3 \end{bmatrix} = \begin{bmatrix} 2 \\ -6 \end{bmatrix}$, so $T(1,2,3) = (2,-6)$.

27. $T(x^2) = x^2$, $T(x) = -x^2$, and $T(1) = 2x$, so $A = \begin{bmatrix} 1 & -1 & 0 \\ 0 & 0 & 2 \\ 0 & 0 & 0 \end{bmatrix}$. $A\begin{bmatrix} 2 \\ -1 \\ 3 \end{bmatrix} = \begin{bmatrix} 3 \\ 6 \\ 0 \end{bmatrix}$, so $T(2x^2 - x + 3) = 3x^2 + 6x$.

28. $T(1,0) = (3,1)$ and $T(0,1) = (0,-1)$, so the matrix of T with respect to the standard basis is $A = \begin{bmatrix} 3 & 0 \\ 1 & -1 \end{bmatrix}$. The transition matrix from the basis {(1,2), (2,3)} to the standard basis is $P = \begin{bmatrix} 1 & 2 \\ 2 & 3 \end{bmatrix}$, so $B = P^{-1}AP = \begin{bmatrix} -3 & 2 \\ 2 & -1 \end{bmatrix}\begin{bmatrix} 3 & 0 \\ 1 & -1 \end{bmatrix}\begin{bmatrix} 1 & 2 \\ 2 & 3 \end{bmatrix} = \begin{bmatrix} -11 & -20 \\ 7 & 13 \end{bmatrix}$ is the matrix of T with respect to the basis {(1,2), (2,3)}.

29. The elementary matrices of T_1 and T_2 are $E_1 = \begin{bmatrix} 1 & 0 & 0 \\ 0 & 1 & 0 \\ 0 & 0 & 2 \end{bmatrix}$ and $E_2 = \begin{bmatrix} 1 & 0 & 0 \\ 0 & 1 & 0 \\ -2 & 0 & 1 \end{bmatrix}$.

30. A is invertible if and only if A is row equivalent to the identity matrix I_n, and if A and I_n are row equivalent matrices, there are elementary matrices $E_1, E_2, \ldots, E_n$ such that $A = E_n \ldots E_2E_1 I_n = E_n \ldots E_2E_1$. Thus A is invertible if and only if A is a product of elementary matrices.

Chapter 8

Exercise Set 8.1, page 357

1. $\begin{vmatrix} 5-\lambda & 4 \\ 1 & 2-\lambda \end{vmatrix} = (5-\lambda)(2-\lambda) - 4 = \lambda^2 - 7\lambda + 6 = (\lambda-6)(\lambda-1)$, so the eigenvalues are $\lambda = 6$ and $\lambda = 1$. For $\lambda = 6$ the eigenvectors are the solutions of $\begin{bmatrix} -1 & 4 \\ 1 & -4 \end{bmatrix}\begin{bmatrix} x_1 \\ x_2 \end{bmatrix} = \mathbf{0}$, so the eigenvectors are vectors of the form $r\begin{bmatrix} 4 \\ 1 \end{bmatrix}$. For $\lambda = 1$ the eigenvectors are the solutions of $\begin{bmatrix} 4 & 4 \\ 1 & 1 \end{bmatrix}\begin{bmatrix} x_1 \\ x_2 \end{bmatrix} = \mathbf{0}$, so the eigenvectors are vectors of the form $s\begin{bmatrix} -1 \\ 1 \end{bmatrix}$.

4. $\begin{vmatrix} 5-\lambda & 2 \\ -8 & -3-\lambda \end{vmatrix} = (5-\lambda)(-3-\lambda) + 16 = \lambda^2 - 2\lambda + 1 = (\lambda-1)(\lambda-1)$, so the only eigenvalue is $\lambda = 1$. The eigenvectors are the solutions of $\begin{bmatrix} 4 & 2 \\ -8 & -4 \end{bmatrix}\begin{bmatrix} x_1 \\ x_2 \end{bmatrix} = \mathbf{0}$, so the eigenvectors are vectors of the form $r\begin{bmatrix} 1 \\ -2 \end{bmatrix}$.

5. $\begin{vmatrix} 1-\lambda & 2 \\ 2 & 1-\lambda \end{vmatrix} = (1-\lambda)(1-\lambda) - 4 = \lambda^2 - 2\lambda - 3 = (\lambda-3)(\lambda+1)$, so the eigenvalues are $\lambda = 3$ and $\lambda = -1$. For $\lambda = 3$ the eigenvectors are the solutions of $\begin{bmatrix} -2 & 2 \\ 2 & -2 \end{bmatrix}\begin{bmatrix} x_1 \\ x_2 \end{bmatrix} = \mathbf{0}$, so the eigenvectors are vectors of the form $r\begin{bmatrix} 1 \\ 1 \end{bmatrix}$. For $\lambda = -1$ the eigenvectors are the solutions of $\begin{bmatrix} 2 & 2 \\ 2 & 2 \end{bmatrix}\begin{bmatrix} x_1 \\ x_2 \end{bmatrix} = \mathbf{0}$, so the eigenvectors are vectors of the form $s\begin{bmatrix} -1 \\ 1 \end{bmatrix}$.

8. $\begin{vmatrix} 2-\lambda & -4 \\ -1 & 2-\lambda \end{vmatrix} = (2-\lambda)(2-\lambda) - 4 = \lambda^2 - 4\lambda = \lambda(\lambda-4)$, so the eigenvalues are $\lambda = 0$ and $\lambda = 4$. For $\lambda = 0$ the eigenvectors are the solutions of $\begin{bmatrix} 2 & -4 \\ -1 & 2 \end{bmatrix}\begin{bmatrix} x_1 \\ x_2 \end{bmatrix} = \mathbf{0}$, so the

eigenvectors are vectors of the form $r\begin{bmatrix} 2 \\ 1 \end{bmatrix}$. For $\lambda = 4$ the eigenvectors are the solutions of $\begin{bmatrix} -2 & -4 \\ -1 & -2 \end{bmatrix}\begin{bmatrix} x_1 \\ x_2 \end{bmatrix} = \mathbf{0}$, so the eigenvectors are vectors of the form $s\begin{bmatrix} -2 \\ 1 \end{bmatrix}$.

9. $\begin{vmatrix} 3-\lambda & 2 & -2 \\ -3 & -1-\lambda & 3 \\ 1 & 2 & -\lambda \end{vmatrix} = (3-\lambda)(-1-\lambda)(-\lambda) + 18 + 2(-1-\lambda) - 6(3-\lambda) - 6\lambda = -\lambda^3 + 2\lambda^2 + \lambda - 2$

$= (1-\lambda^2)(\lambda-2)$, so the eigenvalues are $\lambda = 1$, $\lambda = -1$, and $\lambda = 2$. For $\lambda = 1$ the eigenvectors are the solutions of $\begin{bmatrix} 2 & 2 & -2 \\ -3 & -2 & 3 \\ 1 & 2 & -1 \end{bmatrix}\begin{bmatrix} x_1 \\ x_2 \\ x_3 \end{bmatrix} = \mathbf{0}$, so the eigenvectors are vectors of the form $r\begin{bmatrix} 1 \\ 0 \\ 1 \end{bmatrix}$. For $\lambda = -1$ the eigenvectors are the solutions of

$\begin{bmatrix} 4 & 2 & -2 \\ -3 & 0 & 3 \\ 1 & 2 & 1 \end{bmatrix}\begin{bmatrix} x_1 \\ x_2 \\ x_3 \end{bmatrix} = \mathbf{0}$, so the eigenvectors are vectors of the form $s\begin{bmatrix} 1 \\ -1 \\ 1 \end{bmatrix}$.

For $\lambda = 2$ the eigenvectors are the solutions of $\begin{bmatrix} 1 & 2 & -2 \\ -3 & -3 & 3 \\ 1 & 2 & -2 \end{bmatrix}\begin{bmatrix} x_1 \\ x_2 \\ x_3 \end{bmatrix} = \mathbf{0}$, so the eigenvectors are vectors of the form $t\begin{bmatrix} 0 \\ 1 \\ 1 \end{bmatrix}$.

10. $\begin{vmatrix} 1-\lambda & -2 & 2 \\ -2 & 1-\lambda & 2 \\ -2 & 0 & 3-\lambda \end{vmatrix} = (1-\lambda)^2(3-\lambda)$, so the eigenvalues are $\lambda = 1$ and $\lambda = 3$. For $\lambda = 1$ the eigenvectors are the solutions of $\begin{bmatrix} 0 & -2 & 2 \\ -2 & 0 & 2 \\ -2 & 0 & 2 \end{bmatrix}\begin{bmatrix} x_1 \\ x_2 \\ x_3 \end{bmatrix} = \mathbf{0}$, so the eigenvectors are

vectors of the form $r\begin{bmatrix}1\\1\\1\end{bmatrix}$. For $\lambda = 3$ the eigenvectors are the solutions of

$$\begin{bmatrix}-2 & -2 & 2\\-2 & -2 & 2\\-2 & 0 & 0\end{bmatrix}\begin{bmatrix}x_1\\x_2\\x_3\end{bmatrix} = \mathbf{0},$$ so the eigenvectors are vectors of the form $s\begin{bmatrix}0\\1\\1\end{bmatrix}$.

13. $\begin{vmatrix}15-\lambda & 7 & -7\\-1 & 1-\lambda & 1\\13 & 7 & -5-\lambda\end{vmatrix} = (1-\lambda)(16-10\lambda+\lambda^2) = (1-\lambda)(2-\lambda)(8-\lambda)$, so the eigenvalues are

$\lambda = 1$, $\lambda = 2$, and $\lambda = 8$. For $\lambda = 1$ the eigenvectors are the solutions of

$$\begin{bmatrix}14 & 7 & -7\\-1 & 0 & 1\\13 & 7 & -6\end{bmatrix}\begin{bmatrix}x_1\\x_2\\x_3\end{bmatrix} = \mathbf{0},$$ so the eigenvectors are vectors of the form $r\begin{bmatrix}1\\-1\\1\end{bmatrix}$. For $\lambda = 2$

the eigenvectors are the solutions of $\begin{bmatrix}13 & 7 & -7\\-1 & -1 & 1\\13 & 7 & -7\end{bmatrix}\begin{bmatrix}x_1\\x_2\\x_3\end{bmatrix} = \mathbf{0}$, so the eigenvectors are

vectors of the form $s\begin{bmatrix}0\\1\\1\end{bmatrix}$. For $\lambda = 8$ the eigenvectors are the solutions of

$$\begin{bmatrix}7 & 7 & -7\\-1 & -7 & 1\\13 & 7 & -13\end{bmatrix}\begin{bmatrix}x_1\\x_2\\x_3\end{bmatrix} = \mathbf{0},$$ so the eigenvectors are vectors of the form $t\begin{bmatrix}1\\0\\1\end{bmatrix}$.

15. $\begin{vmatrix}4-\lambda & 2 & -2 & 2\\1 & 3-\lambda & 1 & -1\\0 & 0 & 2-\lambda & 0\\1 & 1 & -3 & 5-\lambda\end{vmatrix} = (2-\lambda)\begin{vmatrix}4-\lambda & 2 & 2\\1 & 3-\lambda & -1\\1 & 1 & 5-\lambda\end{vmatrix} = (2-\lambda)(4-\lambda)(2-\lambda)(6-\lambda)$, so the

eigenvalues are $\lambda = 2$, $\lambda = 4$, and $\lambda = 6$. For $\lambda = 2$ the eigenvectors are the solutions of

$$\begin{bmatrix} 2 & 2 & -2 & 2 \\ 1 & 1 & 1 & -1 \\ 0 & 0 & 0 & 0 \\ 1 & 1 & -3 & 3 \end{bmatrix}\begin{bmatrix} x_1 \\ x_2 \\ x_3 \\ x_4 \end{bmatrix} = \mathbf{0}$$, so the eigenvectors are vectors of the form

$$r\begin{bmatrix} 1 \\ -1 \\ 0 \\ 0 \end{bmatrix} + s\begin{bmatrix} 0 \\ 0 \\ 1 \\ 1 \end{bmatrix}$$. For $\lambda = 4$ the eigenvectors are the solutions of

$$\begin{bmatrix} 0 & 2 & -2 & 2 \\ 1 & -1 & 1 & -1 \\ 0 & 0 & -2 & 0 \\ 1 & 1 & -3 & 1 \end{bmatrix}\begin{bmatrix} x_1 \\ x_2 \\ x_3 \\ x_4 \end{bmatrix} = \mathbf{0}$$, so the eigenvectors are vectors of the form $t\begin{bmatrix} 0 \\ 1 \\ 0 \\ -1 \end{bmatrix}$.

For $\lambda = 6$ the eigenvectors are the solutions of $\begin{bmatrix} -2 & 2 & -2 & 2 \\ 1 & -3 & 1 & -1 \\ 0 & 0 & -4 & 0 \\ 1 & 1 & -3 & -1 \end{bmatrix}\begin{bmatrix} x_1 \\ x_2 \\ x_3 \\ x_4 \end{bmatrix} = \mathbf{0}$, so the

eigenvectors are vectors of the form $p\begin{bmatrix} 1 \\ 0 \\ 0 \\ 1 \end{bmatrix}$.

17. $\begin{vmatrix} 1-\lambda & 0 \\ 0 & 1-\lambda \end{vmatrix} = (1-\lambda)^2$, so the only eigenvalue is $\lambda = 1$. The eigenvectors are the

solutions of $\begin{bmatrix} 0 & 0 \\ 0 & 0 \end{bmatrix}\begin{bmatrix} x_1 \\ x_2 \end{bmatrix} = \mathbf{0}$, so the eigenvectors are all the vectors in $\mathbf{R}^2$. The

transformation represented by the identity matrix is the identity transformation which maps each vector in $\mathbf{R}^2$ into itself.

19. $\begin{vmatrix} -2-\lambda & 0 \\ 0 & -2-\lambda \end{vmatrix} = (-2-\lambda)^2$, so the only eigenvalue is $\lambda = -2$. The eigenvectors are the

solutions of $\begin{bmatrix} 0 & 0 \\ 0 & 0 \end{bmatrix}\begin{bmatrix} x_1 \\ x_2 \end{bmatrix} = \mathbf{0}$, so the eigenvectors are all the vectors in $\mathbf{R}^2$. The transformation represented by the given matrix maps each vector **v** in $\mathbf{R}^2$ into the vector $-2\mathbf{v}$. Thus each vector in $\mathbf{R}^2$ has the direction opposite the direction of its image.

20. $\begin{vmatrix} 1-\lambda & 1 \\ -2 & -1-\lambda \end{vmatrix} = (1-\lambda)(-1-\lambda) + 2 = \lambda^2 + 1 \neq 0$ for any real value of λ, so there are no real eigenvalues.

22. If A is a diagonal matrix with diagonal elements a_{ii} then $A - \lambda I_n$ is also a diagonal matrix with diagonal elements $a_{ii} - \lambda$. Thus $|A - \lambda I_n|$ is the product of the terms $a_{ii} - \lambda$, and the solutions of the equation $|A - \lambda I_n| = 0$ are the values $\lambda = a_{ii}$, the diagonal elements of A.

24. $(A - \lambda I_n)^t = A^t - (\lambda I_n)^t = A^t - \lambda I_n$, so $|A - \lambda I_n| = |(A - \lambda I_n)^t| = |A^t - \lambda I_n|$; i.e., A and A^t have the same characteristic polynomial and therefore the same eigenvalues.

27. $|A - 0I_n| = |A|$, so $|A - 0I_n| = 0$ if and only if $|A| = 0$.

29. The characteristic polynomial of A is $|A - \lambda I_n| = \lambda^n + c_{n-1}\lambda^{n-1} + \ldots + c_1\lambda + c_0$. Substituting $\lambda = 0$, this equation becomes $|A| = c_0$.

31. (a) $\begin{vmatrix} -\lambda & 2 \\ -1 & 3-\lambda \end{vmatrix} = (-\lambda)(3-\lambda) + 2 = \lambda^2 - 3\lambda + 2.$

$$\begin{bmatrix} 0 & 2 \\ -1 & 3 \end{bmatrix}^2 - 3\begin{bmatrix} 0 & 2 \\ -1 & 3 \end{bmatrix} + 2\begin{bmatrix} 1 & 0 \\ 0 & 1 \end{bmatrix} = \begin{bmatrix} -2 & 6 \\ -3 & 7 \end{bmatrix} + \begin{bmatrix} 2 & -6 \\ 3 & -7 \end{bmatrix} = \begin{bmatrix} 0 & 0 \\ 0 & 0 \end{bmatrix}.$$

(c) $\begin{vmatrix} 6-\lambda & -8 \\ 4 & -6-\lambda \end{vmatrix} = (6-\lambda)(-6-\lambda) + 32 = \lambda^2 - 4.$

$$\begin{bmatrix} 6 & -8 \\ 4 & -6 \end{bmatrix}^2 - 4\begin{bmatrix} 1 & 0 \\ 0 & 1 \end{bmatrix} = \begin{bmatrix} 4 & 0 \\ 0 & 4 \end{bmatrix} - \begin{bmatrix} 4 & 0 \\ 0 & 4 \end{bmatrix} = \begin{bmatrix} 0 & 0 \\ 0 & 0 \end{bmatrix}.$$

Exercise Set 8.2, page 367

1. (a) $C^{-1}AC = \begin{bmatrix} 3 & -5 \\ -1 & 2 \end{bmatrix}\begin{bmatrix} 1 & 2 \\ -1 & 3 \end{bmatrix}\begin{bmatrix} 2 & 5 \\ 1 & 3 \end{bmatrix} = \begin{bmatrix} 7 & 13 \\ -2 & -3 \end{bmatrix}$.

(c) $C^{-1}AC = \begin{bmatrix} 4 & -1 \\ -7 & 2 \end{bmatrix}\begin{bmatrix} 0 & 4 \\ 3 & 2 \end{bmatrix}\begin{bmatrix} 2 & 1 \\ 7 & 4 \end{bmatrix} = \begin{bmatrix} 92 & 53 \\ -156 & -90 \end{bmatrix}$.

2. (a) $C^{-1}AC = \begin{bmatrix} -1 & 2 & 2 \\ 0 & 1 & 1 \\ 2 & 0 & -1 \end{bmatrix}\begin{bmatrix} 2 & 0 & 0 \\ -2 & 2 & 1 \\ 2 & 0 & 1 \end{bmatrix}\begin{bmatrix} -1 & 2 & 0 \\ 2 & -3 & 1 \\ -2 & 4 & -1 \end{bmatrix} = \begin{bmatrix} 2 & 0 & 0 \\ 0 & 2 & 0 \\ 0 & 0 & 1 \end{bmatrix}$.

3. (a) The eigenvalues and eigenvectors of this matrix were found in exercise 1 in Section 8.1. They are $\lambda = 6$ with eigenvectors $r\begin{bmatrix} 4 \\ 1 \end{bmatrix}$ and $\lambda = 1$ with eigenvectors $s\begin{bmatrix} -1 \\ 1 \end{bmatrix}$. $\frac{1}{5}\begin{bmatrix} 1 & 1 \\ -1 & 4 \end{bmatrix}\begin{bmatrix} 5 & 4 \\ 1 & 2 \end{bmatrix}\begin{bmatrix} 4 & -1 \\ 1 & 1 \end{bmatrix} = \begin{bmatrix} 6 & 0 \\ 0 & 1 \end{bmatrix}$.

(c) $\begin{vmatrix} 1-\lambda & 1 \\ 0 & 1-\lambda \end{vmatrix} = (1-\lambda)(1-\lambda)$, so the only eigenvalue is $\lambda = 1$. The eigenvectors are the solutions of $\begin{bmatrix} 0 & 1 \\ 0 & 0 \end{bmatrix}\begin{bmatrix} x_1 \\ x_2 \end{bmatrix} = \mathbf{0}$, so the eigenvectors are vectors of the form $r\begin{bmatrix} 1 \\ 0 \end{bmatrix}$.

There are not two linearly independent eigenvectors, so the matrix cannot be diagonalized.

(d) $\begin{vmatrix} 4-\lambda & -1 \\ 2 & 1-\lambda \end{vmatrix} = (4-\lambda)(1-\lambda) + 2 = \lambda^2 - 5\lambda + 6 = (\lambda-2)(\lambda-3)$, so the eigenvalues are $\lambda = 2$ and $\lambda = 3$. For $\lambda = 2$ the eigenvectors are the solutions of $\begin{bmatrix} 2 & -1 \\ 2 & -1 \end{bmatrix}\begin{bmatrix} x_1 \\ x_2 \end{bmatrix} = \mathbf{0}$, so the eigenvectors are vectors of the form $r\begin{bmatrix} 1 \\ 2 \end{bmatrix}$. For $\lambda = 3$ the eigenvectors are the solutions of $\begin{bmatrix} 1 & -1 \\ 2 & -2 \end{bmatrix}\begin{bmatrix} x_1 \\ x_2 \end{bmatrix} = \mathbf{0}$, so the eigenvectors are vectors of the form

$s\begin{bmatrix}1\\1\end{bmatrix}$. $\begin{bmatrix}-1 & 1\\ 2 & -1\end{bmatrix}\begin{bmatrix}4 & -1\\ 2 & 1\end{bmatrix}\begin{bmatrix}1 & 1\\ 2 & 1\end{bmatrix} = \begin{bmatrix}2 & 0\\ 0 & 3\end{bmatrix}$.

4. (a) $\begin{vmatrix}-7-\lambda & 10\\ -5 & 8-\lambda\end{vmatrix} = (-7-\lambda)(8-\lambda) + 50 = \lambda^2 - \lambda - 6 = (\lambda+2)(\lambda-3)$, so the eigenvalues are $\lambda = -2$ and $\lambda = 3$. For $\lambda = -2$ the eigenvectors are the solutions of $\begin{bmatrix}-5 & 10\\ -5 & 10\end{bmatrix}\begin{bmatrix}x_1\\ x_2\end{bmatrix} = \mathbf{0}$, so the eigenvectors are vectors of the form $r\begin{bmatrix}2\\1\end{bmatrix}$. For $\lambda = 3$ the eigenvectors are the solutions of $\begin{bmatrix}-10 & 10\\ -5 & 5\end{bmatrix}\begin{bmatrix}x_1\\ x_2\end{bmatrix} = \mathbf{0}$, so the eigenvectors are vectors of the form $s\begin{bmatrix}1\\1\end{bmatrix}$. $\begin{bmatrix}1 & -1\\ -1 & 2\end{bmatrix}\begin{bmatrix}-7 & 10\\ -5 & 8\end{bmatrix}\begin{bmatrix}2 & 1\\ 1 & 1\end{bmatrix} = \begin{bmatrix}-2 & 0\\ 0 & 3\end{bmatrix}$.

(c) $\begin{vmatrix}1-\lambda & -2\\ 2 & -3-\lambda\end{vmatrix} = (1-\lambda)(-3-\lambda) + 4 = \lambda^2 + 2\lambda + 1 = (\lambda+1)(\lambda+1)$, so the only eigenvalue is $\lambda = -1$. The eigenvectors are the solutions of $\begin{bmatrix}2 & -2\\ 2 & -2\end{bmatrix}\begin{bmatrix}x_1\\ x_2\end{bmatrix} = \mathbf{0}$, so the eigenvectors are vectors of the form $r\begin{bmatrix}1\\-1\end{bmatrix}$. Since there are not two linearly independent eigenvectors the matrix cannot be diagonalized.

(e) $\begin{vmatrix}a-\lambda & b\\ 0 & a-\lambda\end{vmatrix} = (a-\lambda)(a-\lambda)$, so the only eigenvalue is $\lambda = a$. The eigenvectors are the solutions of $\begin{bmatrix}0 & b\\ 0 & 0\end{bmatrix}\begin{bmatrix}x_1\\ x_2\end{bmatrix} = \mathbf{0}$, so the eigenvectors are vectors of the form $r\begin{bmatrix}1\\0\end{bmatrix}$. There are not two linearly independent eigenvectors, so the matrix cannot be diagonalized.

5. (a) The eigenvalues and eigenvectors of this matrix were found in exercise 13 in Section 8.1. They are $\lambda = 1$, $\lambda = 2$, $\lambda = 8$, with corresponding eigenvectors

$$r\begin{bmatrix}1\\-1\\1\end{bmatrix}, s\begin{bmatrix}0\\1\\1\end{bmatrix}, t\begin{bmatrix}1\\0\\1\end{bmatrix}. \quad \begin{bmatrix}-1 & -1 & 1\\-1 & 0 & 1\\2 & 1 & -1\end{bmatrix}\begin{bmatrix}15 & 7 & -7\\-1 & 1 & 1\\13 & 7 & -5\end{bmatrix}\begin{bmatrix}1 & 0 & 1\\-1 & 1 & 0\\1 & 1 & 1\end{bmatrix}=\begin{bmatrix}1 & 0 & 0\\0 & 2 & 0\\0 & 0 & 8\end{bmatrix}.$$

(c) $\begin{vmatrix}1-\lambda & 0 & 0\\-2 & 1-\lambda & 2\\-2 & 0 & 3-\lambda\end{vmatrix} = (1-\lambda)^2(3-\lambda)$, so the eigenvalues are $\lambda = 1$ and $\lambda = 3$.

For $\lambda = 1$ the eigenvectors are $r\begin{bmatrix}1\\0\\1\end{bmatrix} + s\begin{bmatrix}0\\1\\0\end{bmatrix}$. For $\lambda = 3$ the eigenvectors are

$$t\begin{bmatrix}0\\1\\1\end{bmatrix}. \quad \begin{bmatrix}1 & 0 & 0\\1 & 1 & -1\\-1 & 0 & 1\end{bmatrix}\begin{bmatrix}1 & 0 & 0\\-2 & 1 & 2\\-2 & 0 & 3\end{bmatrix}\begin{bmatrix}1 & 0 & 0\\0 & 1 & 1\\1 & 0 & 1\end{bmatrix}=\begin{bmatrix}1 & 0 & 0\\0 & 1 & 0\\0 & 0 & 3\end{bmatrix}.$$

6. (a) The eigenvalues (found in exercise 5, Section 8.1) are $\lambda = 3$ and $\lambda = -1$, with corresponding eigenvectors $r\begin{bmatrix}1\\1\end{bmatrix}$ and $s\begin{bmatrix}-1\\1\end{bmatrix}$. The set $\left\{\begin{bmatrix}\frac{1}{\sqrt{2}}\\\frac{1}{\sqrt{2}}\end{bmatrix}, \begin{bmatrix}\frac{-1}{\sqrt{2}}\\\frac{1}{\sqrt{2}}\end{bmatrix}\right\}$

is an orthonormal basis. $\begin{bmatrix}\frac{1}{\sqrt{2}} & \frac{1}{\sqrt{2}}\\\frac{-1}{\sqrt{2}} & \frac{1}{\sqrt{2}}\end{bmatrix}\begin{bmatrix}1 & 2\\2 & 1\end{bmatrix}\begin{bmatrix}\frac{1}{\sqrt{2}} & \frac{-1}{\sqrt{2}}\\\frac{1}{\sqrt{2}} & \frac{1}{\sqrt{2}}\end{bmatrix} = \begin{bmatrix}3 & 0\\0 & -1\end{bmatrix}$.

(b) $\begin{vmatrix}11-\lambda & 2\\2 & 14-\lambda\end{vmatrix} = (11-\lambda)(14-\lambda) - 4 = \lambda^2 - 25\lambda + 150 = (\lambda-15)(\lambda-10)$, so the

eigenvalues are $\lambda = 15$ and $\lambda = 10$, with corresponding eigenvectors

$r\begin{bmatrix}1\\2\end{bmatrix}$ and $s\begin{bmatrix}-2\\1\end{bmatrix}$. The set $\left\{\begin{bmatrix}\frac{1}{\sqrt{5}}\\\frac{2}{\sqrt{5}}\end{bmatrix}, \begin{bmatrix}\frac{-2}{\sqrt{5}}\\\frac{1}{\sqrt{5}}\end{bmatrix}\right\}$ is an orthonormal basis.

$$\begin{bmatrix} \frac{1}{\sqrt{5}} & \frac{2}{\sqrt{5}} \\ \frac{-2}{\sqrt{5}} & \frac{1}{\sqrt{5}} \end{bmatrix} \begin{bmatrix} 11 & 2 \\ 2 & 14 \end{bmatrix} \begin{bmatrix} \frac{1}{\sqrt{5}} & \frac{-2}{\sqrt{5}} \\ \frac{2}{\sqrt{5}} & \frac{1}{\sqrt{5}} \end{bmatrix} = \begin{bmatrix} 15 & 0 \\ 0 & 10 \end{bmatrix}.$$

7. (a) $\begin{vmatrix} 1-\lambda & 5 \\ 5 & 1-\lambda \end{vmatrix} = (1-\lambda)(1-\lambda) - 25 = \lambda^2 - 2\lambda - 24 = (\lambda-6)(\lambda+4)$, so the eigenvalues are $\lambda = 6$ and $\lambda = -4$, with corresponding eigenvectors $r\begin{bmatrix} 1 \\ 1 \end{bmatrix}$ and $s\begin{bmatrix} -1 \\ 1 \end{bmatrix}$. The set $\left\{ \begin{bmatrix} \frac{1}{\sqrt{2}} \\ \frac{1}{\sqrt{2}} \end{bmatrix}, \begin{bmatrix} \frac{-1}{\sqrt{2}} \\ \frac{1}{\sqrt{2}} \end{bmatrix} \right\}$ is an orthonormal basis.

$$\begin{bmatrix} \frac{1}{\sqrt{2}} & \frac{1}{\sqrt{2}} \\ \frac{-1}{\sqrt{2}} & \frac{1}{\sqrt{2}} \end{bmatrix} \begin{bmatrix} 1 & 5 \\ 5 & 1 \end{bmatrix} \begin{bmatrix} \frac{1}{\sqrt{2}} & \frac{-1}{\sqrt{2}} \\ \frac{1}{\sqrt{2}} & \frac{1}{\sqrt{2}} \end{bmatrix} = \begin{bmatrix} 6 & 0 \\ 0 & -4 \end{bmatrix}.$$

(c) $\begin{vmatrix} 1-\lambda & 3 \\ 3 & 9-\lambda \end{vmatrix} = (1-\lambda)(9-\lambda) - 9 = \lambda^2 - 10\lambda = (\lambda-10)\lambda$, so the eigenvalues are $\lambda = 10$ and $\lambda = 0$, with corresponding eigenvectors $r\begin{bmatrix} 1 \\ 3 \end{bmatrix}$ and $s\begin{bmatrix} -3 \\ 1 \end{bmatrix}$.

The set $\left\{ \begin{bmatrix} \frac{1}{\sqrt{10}} \\ \frac{3}{\sqrt{10}} \end{bmatrix}, \begin{bmatrix} \frac{-3}{\sqrt{10}} \\ \frac{1}{\sqrt{10}} \end{bmatrix} \right\}$ is an orthonormal basis.

$$\begin{bmatrix} \frac{1}{\sqrt{10}} & \frac{3}{\sqrt{10}} \\ \frac{-3}{\sqrt{10}} & \frac{1}{\sqrt{10}} \end{bmatrix} \begin{bmatrix} 1 & 3 \\ 3 & 9 \end{bmatrix} \begin{bmatrix} \frac{1}{\sqrt{10}} & \frac{-3}{\sqrt{10}} \\ \frac{3}{\sqrt{10}} & \frac{1}{\sqrt{10}} \end{bmatrix} = \begin{bmatrix} 10 & 0 \\ 0 & 0 \end{bmatrix}.$$

8. (a) $\begin{vmatrix} -\lambda & 2 & 0 \\ 2 & -\lambda & 0 \\ 0 & 0 & 1-\lambda \end{vmatrix} = \lambda^2(1-\lambda) - 4(1-\lambda)$, so the eigenvalues are $\lambda = 1$ and $\lambda = \pm 2$.

For $\lambda = 1$ the eigenvectors are $r\begin{bmatrix} 0 \\ 0 \\ 1 \end{bmatrix}$, for $\lambda = 2$ the eigenvectors are $s\begin{bmatrix} 1 \\ 1 \\ 0 \end{bmatrix}$, and for $\lambda = -2$ the eigenvectors are $t\begin{bmatrix} 1 \\ -1 \\ 0 \end{bmatrix}$.

$$\begin{bmatrix} 0 & 0 & 1 \\ 1/\sqrt{2} & 1/\sqrt{2} & 0 \\ 1/\sqrt{2} & -1/\sqrt{2} & 0 \end{bmatrix} \begin{bmatrix} 0 & 2 & 0 \\ 2 & 0 & 0 \\ 0 & 0 & 1 \end{bmatrix} \begin{bmatrix} 0 & 1/\sqrt{2} & 1/\sqrt{2} \\ 0 & 1/\sqrt{2} & -1/\sqrt{2} \\ 1 & 0 & 0 \end{bmatrix} = \begin{bmatrix} 1 & 0 & 0 \\ 0 & 2 & 0 \\ 0 & 0 & -2 \end{bmatrix}.$$

(b) $\begin{vmatrix} 9-\lambda & -3 & 3 \\ -3 & 6-\lambda & -6 \\ 3 & -6 & 6-\lambda \end{vmatrix} = (9-\lambda)(6-\lambda)^2 + 54\lambda - 324 = -\lambda(15-\lambda)(6-\lambda)$, so the eigenvalues are $\lambda = 0$, $\lambda = 15$, and $\lambda = 6$. For $\lambda = 0$ the eigenvectors are $r\begin{bmatrix} 0 \\ 1 \\ 1 \end{bmatrix}$, for $\lambda = 15$ the eigenvectors are $s\begin{bmatrix} 1 \\ -1 \\ 1 \end{bmatrix}$, and for $\lambda = 6$ the eigenvectors are $t\begin{bmatrix} 2 \\ 1 \\ -1 \end{bmatrix}$.

$$\begin{bmatrix} 0 & 1/\sqrt{2} & 1/\sqrt{2} \\ 1/\sqrt{3} & -1/\sqrt{3} & 1/\sqrt{3} \\ 2/\sqrt{6} & 1/\sqrt{6} & -1/\sqrt{6} \end{bmatrix} \begin{bmatrix} 9 & -3 & 3 \\ -3 & 6 & -6 \\ 3 & -6 & 6 \end{bmatrix} \begin{bmatrix} 0 & 1/\sqrt{3} & 2/\sqrt{6} \\ 1/\sqrt{2} & -1/\sqrt{3} & 1/\sqrt{6} \\ 1/\sqrt{2} & 1/\sqrt{3} & -1/\sqrt{6} \end{bmatrix}$$

$$= \begin{bmatrix} 0 & 0 & 0 \\ 0 & 15 & 0 \\ 0 & 0 & 6 \end{bmatrix}.$$

9. (a) $\begin{bmatrix} 1 & 5 \\ 5 & 1 \end{bmatrix}^8 = \begin{bmatrix} \frac{1}{\sqrt{2}} & \frac{-1}{\sqrt{2}} \\ \frac{1}{\sqrt{2}} & \frac{1}{\sqrt{2}} \end{bmatrix} \begin{bmatrix} 6 & 0 \\ 0 & -4 \end{bmatrix}^8 \begin{bmatrix} \frac{1}{\sqrt{2}} & \frac{1}{\sqrt{2}} \\ \frac{-1}{\sqrt{2}} & \frac{1}{\sqrt{2}} \end{bmatrix}$

$$= \frac{1}{2}\begin{bmatrix} 1 & -1 \\ 1 & 1 \end{bmatrix}\begin{bmatrix} 1679616 & 0 \\ 0 & 65536 \end{bmatrix}\begin{bmatrix} 1 & 1 \\ -1 & 1 \end{bmatrix} = \begin{bmatrix} 872576 & 807040 \\ 807040 & 872576 \end{bmatrix}.$$

(d) $$\begin{bmatrix} 1.5 & -.5 \\ -.5 & 1.5 \end{bmatrix}^{16} = \begin{bmatrix} \frac{1}{\sqrt{2}} & \frac{-1}{\sqrt{2}} \\ \frac{1}{\sqrt{2}} & \frac{1}{\sqrt{2}} \end{bmatrix}\begin{bmatrix} 1 & 0 \\ 0 & 2 \end{bmatrix}^{16}\begin{bmatrix} \frac{1}{\sqrt{2}} & \frac{1}{\sqrt{2}} \\ \frac{-1}{\sqrt{2}} & \frac{1}{\sqrt{2}} \end{bmatrix}$$

$$= \frac{1}{2}\begin{bmatrix} 1 & -1 \\ 1 & 1 \end{bmatrix}\begin{bmatrix} 1 & 0 \\ 0 & 65536 \end{bmatrix}\begin{bmatrix} 1 & 1 \\ -1 & 1 \end{bmatrix} = \begin{bmatrix} 32768.5 & -32767.5 \\ -32767.5 & 32768.5 \end{bmatrix}.$$

10. (a) $$\begin{bmatrix} 0 & 2 & 0 \\ 2 & 0 & 0 \\ 0 & 0 & 1 \end{bmatrix}^{6} = \begin{bmatrix} 0 & 1/\sqrt{2} & 1/\sqrt{2} \\ 0 & 1/\sqrt{2} & -1/\sqrt{2} \\ 1 & 0 & 0 \end{bmatrix}\begin{bmatrix} 1 & 0 & 0 \\ 0 & 2 & 0 \\ 0 & 0 & -2 \end{bmatrix}^{6}\begin{bmatrix} 0 & 0 & 1 \\ 1/\sqrt{2} & 1/\sqrt{2} & 0 \\ 1/\sqrt{2} & -1/\sqrt{2} & 0 \end{bmatrix}$$

$$= \frac{1}{2}\begin{bmatrix} 0 & 1 & 1 \\ 0 & 1 & -1 \\ \sqrt{2} & 0 & 0 \end{bmatrix}\begin{bmatrix} 1 & 0 & 0 \\ 0 & 64 & 0 \\ 0 & 0 & 64 \end{bmatrix}\begin{bmatrix} 0 & 0 & \sqrt{2} \\ 1 & 1 & 0 \\ 1 & -1 & 0 \end{bmatrix} = \begin{bmatrix} 64 & 0 & 0 \\ 0 & 64 & 0 \\ 0 & 0 & 1 \end{bmatrix}.$$

(b) $$\begin{bmatrix} 9 & -3 & 3 \\ -3 & 6 & -6 \\ 3 & -6 & 6 \end{bmatrix}^{5} = \begin{bmatrix} 0 & 1/\sqrt{3} & 2/\sqrt{6} \\ 1/\sqrt{2} & -1/\sqrt{3} & 1/\sqrt{6} \\ 1/\sqrt{2} & 1/\sqrt{3} & -1/\sqrt{6} \end{bmatrix}\begin{bmatrix} 0 & 0 & 0 \\ 0 & 15 & 0 \\ 0 & 0 & 6 \end{bmatrix}^{5}\begin{bmatrix} 0 & 1/\sqrt{2} & 1/\sqrt{2} \\ 1/\sqrt{3} & -1/\sqrt{3} & 1/\sqrt{3} \\ 2/\sqrt{6} & 1/\sqrt{6} & -1/\sqrt{6} \end{bmatrix}$$

$$= \frac{1}{6}\begin{bmatrix} 0 & \sqrt{2} & 2 \\ \sqrt{3} & -\sqrt{2} & 1 \\ \sqrt{3} & \sqrt{2} & -1 \end{bmatrix}\begin{bmatrix} 0 & 0 & 0 \\ 0 & 759375 & 0 \\ 0 & 0 & 7776 \end{bmatrix}\begin{bmatrix} 0 & \sqrt{3} & \sqrt{3} \\ \sqrt{2} & -\sqrt{2} & \sqrt{2} \\ 2 & 1 & -1 \end{bmatrix}$$

$$= \begin{bmatrix} 258309 & -250533 & 250533 \\ -250533 & 254421 & -254421 \\ 250533 & -254421 & 254421 \end{bmatrix}.$$

11. (a) $T(1,0) = (4,2) = 4(1,0) + 2(0,1)$ and $T(0,1) = (2,4) = 2(1,0) + 4(0,1)$, so the matrix representation of T relative to the standard basis is $A = \begin{bmatrix} 4 & 2 \\ 2 & 4 \end{bmatrix}$.

$$\begin{vmatrix} 4-\lambda & 2 \\ 2 & 4-\lambda \end{vmatrix} = (4-\lambda)(4-\lambda) - 4 = \lambda^2 - 8\lambda + 12 = (\lambda-6)(\lambda-2)$$, so the eigenvalues are

$\lambda = 6$ and $\lambda = 2$ with corresponding eigenvectors (written as row vectors) r [1 1] and

s [-1 1]. Orthonormal eigenvectors are $\left[\frac{1}{\sqrt{2}}\ \frac{1}{\sqrt{2}}\right]$ and $\left[\frac{-1}{\sqrt{2}}\ \frac{1}{\sqrt{2}}\right]$. Let B' be the

basis $\left\{ \left(\frac{1}{\sqrt{2}}, \frac{1}{\sqrt{2}}\right), \left(\frac{-1}{\sqrt{2}}, \frac{1}{\sqrt{2}}\right) \right\}$. The matrix representation of T relative to B' is

$A' = \begin{bmatrix} 6 & 0 \\ 0 & 2 \end{bmatrix}$. The transition matrix from the standard basis to B' is $C = \begin{bmatrix} \frac{1}{\sqrt{2}} & \frac{-1}{\sqrt{2}} \\ \frac{1}{\sqrt{2}} & \frac{1}{\sqrt{2}} \end{bmatrix}$

and $C^{-1}AC = A'$.

The standard basis defines an xy coordinate system and the basis B' defines an x'y' coordinate system, rotated 45° counterclockwise from the xy system. The transformation T is a scaling in the x'y' system with factor 6 in the x' direction and factor 2 in the y' direction.

(c) T(1,0) = (9,2) = 9(1,0) + 2(0,1) and T(0,1) = (2,4) = 2(1,0) + 6(0,1), so the matrix

representation of T relative to the standard basis is $A = \begin{bmatrix} 9 & 2 \\ 2 & 6 \end{bmatrix}$.

A has eigenvalues $\lambda = 10$ and $\lambda = 5$ with corresponding eigenvectors r [2 1]

and s [-1 2]. Orthonormal eigenvectors are $\left[\frac{2}{\sqrt{5}}\ \frac{1}{\sqrt{5}}\right]$ and $\left[\frac{-1}{\sqrt{5}}\ \frac{2}{\sqrt{5}}\right]$.

Let B' be the basis $\left\{ \left(\frac{2}{\sqrt{5}}, \frac{1}{\sqrt{5}}\right), \left(\frac{-1}{\sqrt{5}}, \frac{2}{\sqrt{5}}\right) \right\}$. The matrix representation of T

relative to B' is $A' = \begin{bmatrix} 10 & 0 \\ 0 & 5 \end{bmatrix}$. The transition matrix from the standard basis to B' is

$C = \begin{bmatrix} \frac{2}{\sqrt{5}} & \frac{-1}{\sqrt{5}} \\ \frac{1}{\sqrt{5}} & \frac{2}{\sqrt{5}} \end{bmatrix}$ and $C^{-1}AC = A'$.

The standard basis defines an xy coordinate system and the basis B' defines an x'y' coordinate system, rotated counterclockwise through an angle θ from the xy system, where $\cos \theta = 2/\sqrt{5}$ and $\sin \theta = 1/\sqrt{5}$. The transformation T is a scaling in the x'y' system with factor 10 in the x' direction and factor 5 in the y' direction.

12. (a) $T(1,0) = (8,9) = 8(1,0) + 9(0,1)$ and $T(0,1) = (-6,-7) = -6(1,0) - 7(0,1)$, so the matrix representation of T relative to the standard basis is $A = \begin{bmatrix} 8 & -6 \\ 9 & -7 \end{bmatrix}$.

$$\begin{vmatrix} 8-\lambda & -6 \\ 9 & -7-\lambda \end{vmatrix} = (8-\lambda)(-7-\lambda) + 54 = \lambda^2 - \lambda - 2 = (\lambda-2)(\lambda+1), \text{ so the}$$

eigenvalues are $\lambda = 2$ and $\lambda = -1$ with corresponding eigenvectors r [1 1] and s [2 3]. Let B' be the basis {(1,1), (2,3)}. The matrix representation of T relative to B' is $A' = \begin{bmatrix} 2 & 0 \\ 0 & -1 \end{bmatrix}$. The transition matrix from the standard basis to B' is

$C = \begin{bmatrix} 1 & 2 \\ 1 & 3 \end{bmatrix}$ and $C^{-1}AC = A'$.

The standard basis defines an xy coordinate system and the basis B' defines an x'y' coordinate system, which is not rectangular. The transformation T is a scaling in the x'y' system with factor 2 in the x' direction and factor 1 in the y' direction followed by a reflection about the x' axis.

(c) $T(1,0) = (3,2) = 3(1,0) + 2(0,1)$ and $T(0,1) = (-4,-3) = -4(1,0) - 3(0,1)$, so the matrix representation of T relative to the standard basis is $A = \begin{bmatrix} 3 & -4 \\ 2 & -3 \end{bmatrix}$.

$$\begin{vmatrix} 3-\lambda & -4 \\ 2 & -3-\lambda \end{vmatrix} = (3-\lambda)(-3-\lambda) + 8 = \lambda^2 - 1 = (\lambda-1)(\lambda+1), \text{ so the}$$

eigenvalues are $\lambda = 1$ and $\lambda = -1$ with corresponding eigenvectors r [2 1] and s [1 1]. Let B' be the basis {(2,1), (1,1)}. The matrix representation of T relative to B' is $A' = \begin{bmatrix} 1 & 0 \\ 0 & -1 \end{bmatrix}$. The transition matrix from the standard basis to B' is

$C = \begin{bmatrix} 2 & 1 \\ 1 & 1 \end{bmatrix}$ and $C^{-1}AC = A'$.

The standard basis defines an xy coordinate system and the basis B' defines an x'y' coordinate system, which is not rectangular. The transformation T is a reflection about the x' axis.

13. (a) If A and B are similar then $B = C^{-1}AC$, so that $|B| = |C^{-1}AC| = |C^{-1}||A||C| = |A|$.

(c) We know from exercise 17, Section 2.3 that for nxn matrices E and F, tr(EF) = tr(FE). Thus if $B = C^{-1}AC$ then $tr(B) = tr(C^{-1}(AC)) = tr((AC)C^{-1}) = tr(A(CC^{-1})) = tr(A)$.

(f) If $B = C^{-1}AC$ and A is nonsingular then B is also nonsingular and $B^{-1} = (C^{-1}AC)^{-1} = C^{-1}A^{-1}C$, so B^{-1} is similar to A^{-1}.

15. (b) If $A = C^{-1}BC$ then $A + kI = C^{-1}BC + kC^{-1}IC = C^{-1}BC + C^{-1}kIC = C^{-1}(B+kI)C$, so $B + kI$ and $A + kI$ are similar for any scalar k.

17. The matrix $\begin{bmatrix} 4 & -1 \\ 2 & 1 \end{bmatrix}$ has eigenvectors $\lambda = 2$ and $\lambda = 3$ with corresponding eigenvectors

$r\begin{bmatrix} 1 \\ 2 \end{bmatrix}$ and $s\begin{bmatrix} 1 \\ 1 \end{bmatrix}$. If $C = \begin{bmatrix} 1 & 1 \\ 2 & 1 \end{bmatrix}$ then $C^{-1}AC = \begin{bmatrix} 2 & 0 \\ 0 & 3 \end{bmatrix}$, but if $C = \begin{bmatrix} 1 & 1 \\ 1 & 2 \end{bmatrix}$ then

$C^{-1}AC = \begin{bmatrix} 3 & 0 \\ 0 & 2 \end{bmatrix}$. Thus the diagonal matrix is not unique. An eigenvalue λ will occupy

the ith diagonal position in the diagonal matrix if the ith column of C is an eigenvector corresponding to λ.

19. If $B = C^{-1}AC = C^tAC$, then $B^t = (C^tAC)^t = C^tA^t(C^t)^t = C^tA^tC = C^tAC = B$. Thus B is symmetric.

Exercise Set 8.3, page 380

1 (a) $x^2 + 4xy + 2y^2 = [x\ y]\begin{bmatrix} 1 & 2 \\ 2 & 2 \end{bmatrix}\begin{bmatrix} x \\ y \end{bmatrix}$. (c) $7x^2 - 6xy - y^2 = [x\ y]\begin{bmatrix} 7 & -3 \\ -3 & -1 \end{bmatrix}\begin{bmatrix} x \\ y \end{bmatrix}$.

(e) $-3x^2 - 7xy + 4y^2 = [x\ \ y]\begin{bmatrix} -3 & -7/2 \\ -7/2 & 4 \end{bmatrix}\begin{bmatrix} x \\ y \end{bmatrix}$.

2. (a) $11x^2 + 4xy + 14y^2 - 60 = [x\ \ y]\begin{bmatrix} 11 & 2 \\ 2 & 14 \end{bmatrix}\begin{bmatrix} x \\ y \end{bmatrix} - 60$. From exercise 6(b) in Section 8.2, $C^t\begin{bmatrix} 11 & 2 \\ 2 & 14 \end{bmatrix}C = \begin{bmatrix} 15 & 0 \\ 0 & 10 \end{bmatrix}$, where $C = \begin{bmatrix} \frac{1}{\sqrt{5}} & \frac{-2}{\sqrt{5}} \\ \frac{2}{\sqrt{5}} & \frac{1}{\sqrt{5}} \end{bmatrix}$, so the given equation becomes $[x\ \ y]\,C\begin{bmatrix} 15 & 0 \\ 0 & 10 \end{bmatrix}C^t\begin{bmatrix} x \\ y \end{bmatrix} - 60 = 0$, or $[x'\ \ y']\begin{bmatrix} 15 & 0 \\ 0 & 10 \end{bmatrix}\begin{bmatrix} x' \\ y' \end{bmatrix} - 60 = 0$, where $[x'\ \ y'] = [x\ \ y]\,C$. Thus $15x'^2 + 10y'^2 = 60$; i.e., $\frac{x'^2}{4} + \frac{y'^2}{6} = 1$. The graph is an ellipse with the lines $y = 2x$ and $x = -2y$ as axes.

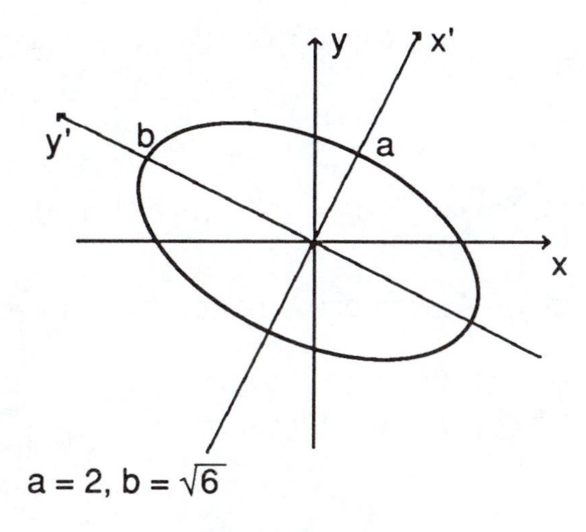

$a = 2$, $b = \sqrt{6}$

figure for 2(a)

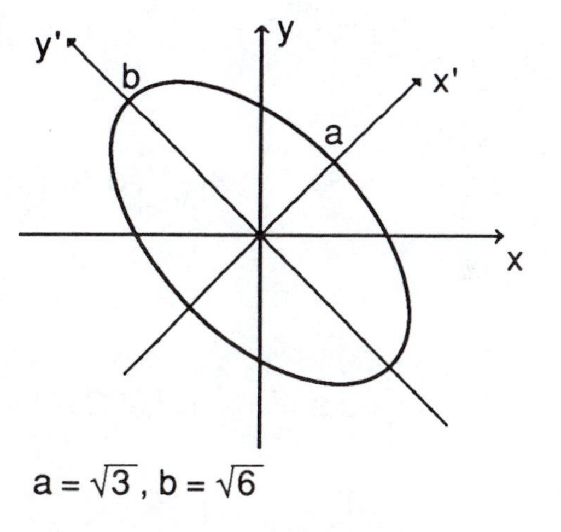

$a = \sqrt{3}$, $b = \sqrt{6}$

figure for 2(b)

(b) $3x^2 + 2xy + 3y^2 - 12 = [x\ \ y]\begin{bmatrix} 3 & 1 \\ 1 & 3 \end{bmatrix}\begin{bmatrix} x \\ y \end{bmatrix} - 12$. From exercise 6(c) in Section 8.2, $C^t\begin{bmatrix} 3 & 1 \\ 1 & 3 \end{bmatrix}C = \begin{bmatrix} 4 & 0 \\ 0 & 2 \end{bmatrix}$, where $C = \begin{bmatrix} \frac{1}{\sqrt{2}} & \frac{-1}{\sqrt{2}} \\ \frac{1}{\sqrt{2}} & \frac{1}{\sqrt{2}} \end{bmatrix}$, so the given equation becomes $[x\ \ y]\, C \begin{bmatrix} 4 & 0 \\ 0 & 2 \end{bmatrix} C^t \begin{bmatrix} x \\ y \end{bmatrix} - 12 = 0$, or $[x'\ \ y'] \begin{bmatrix} 4 & 0 \\ 0 & 2 \end{bmatrix} \begin{bmatrix} x' \\ y' \end{bmatrix} - 12 = 0$,

where $[x'\ \ y'] = [x\ \ y]\, C$. Thus $4x'^2 + 2y'^2 = 12$, i.e. $\frac{x'^2}{3} + \frac{y'^2}{6} = 1$. The graph is an ellipse with the lines $y = x$ and $y = -x$ as axes.

3. (a) $a_n = a_{n-1} + 2a_{n-2}$, $a_1 = 1$, and $a_2 = 2$. Let $b_n = a_{n-1}$. Thus $\begin{bmatrix} a_n \\ b_n \end{bmatrix} = \begin{bmatrix} 1 & 2 \\ 1 & 0 \end{bmatrix} \begin{bmatrix} a_{n-1} \\ b_{n-1} \end{bmatrix}$.

The matrix has eigenvalues $\lambda = -1$ and $\lambda = 2$, with corresponding eigenvectors

$\begin{bmatrix} 1 \\ -1 \end{bmatrix}$ and $\begin{bmatrix} 2 \\ 1 \end{bmatrix}$. Let $C = \begin{bmatrix} 1 & 2 \\ -1 & 1 \end{bmatrix}$. $C^{-1} = \frac{1}{3} \begin{bmatrix} 1 & -2 \\ 1 & 1 \end{bmatrix}$, and

$\begin{bmatrix} a_n \\ b_n \end{bmatrix} = C \begin{bmatrix} (-1)^{n-2} & 0 \\ 0 & 2^{n-2} \end{bmatrix} C^{-1} \begin{bmatrix} 2 \\ 1 \end{bmatrix} = \frac{1}{3} \begin{bmatrix} 2^n + 2^{n-1} \\ 2^{n-1} + 2^{n-2} \end{bmatrix}$, so $a_n = \frac{1}{3}(2^n + 2^{n-1}) = 2^{n-1}$.

$a_{10} = 2^9 = 512$.

(c) $a_n = 3a_{n-1} + 4a_{n-2}$, $a_1 = 1$, and $a_2 = -1$. Let $b_n = a_{n-1}$. Thus $\begin{bmatrix} a_n \\ b_n \end{bmatrix} = \begin{bmatrix} 3 & 4 \\ 1 & 0 \end{bmatrix} \begin{bmatrix} a_{n-1} \\ b_{n-1} \end{bmatrix}$.

The matrix has eigenvalues $\lambda = -1$ and $\lambda = 4$, with corresponding eigenvectors

$\begin{bmatrix} 1 \\ -1 \end{bmatrix}$ and $\begin{bmatrix} 4 \\ 1 \end{bmatrix}$. Let $C = \begin{bmatrix} 1 & 4 \\ -1 & 1 \end{bmatrix}$. $C^{-1} = \frac{1}{5} \begin{bmatrix} 1 & -4 \\ 1 & 1 \end{bmatrix}$, and

$\begin{bmatrix} a_n \\ b_n \end{bmatrix} = C \begin{bmatrix} (-1)^{n-2} & 0 \\ 0 & 4^{n-2} \end{bmatrix} C^{-1} \begin{bmatrix} -1 \\ 1 \end{bmatrix} = \begin{bmatrix} (-1)^{n-1} \\ (-1)^{n-2} \end{bmatrix}$, so $a_n = (-1)^{n-1}$.

$a_{12} = (-1)^{11} = -1$.

5. $m\ddot{x}_1 = \frac{1}{a}(-2Tx_1 + Tx_2)$ and $m\ddot{x}_2 = \frac{1}{a}(Tx_1 - 2Tx_2)$, so $\begin{bmatrix} \ddot{x}_1 \\ \ddot{x}_2 \end{bmatrix} = \frac{T}{ma} \begin{bmatrix} -2 & 1 \\ 1 & -2 \end{bmatrix} \begin{bmatrix} x_1 \\ x_2 \end{bmatrix}$.

The matrix has eigenvalues $\lambda = -3$ and $\lambda = -1$ with corresponding eigenvectors $\begin{bmatrix} 1 \\ -1 \end{bmatrix}$

and $\begin{bmatrix} 1 \\ 1 \end{bmatrix}$. $C^{-1}AC = \begin{bmatrix} -3 & 0 \\ 0 & -1 \end{bmatrix}$, where $C = \begin{bmatrix} 1 & 1 \\ -1 & 1 \end{bmatrix}$. Let $\begin{bmatrix} x_1' \\ x_2' \end{bmatrix} = C^{-1}\begin{bmatrix} x_1 \\ x_2 \end{bmatrix}$.

$\begin{bmatrix} \ddot{x}_1' \\ \ddot{x}_2' \end{bmatrix} = \frac{T}{ma}\begin{bmatrix} -3 & 0 \\ 0 & -1 \end{bmatrix}\begin{bmatrix} x_1' \\ x_2' \end{bmatrix}$, so that $\ddot{x}_1' = \frac{-3T}{ma}x_1'$ and $\ddot{x}_2' = \frac{-T}{ma}x_2'$.

Solutions of these equations are $x_1' = b_1\cos(\alpha_1 t + \gamma_1)$ and $x_2' = b_2\cos(\alpha_2 t + \gamma_2)$, where $\alpha_1 = (3T/ma)^{1/2}$ and $\alpha_2 = (T/ma)^{1/2}$. We now must solve for x_1 and x_2.

$\begin{bmatrix} x_1 \\ x_2 \end{bmatrix} = C\begin{bmatrix} x_1' \\ x_2' \end{bmatrix} = \begin{bmatrix} 1 & 1 \\ -1 & 1 \end{bmatrix}\begin{bmatrix} x_1' \\ x_2' \end{bmatrix} = x_1'\begin{bmatrix} 1 \\ -1 \end{bmatrix} + x_2'\begin{bmatrix} 1 \\ 1 \end{bmatrix}$

$= b_1\cos(\alpha_1 t + \gamma_1)\begin{bmatrix} 1 \\ -1 \end{bmatrix} + b_2\cos(\alpha_2 t + \gamma_2)\begin{bmatrix} 1 \\ 1 \end{bmatrix}$. Thus the normal modes are

mode 1: $\begin{bmatrix} x_1 \\ x_2 \end{bmatrix} = \cos(\alpha_1 t + \gamma_1)\begin{bmatrix} 1 \\ -1 \end{bmatrix}$, where $\alpha_1 = \left(\frac{3T}{ma}\right)^{1/2}$, and

mode 2: $\begin{bmatrix} x_1 \\ x_2 \end{bmatrix} = \cos(\alpha_2 t + \gamma_2)\begin{bmatrix} 1 \\ 1 \end{bmatrix}$, where $\alpha_2 = \left(\frac{T}{ma}\right)^{1/2}$.

6. $M\ddot{x}_1 = -5x_1 + 2x_2$ and $M\ddot{x}_2 = 2x_1 - 2x_2$, so $\begin{bmatrix} \ddot{x}_1 \\ \ddot{x}_2 \end{bmatrix} = \frac{1}{M}\begin{bmatrix} -5 & 2 \\ 2 & -2 \end{bmatrix}\begin{bmatrix} x_1 \\ x_2 \end{bmatrix}$.

The matrix has eigenvalues $\lambda = -1$ and $\lambda = -6$ with corresponding eigenvectors $\begin{bmatrix} 1 \\ 2 \end{bmatrix}$

and $\begin{bmatrix} 2 \\ -1 \end{bmatrix}$. $C^{-1}AC = \begin{bmatrix} -1 & 0 \\ 0 & -6 \end{bmatrix}$, where $C = \begin{bmatrix} 1 & 2 \\ 2 & -1 \end{bmatrix}$. Let $\begin{bmatrix} x_1' \\ x_2' \end{bmatrix} = C^{-1}\begin{bmatrix} x_1 \\ x_2 \end{bmatrix}$.

$\begin{bmatrix} \ddot{x}_1' \\ \ddot{x}_2' \end{bmatrix} = \frac{1}{M}\begin{bmatrix} -1 & 0 \\ 0 & -6 \end{bmatrix}\begin{bmatrix} x_1' \\ x_2' \end{bmatrix}$, so that $\ddot{x}_1' = \frac{-1}{M}x_1'$ and $\ddot{x}_2' = \frac{-6}{M}x_2'$.

Solutions of these equations are $x_1' = b_1 \cos(\alpha_1 t + \gamma_1)$ and $x_2' = b_2 \cos(\alpha_2 t + \gamma_2)$, where $\alpha_1 = (1/M)^{1/2}$ and $\alpha_2 = (6/M)^{1/2}$. We now must solve for x_1 and x_2.

$$\begin{bmatrix} x_1 \\ x_2 \end{bmatrix} = C\begin{bmatrix} x_1' \\ x_2' \end{bmatrix} = \begin{bmatrix} 1 & 2 \\ 2 & -1 \end{bmatrix}\begin{bmatrix} x_1' \\ x_2' \end{bmatrix} = x_1'\begin{bmatrix} 1 \\ 2 \end{bmatrix} + x_2'\begin{bmatrix} 2 \\ -1 \end{bmatrix}$$

$$= b_1 \cos(\alpha_1 t + \gamma_1)\begin{bmatrix} 1 \\ 2 \end{bmatrix} + b_2 \cos(\alpha_2 t + \gamma_2)\begin{bmatrix} 2 \\ -1 \end{bmatrix}.$$

Exercise Set 8.4, page 386

1. The eigenvectors of $\lambda = 1$ are vectors of the form $r\begin{bmatrix} 2 \\ 1 \end{bmatrix}$. If there is no change in total population $2r + r = 185 + 55 = 240$, so $r = 80$. Thus the long-term prediction is that population in metropolitan areas will be 160 million and population in nonmetropolitan areas will be 80 million.

3. $P^2 = \begin{bmatrix} .375 & .25 & .125 \\ .5 & .5 & .5 \\ .125 & .25 & .375 \end{bmatrix}$, and since all terms are positive, P is regular. The eigenvectors of $\lambda = 1$ are vectors of the form $r\begin{bmatrix} 1 \\ 2 \\ 1 \end{bmatrix}$. The powers of P approach the stochastic matrix $Q = \begin{bmatrix} s & s & s \\ 2s & 2s & 2s \\ s & s & s \end{bmatrix}$, so $s = .25$ and $Q = \begin{bmatrix} .25 & .25 & .25 \\ .5 & .5 & .5 \\ .25 & .25 & .25 \end{bmatrix}$.

The columns of Q indicate that when guinea pigs are bred with hybrids only the long-term distribution of types AA, Aa, and aa will be 1:2:1. That is, the long-term probabilities of Types AA, Aa, and aa are .25, .5, and .25.

4.

$$P = \begin{array}{c} \text{wet} \quad \text{dry} \\ \begin{bmatrix} .65 & .23 \\ .35 & .77 \end{bmatrix} \end{array} \begin{array}{l} \text{wet} \\ \text{dry} \end{array}$$

(a) $P^2 = \begin{bmatrix} .5 & .33 \\ .5 & .67 \end{bmatrix}$. If Thursday is dry, the probability that Saturday will also be dry is

.67, the (2,2) term in P^2.

(b) The eigenvectors of P corresponding to $\lambda = 1$ are vectors of the form $r\begin{bmatrix} 23 \\ 35 \end{bmatrix}$.

Thus the powers of P approach the stochastic matrix $Q = \begin{bmatrix} 23s & 23s \\ 35s & 35s \end{bmatrix}$, so

$(23+35)s = 1$ and $s = \frac{1}{58}$. $\frac{23}{58} = .4$, $\frac{35}{58} = .6$, and $Q = \begin{bmatrix} .4 & .4 \\ .6 & .6 \end{bmatrix}$, so the long-term

probability for a wet day in December is .4 and for a dry day is .6.

6. room 1 2 3 4

$$P = \begin{bmatrix} 0 & 1/3 & 0 & 1/4 \\ 1/2 & 0 & 1/3 & 1/4 \\ 0 & 1/3 & 0 & 1/2 \\ 1/2 & 1/3 & 2/3 & 0 \end{bmatrix} \begin{matrix} 1 \\ 2 \\ 3 \\ 4 \end{matrix}$$

P is regular since every term in P^2 is positive.

The eigenvectors of P corresponding to $\lambda = 1$ are vectors of the form $r\begin{bmatrix} 2 \\ 3 \\ 3 \\ 4 \end{bmatrix}$; thus

the distribution of rats in rooms 1, 2, 3, and 4 is 2:3:3:4. The powers of P approach

the stochastic matrix $Q = \begin{bmatrix} 2s & 2s & 2s & 2s \\ 3s & 3s & 3s & 3s \\ 3s & 3s & 3s & 3s \\ 4s & 4s & 4s & 4s \end{bmatrix}$, so $2s + 3s + 3s + 4s = 1$ and $s = \frac{1}{12}$.

The long-term probability that a given rat will be in room 4 is therefore 4/12 = 1/3.

7. $P = \begin{bmatrix} .75 & .20 \\ .25 & .80 \end{bmatrix}$. The eigenvectors of $\lambda = 1$ are vectors of the form $r\begin{bmatrix} 4 \\ 5 \end{bmatrix}$. The powers of

P approach $Q = \begin{bmatrix} 4s & 4s \\ 5s & 5s \end{bmatrix}$, so $4s + 5s = 1$ and $s = \frac{1}{9}$. If current trends continue, the

eventual distribution will be 4/9 = 44.4% using company A and 5/9 = 55.6% using

company B.

9. The sum of the terms in each column of a stochastic matrix A is 1, so the sum of the terms in each column of A – I is zero. It has previously been proved (exercise 15, Section 3.2) that if the sum of the terms in each column of a matrix is zero the determinant of the matrix is zero. Thus $|A - 1I| = |A - I| = 0$, and 1 is an eigenvalue of A.

Chapter 8 Review Exercises, page 388

1. $\begin{vmatrix} 5-\lambda & -7 & 7 \\ 4 & -3-\lambda & 4 \\ 4 & -1 & 2-\lambda \end{vmatrix} = (5-\lambda)(1-\lambda)(-2-\lambda)$, so the eigenvalues are $\lambda = 5$, $\lambda = 1$, and $\lambda = -2$. For $\lambda = 5$ the eigenvectors are the solutions of $\begin{bmatrix} 0 & -7 & 7 \\ 4 & -8 & 4 \\ 4 & -1 & -3 \end{bmatrix}\begin{bmatrix} x_1 \\ x_2 \\ x_3 \end{bmatrix} = \mathbf{0}$, so the eigenvectors are vectors of the form $r\begin{bmatrix} 1 \\ 1 \\ 1 \end{bmatrix}$. For $\lambda = 1$ the eigenvectors are the solutions of $\begin{bmatrix} 4 & -7 & 7 \\ 4 & -4 & 4 \\ 4 & -1 & 1 \end{bmatrix}\begin{bmatrix} x_1 \\ x_2 \\ x_3 \end{bmatrix} = \mathbf{0}$, so the eigenvectors are vectors of the form $s\begin{bmatrix} 0 \\ 1 \\ 1 \end{bmatrix}$. For $\lambda = -2$ the eigenvectors are the solutions of $\begin{bmatrix} 7 & -7 & 7 \\ 4 & -1 & 4 \\ 4 & -1 & 4 \end{bmatrix}\begin{bmatrix} x_1 \\ x_2 \\ x_3 \end{bmatrix} = \mathbf{0}$, so the eigenvectors are vectors of the form $t\begin{bmatrix} 1 \\ 0 \\ -1 \end{bmatrix}$.

2. Let λ be an eigenvalue of A with eigenvector $\mathbf{x}$. A is invertible, so $\lambda \neq 0$. $A\mathbf{x} = \lambda\mathbf{x}$, so $\mathbf{x} = A^{-1}A\mathbf{x} = A^{-1}\lambda\mathbf{x} = \lambda A^{-1}\mathbf{x}$. Thus $\frac{1}{\lambda}\mathbf{x} = A^{-1}\mathbf{x}$, so the eigenvalues for A^{-1} are the inverses of the eigenvalues for A and the corresponding eigenvectors are the same.

3. $A\mathbf{x} = \lambda\mathbf{x}$, so $A\mathbf{x} - kI\mathbf{x} = \lambda\mathbf{x} - kI\mathbf{x} = \lambda\mathbf{x} - k\mathbf{x}$, so $(A - kI)\mathbf{x} = (\lambda - k)\mathbf{x}$. Thus $\lambda - k$ is an eigenvalue for $A - kI$ with corresponding eigenvector $\mathbf{x}$.

4. $C^{-1}AC = \begin{bmatrix} 1 & -1 \\ -1 & 2 \end{bmatrix}\begin{bmatrix} 4 & -2 \\ 1 & 1 \end{bmatrix}\begin{bmatrix} 2 & 1 \\ 1 & 1 \end{bmatrix} = \begin{bmatrix} 3 & 0 \\ 0 & 2 \end{bmatrix}$.

5. $\begin{vmatrix} 1-\lambda & 1 \\ -2 & 4-\lambda \end{vmatrix} = (3-\lambda)(2-\lambda)$, so the eigenvalues are $\lambda = 3$ and $\lambda = 2$. For $\lambda = 3$ the eigenvectors are $r\begin{bmatrix} 1 \\ 2 \end{bmatrix}$ and for $\lambda = 2$ the eigenvectors are $s\begin{bmatrix} 1 \\ 1 \end{bmatrix}$. Let $C = \begin{bmatrix} 1 & 1 \\ 2 & 1 \end{bmatrix}$.

$C^{-1}AC = \begin{bmatrix} -1 & 1 \\ 2 & -1 \end{bmatrix}\begin{bmatrix} 1 & 1 \\ -2 & 4 \end{bmatrix}\begin{bmatrix} 1 & 1 \\ 2 & 1 \end{bmatrix} = \begin{bmatrix} 3 & 0 \\ 0 & 2 \end{bmatrix}$.

6. $\begin{vmatrix} 7-\lambda & -2 & 1 \\ -2 & 10-\lambda & -2 \\ 1 & -2 & 7-\lambda \end{vmatrix} = (6-\lambda)(6-\lambda)(12-\lambda)$, so the eigenvalues are $\lambda = 6$ and $\lambda = 12$.

For $\lambda = 6$ the eigenvectors are vectors of the form $r\begin{bmatrix} 1 \\ 0 \\ -1 \end{bmatrix} + s\begin{bmatrix} 1 \\ 1 \\ 1 \end{bmatrix}$. For $\lambda = 12$ the eigenvectors are vectors of the form $t\begin{bmatrix} 1 \\ -2 \\ 1 \end{bmatrix}$. Orthonormal eigenvectors are

$\begin{bmatrix} 1/\sqrt{2} \\ 0 \\ -1/\sqrt{2} \end{bmatrix}, \begin{bmatrix} 1/\sqrt{3} \\ 1/\sqrt{3} \\ 1/\sqrt{3} \end{bmatrix}, \begin{bmatrix} 1/\sqrt{6} \\ -2/\sqrt{6} \\ 1/\sqrt{6} \end{bmatrix}$. Let $C = \begin{bmatrix} 1/\sqrt{2} & 1/\sqrt{3} & 1/\sqrt{6} \\ 0 & 1/\sqrt{3} & -2/\sqrt{6} \\ -1/\sqrt{2} & 1/\sqrt{3} & 1/\sqrt{6} \end{bmatrix}$.

$$\begin{bmatrix} 1/\sqrt{2} & 0 & -1/\sqrt{2} \\ 1/\sqrt{3} & 1/\sqrt{3} & 1/\sqrt{3} \\ 1/\sqrt{6} & -2/\sqrt{6} & 1/\sqrt{6} \end{bmatrix}\begin{bmatrix} 7 & -2 & 1 \\ -2 & 10 & -2 \\ 1 & -2 & 7 \end{bmatrix}\begin{bmatrix} 1/\sqrt{2} & 1/\sqrt{3} & 1/\sqrt{6} \\ 0 & 1/\sqrt{3} & -2/\sqrt{6} \\ -1/\sqrt{2} & 1/\sqrt{3} & 1/\sqrt{6} \end{bmatrix} = \begin{bmatrix} 6 & 0 & 0 \\ 0 & 6 & 0 \\ 0 & 0 & 12 \end{bmatrix}.$$

7. $\begin{vmatrix} a-\lambda & b \\ b & c-\lambda \end{vmatrix} = (a-\lambda)(c-\lambda) - b^2 = \lambda^2 - (a+c)\lambda + ac - b^2$. The characteristic equation $\lambda^2 - (a+c)\lambda + ac - b^2 = 0$ has roots $\lambda = (a+c \pm \sqrt{D})/2$ where $D = (a-c)^2 + 4b^2$ (from the quadratic formula). D is nonnegative for all values of a, b, and c, so the roots are real.

8. If A is symmetric then A can be diagonalized, $D = C^{-1}AC$, where the diagonal elements of D are the eigenvalues of A. If A has only one eigenvalue, λ, then $D = \lambda I$, so $\lambda I = C^{-1}AC$. Thus $A = CC^{-1}ACC^{-1} = C\lambda IC^{-1} = \lambda CIC^{-1} = \lambda CC^{-1} = \lambda I$.

9. $-x^2 - 16xy + 11y^2 - 30 = [x\ \ y]\begin{bmatrix} -1 & -8 \\ -8 & 11 \end{bmatrix}\begin{bmatrix} x \\ y \end{bmatrix} - 30$. The symmetric matrix has eigenvalues $\lambda = 15$ and $\lambda = -5$ with corresponding orthonormal eigenvectors

$\begin{bmatrix} \frac{1}{\sqrt{5}} \\ \frac{-2}{\sqrt{5}} \end{bmatrix}$ and $\begin{bmatrix} \frac{2}{\sqrt{5}} \\ \frac{1}{\sqrt{5}} \end{bmatrix}$, so $C^t \begin{bmatrix} -1 & -8 \\ -8 & 11 \end{bmatrix} C = \begin{bmatrix} 15 & 0 \\ 0 & -5 \end{bmatrix}$, where $C = \begin{bmatrix} \frac{1}{\sqrt{5}} & \frac{2}{\sqrt{5}} \\ \frac{-2}{\sqrt{5}} & \frac{1}{\sqrt{5}} \end{bmatrix}$, so the

given equation becomes $[x\ \ y]\, C \begin{bmatrix} 15 & 0 \\ 0 & -5 \end{bmatrix} C^t \begin{bmatrix} x \\ y \end{bmatrix} - 30 = 0$, or

$[x'\ \ y'] \begin{bmatrix} 15 & 0 \\ 0 & -5 \end{bmatrix} \begin{bmatrix} x' \\ y' \end{bmatrix} - 30 = 0$, where $[x'\ \ y'] = [x\ \ y]\, C$. Thus $15x'^2 - 5y'^2 = 30$;

i.e., $\frac{x'^2}{2} - \frac{y'^2}{6} = 1$. The graph is a hyperbola with axes $y = -2x$ and $x = 2y$.

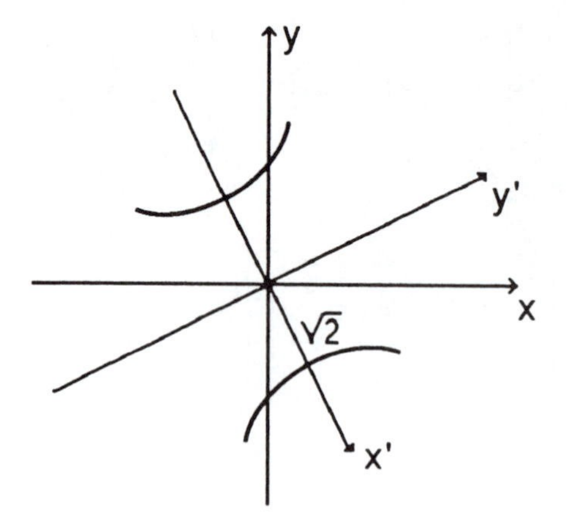

10. $a_n = 4a_{n-1} + 5a_{n-2}$, $a_1 = 3$, and $a_2 = 2$. Let $b_n = a_{n-1}$. Thus $\begin{bmatrix} a_n \\ b_n \end{bmatrix} = \begin{bmatrix} 4 & 5 \\ 1 & 0 \end{bmatrix} \begin{bmatrix} a_{n-1} \\ b_{n-1} \end{bmatrix}$.

The matrix has eigenvalues $\lambda = -1$ and $\lambda = 5$ with corresponding eigenvectors

$\begin{bmatrix} 1 \\ -1 \end{bmatrix}$ and $\begin{bmatrix} 5 \\ 1 \end{bmatrix}$. Let $C = \begin{bmatrix} 1 & 5 \\ -1 & 1 \end{bmatrix}$. $C^{-1} = \frac{1}{6}\begin{bmatrix} 1 & -5 \\ 1 & 1 \end{bmatrix}$, and

$\begin{bmatrix} a_n \\ b_n \end{bmatrix} = C\begin{bmatrix} (-1)^{n-2} & 0 \\ 0 & 5^{n-2} \end{bmatrix} C^{-1} \begin{bmatrix} 2 \\ 3 \end{bmatrix} = \frac{1}{6}\begin{bmatrix} 13(-1)^{n-1} + 5^n \\ 13(-1)^{n-2} + 5^{n-1} \end{bmatrix}$, so

$a_n = \frac{1}{6}(13(-1)^{n-1} + 5^n)$. $a_{12} = \frac{1}{6}(13(-1)^{11} + 5^{12}) = \frac{1}{6}(-13 + 244140625)$

$= 40{,}690{,}102$.

Chapter 9

Exercise Set 9.1, page 395

1. From the first equation $x_1 = 1$, and from the second equation $2 - x_2 = -2$, so $x_2 = 4$.

   From the third equation $3 + 4 - x_3 = 8$, so $x_3 = -1$.

3. From the first equation $x_1 = -2$, and from the second equation $-6 + x_2 = -5$, so $x_2 = 1$.

   From the third equation $-2 + 4 + 2x_3 = 4$, so $x_3 = 1$.

4. From the third equation $x_3 = 2$, and from the second equation $2x_2 + 2 = 3$, so $x_2 = 1/2$.

   From the first equation $x_1 + 2 + 1/2 = 3$, so $x_1 = 1/2$.

6. From the third equation $x_3 = 7$, and from the second equation $2x_2 + 28 = 26$, so $x_2 = -1$.

   From the first equation $3x_1 - 2 - 7 = -6$, so $x_1 = 1$.

7. The "rule" is $Ri - aRj$ causes $a_{ij} = a$, where a_{ij} is the (i,j)th position in L.

   (a) $L = \begin{bmatrix} 1 & 0 & 0 \\ 1 & 1 & 0 \\ -1 & -1 & 1 \end{bmatrix}$. (c) $L = \begin{bmatrix} 1 & 0 & 0 \\ -1/2 & 1 & 0 \\ -1/5 & 1 & 1 \end{bmatrix}$.

8. (a) $R2 - 2R1$, $R3 + 3R1$, $R3 - 5R2$ (c) $R3 - \frac{1}{7}R1$, $R3 + \frac{3}{4}R2$

9. The matrix equation $\mathbf{A}\mathbf{x} = \mathbf{b}$ is $\begin{bmatrix} 1 & 2 & -1 \\ -2 & -1 & 3 \\ 1 & -1 & -4 \end{bmatrix}\begin{bmatrix} x_1 \\ x_2 \\ x_3 \end{bmatrix} = \begin{bmatrix} 2 \\ 3 \\ -7 \end{bmatrix}$.

$$\begin{bmatrix} 1 & 2 & -1 \\ -2 & -1 & 3 \\ 1 & -1 & -4 \end{bmatrix} \underset{\substack{R2+2R1 \\ R3+(-1)R1}}{\approx} \begin{bmatrix} 1 & 2 & -1 \\ 0 & 3 & 1 \\ 0 & -3 & -3 \end{bmatrix} \underset{R3+R2}{\approx} \begin{bmatrix} 1 & 2 & -1 \\ 0 & 3 & 1 \\ 0 & 0 & -2 \end{bmatrix} = U, \text{ and}$$

$L = \begin{bmatrix} 1 & 0 & 0 \\ -2 & 1 & 0 \\ 1 & -1 & 1 \end{bmatrix}$. $L\mathbf{y} = \begin{bmatrix} 2 \\ 3 \\ -7 \end{bmatrix}$ gives $\mathbf{y} = \begin{bmatrix} 2 \\ 7 \\ -2 \end{bmatrix}$, and $U\mathbf{x} = \mathbf{y}$ gives $\mathbf{x} = \begin{bmatrix} -1 \\ 2 \\ 1 \end{bmatrix}$.

11. The matrix equation $\mathbf{A}\mathbf{x} = \mathbf{b}$ is $\begin{bmatrix} 3 & -1 & 1 \\ -3 & 2 & 1 \\ 9 & 5 & -3 \end{bmatrix}\begin{bmatrix} x_1 \\ x_2 \\ x_3 \end{bmatrix} = \begin{bmatrix} 10 \\ -8 \\ 24 \end{bmatrix}$.

$$\begin{bmatrix} 3 & -1 & 1 \\ -3 & 2 & 1 \\ 9 & 5 & -3 \end{bmatrix} \underset{\substack{R2+R1 \\ R3+(-3)R1}}{\approx} \begin{bmatrix} 3 & -1 & 1 \\ 0 & 1 & 2 \\ 0 & 8 & -6 \end{bmatrix} \underset{R3+(-8)R2}{\approx} \begin{bmatrix} 3 & -1 & 1 \\ 0 & 1 & 2 \\ 0 & 0 & -22 \end{bmatrix} = U, \text{ and}$$

$L = \begin{bmatrix} 1 & 0 & 0 \\ -1 & 1 & 0 \\ 3 & 8 & 1 \end{bmatrix}$. $L\mathbf{y} = \begin{bmatrix} 10 \\ -8 \\ 24 \end{bmatrix}$ gives $\mathbf{y} = \begin{bmatrix} 10 \\ 2 \\ -22 \end{bmatrix}$, and $U\mathbf{x} = \mathbf{y}$ gives $\mathbf{x} = \begin{bmatrix} 3 \\ 0 \\ 1 \end{bmatrix}$.

14. The matrix equation $\mathbf{A}\mathbf{x} = \mathbf{b}$ is $\begin{bmatrix} -2 & 0 & 3 \\ -4 & 3 & 5 \\ 8 & 9 & -11 \end{bmatrix}\begin{bmatrix} x_1 \\ x_2 \\ x_3 \end{bmatrix} = \begin{bmatrix} 3 \\ 11 \\ 7 \end{bmatrix}$.

$$\begin{bmatrix} -2 & 0 & 3 \\ -4 & 3 & 5 \\ 8 & 9 & -11 \end{bmatrix} \underset{\substack{R2+(-2)R1 \\ R3+4R1}}{\approx} \begin{bmatrix} -2 & 0 & 3 \\ 0 & 3 & -1 \\ 0 & 9 & 1 \end{bmatrix} \underset{R3+(-3)R2}{\approx} \begin{bmatrix} -2 & 0 & 3 \\ 0 & 3 & -1 \\ 0 & 0 & 4 \end{bmatrix} = U, \text{ and}$$

$L = \begin{bmatrix} 1 & 0 & 0 \\ 2 & 1 & 0 \\ -4 & 3 & 1 \end{bmatrix}$. $L\mathbf{y} = \begin{bmatrix} 3 \\ 11 \\ 7 \end{bmatrix}$ gives $\mathbf{y} = \begin{bmatrix} 3 \\ 5 \\ 4 \end{bmatrix}$, and $U\mathbf{x} = \mathbf{y}$ gives $\mathbf{x} = \begin{bmatrix} 0 \\ 2 \\ 1 \end{bmatrix}$.

16. The matrix equation $\mathbf{Ax} = \mathbf{b}$ is $\begin{bmatrix} 2 & -3 & 1 \\ 4 & -5 & 6 \\ -10 & 19 & 9 \end{bmatrix}\begin{bmatrix} x_1 \\ x_2 \\ x_3 \end{bmatrix} = \begin{bmatrix} -5 \\ 2 \\ 55 \end{bmatrix}$.

$$\begin{bmatrix} 2 & -3 & 1 \\ 4 & -5 & 6 \\ -10 & 19 & 9 \end{bmatrix} \underset{\substack{R2+(-2)R1 \\ R3+5R1}}{\approx} \begin{bmatrix} 2 & -3 & 1 \\ 0 & 1 & 4 \\ 0 & 4 & 14 \end{bmatrix} \underset{R3+(-4)R2}{\approx} \begin{bmatrix} 2 & -3 & 1 \\ 0 & 1 & 4 \\ 0 & 0 & -2 \end{bmatrix} = U, \text{ and}$$

$L = \begin{bmatrix} 1 & 0 & 0 \\ 2 & 1 & 0 \\ -5 & 4 & 1 \end{bmatrix}$. $L\mathbf{y} = \begin{bmatrix} -5 \\ 2 \\ 55 \end{bmatrix}$ gives $\mathbf{y} = \begin{bmatrix} -5 \\ 12 \\ -18 \end{bmatrix}$, and $U\mathbf{x} = \mathbf{y}$ gives $\mathbf{x} = \begin{bmatrix} -43 \\ -24 \\ 9 \end{bmatrix}$.

18. The matrix equation $\mathbf{Ax} = \mathbf{b}$ is $\begin{bmatrix} 1 & 2 & -1 \\ 2 & 5 & 1 \\ -1 & -1 & 4 \end{bmatrix}\begin{bmatrix} x_1 \\ x_2 \\ x_3 \end{bmatrix} = \begin{bmatrix} 2 \\ 3 \\ -3 \end{bmatrix}$.

$$\begin{bmatrix} 1 & 2 & -1 \\ 2 & 5 & 1 \\ -1 & -1 & 4 \end{bmatrix} \underset{\substack{R2+(-2)R1 \\ R3+R1}}{\approx} \begin{bmatrix} 1 & 2 & -1 \\ 0 & 1 & 3 \\ 0 & 1 & 3 \end{bmatrix} \underset{R3+(-1)R2}{\approx} \begin{bmatrix} 1 & 2 & -1 \\ 0 & 1 & 3 \\ 0 & 0 & 0 \end{bmatrix} = U, \text{ and}$$

$L = \begin{bmatrix} 1 & 0 & 0 \\ 2 & 1 & 0 \\ -1 & 1 & 1 \end{bmatrix}$. $L\mathbf{y} = \begin{bmatrix} 2 \\ 3 \\ -3 \end{bmatrix}$ gives $\mathbf{y} = \begin{bmatrix} 2 \\ -1 \\ 0 \end{bmatrix}$, and $U\mathbf{x} = \mathbf{y}$ gives $\mathbf{x} = \begin{bmatrix} 4+7r \\ -1-3r \\ r \end{bmatrix}$.

19. The matrix equation $\mathbf{Ax} = \mathbf{b}$ is $\begin{bmatrix} 4 & 1 & -2 \\ -4 & 2 & 3 \\ 8 & -7 & -7 \end{bmatrix}\begin{bmatrix} x_1 \\ x_2 \\ x_3 \end{bmatrix} = \begin{bmatrix} 3 \\ 1 \\ -2 \end{bmatrix}$.

$$\begin{bmatrix} 4 & 1 & -2 \\ -4 & 2 & 3 \\ 8 & -7 & -7 \end{bmatrix} \underset{\substack{R2+R1 \\ R3+(-2)R1}}{\approx} \begin{bmatrix} 4 & 1 & -2 \\ 0 & 3 & 1 \\ 0 & -9 & -3 \end{bmatrix} \underset{R3+3R2}{\approx} \begin{bmatrix} 4 & 1 & -2 \\ 0 & 3 & 1 \\ 0 & 0 & 0 \end{bmatrix} = U, \text{ and}$$

$L = \begin{bmatrix} 1 & 0 & 0 \\ -1 & 1 & 0 \\ 2 & -3 & 1 \end{bmatrix}$. $L\mathbf{y} = \begin{bmatrix} 3 \\ 1 \\ -2 \end{bmatrix}$ gives $\mathbf{y} = \begin{bmatrix} 3 \\ 4 \\ 4 \end{bmatrix}$, and $U\mathbf{x} = \mathbf{y}$ is a system with no solution.

20. The matrix equation $A\mathbf{x} = \mathbf{b}$ is $\begin{bmatrix} 1 & 1 & -1 & 2 \\ 1 & 3 & 2 & 2 \\ -1 & -3 & -4 & 6 \\ 0 & 4 & 7 & -2 \end{bmatrix}\begin{bmatrix} x_1 \\ x_2 \\ x_3 \\ x_4 \end{bmatrix} = \begin{bmatrix} 7 \\ 6 \\ 12 \\ -7 \end{bmatrix}$.

$$\begin{bmatrix} 1 & 1 & -1 & 2 \\ 1 & 3 & 2 & 2 \\ -1 & -3 & -4 & 6 \\ 0 & 4 & 7 & -2 \end{bmatrix} \underset{\substack{R2+(-1)R1 \\ R3+R1}}{\approx} \begin{bmatrix} 1 & 1 & -1 & 2 \\ 0 & 2 & 3 & 0 \\ 0 & -2 & -5 & 8 \\ 0 & 4 & 7 & -2 \end{bmatrix} \underset{\substack{R3+R2 \\ R4+(-2)R2}}{\approx} \begin{bmatrix} 1 & 1 & -1 & 2 \\ 0 & 2 & 3 & 0 \\ 0 & 0 & -2 & 8 \\ 0 & 0 & 1 & -2 \end{bmatrix}$$

$$\underset{R4+(1/2)R3}{\approx} \begin{bmatrix} 1 & 1 & -1 & 2 \\ 0 & 2 & 3 & 0 \\ 0 & 0 & -2 & 8 \\ 0 & 0 & 0 & 2 \end{bmatrix} = U, \text{ and } L = \begin{bmatrix} 1 & 0 & 0 & 0 \\ 1 & 1 & 0 & 0 \\ -1 & -1 & 1 & 0 \\ 0 & 2 & -1/2 & 1 \end{bmatrix}.$$

$$L\mathbf{y} = \begin{bmatrix} 7 \\ 6 \\ 12 \\ -7 \end{bmatrix} \text{ gives } \mathbf{y} = \begin{bmatrix} 7 \\ -1 \\ 18 \\ 4 \end{bmatrix}, \text{ and } U\mathbf{x} = \mathbf{y} \text{ gives } \mathbf{x} = \begin{bmatrix} 1 \\ 1 \\ -1 \\ 2 \end{bmatrix}.$$

24. $\begin{bmatrix} 6 & -2 \\ 12 & 8 \end{bmatrix} = \begin{bmatrix} 1 & 0 \\ 2 & 1 \end{bmatrix}\begin{bmatrix} 6 & -2 \\ 0 & 12 \end{bmatrix} = \begin{bmatrix} 2 & 0 \\ 4 & 2 \end{bmatrix}\begin{bmatrix} 3 & -1 \\ 0 & 6 \end{bmatrix}$.

26. The first pair of row operations below require three multiplications and two additions each. The last row operation requires two multiplications and one addition. Thus a total of thirteen arithmetic operations are required to obtain U.

$$\begin{bmatrix} a & b & c \\ d & e & f \\ g & h & i \end{bmatrix} \underset{\substack{R2-(d/a)R1 \\ R3-(g/a)R1}}{\approx} \begin{bmatrix} a & b & c \\ 0 & j & k \\ 0 & m & n \end{bmatrix} \underset{R3-(m/j)R2}{\approx} \begin{bmatrix} a & b & c \\ 0 & j & k \\ 0 & 0 & s \end{bmatrix} = U. \text{ No arithmetic operations}$$

are required for L since it can be written down with no additional calculation.

$$L = \begin{bmatrix} 1 & 0 & 0 \\ -p & 1 & 0 \\ -q & -r & 0 \end{bmatrix}, \text{ where } p = \frac{d}{a},\ q = \frac{g}{a}, \text{ and } r = \frac{m}{j}.$$

To solve $L\mathbf{y} = \mathbf{b}$ requires six operations as follows: $y_1 = b_1$ requires no operations,

$y_2 = b_2 + pb_1$ requires one multiplication and one addition, and $y_3 = b_3 + qy_1 + ry_2$ requires two multiplications and two additions.

To solve $\mathbf{Ux} = \mathbf{y}$ requires nine operations because there is an additional multiplication at each stage: $x_3 = y_3/s$ requires one multiplication, $x_2 = (y_2 - kx_3)/j$ requires two multiplications and one addition, and $x_1 = (y_1 - bx_2 - cx_3)/a$ requires three multiplications and two additions.

Exercise Set 9.2, page 406

1. (a) max{1+3, 2+4} = max{4,6} = 6

   (c) max{1+3+6, 2+5+1, 0+4+2} = max{10,8,6} = 10

2. (a) $A = \begin{bmatrix} 1 & 2 \\ 3 & 4 \end{bmatrix}$ and $A^{-1} = \begin{bmatrix} -2 & 1 \\ 3/2 & -1/2 \end{bmatrix}$, so $||A||$ = max{4,6} = 6 and

   $||A^{-1}||$ = max{7/2, 3/2} = 7/2. Thus c(A) = 21.

   (c) $A = \begin{bmatrix} 5 & 2 \\ 8 & 3 \end{bmatrix}$ and $A^{-1} = \begin{bmatrix} -3 & 2 \\ 8 & -5 \end{bmatrix}$, so $||A||$ = max{13,5} = 13 and

   $||A^{-1}||$ = max{11,7} = 11. Thus c(A) = 143.

   (e) $A = \begin{bmatrix} 5.2 & 3.7 \\ 3.8 & 2.6 \end{bmatrix}$ and $A^{-1} = \begin{bmatrix} -2.6/.54 & 3.7/.54 \\ 3.8/.54 & -5.2/.54 \end{bmatrix}$, so $||A||$ = max{9,6.3} = 9 and

   $||A^{-1}||$ = max{6.4/.54, 8.9/.54} = 8.9/.54. Thus c(A) = $148\frac{1}{3}$.

3. (a) $A = \begin{bmatrix} 9 & 2 \\ 4 & 7 \end{bmatrix}$ and $A^{-1} = \begin{bmatrix} 7/55 & -2/55 \\ -4/55 & 9/55 \end{bmatrix}$, so $||A||$ = max{13,9} = 13 and

   $||A^{-1}||$ = max{1/5, 1/5} = 1/5. Thus c(A) = 13/5 = 2.6 = $.26\text{x}10^1$.

   The solution of a system of equations AX = B can have one fewer significant digit of accuracy than the elements of A.

(c) $A = \begin{bmatrix} 300 & 1001 \\ 75 & 250 \end{bmatrix}$ and $A^{-1} = \begin{bmatrix} -10/3 & 1001/75 \\ 1 & -4 \end{bmatrix}$, so $||A|| = \max\{375, 1251\}$

$= 1251$ and $||A^{-1}|| = \max\{13/3, 1301/75\} = 1301/75$. Thus $c(A) = 21700.7$
$= .217\text{x}10^5$.

The solution of a system of equations AX = B can have five fewer significant digits of accuracy than the elements of A.

(e) $A = \begin{bmatrix} 5 & 3 \\ 4 & 2 \end{bmatrix}$ and $A^{-1} = \begin{bmatrix} -1 & 3/2 \\ 2 & -5/2 \end{bmatrix}$, so $||A|| = \max\{9,5\} = 9$ and

$||A^{-1}|| = \max\{3,4\} = 4$. Thus $c(A) = 36 = .36\text{x}10^2$.

The solution of a system of equations AX = B can have two fewer significant digits of accuracy than the elements of A.

4. (a) $A = \begin{bmatrix} 1 & 1 & -1 \\ 1 & 0 & 2 \\ 1 & -2 & 0 \end{bmatrix}$ and $A^{-1} = \begin{bmatrix} 1/2 & 1/4 & 1/4 \\ 1/4 & 1/8 & -3/8 \\ -1/4 & 3/8 & -1/8 \end{bmatrix}$, so $||A|| = \max\{3,3,3\} = 3$ and

$||A^{-1}|| = \max\{1, 3/4, 3/4\} = 1$. Thus $c(A) = 3 = .3\text{x}10^1$.

The solution of a system of equations AX = B can have one fewer significant digit of accuracy than the elements of A.

5. $A = \begin{bmatrix} 1/3 & 1/4 & 1/5 \\ 1/4 & 1/5 & 1/6 \\ 1/5 & 1/6 & 1/7 \end{bmatrix}$ and $A^{-1} = \begin{bmatrix} 300 & -900 & 630 \\ -900 & 2880 & -2100 \\ 630 & -2100 & 1575 \end{bmatrix}$, so $||A|| = \frac{47}{60}$ and

$||A^{-1}|| = 5880$. Thus $c(A) = 4606$.

6. $A = \begin{bmatrix} 1 & k \\ 1 & 1 \end{bmatrix}$, $k \neq 1$, and $A^{-1} = \begin{bmatrix} 1/(1-k) & -k/(1-k) \\ -1/(1-k) & 1/(1-k) \end{bmatrix}$, so $||A|| = \max\{2, 1+|k|\}$ and

$||A^{-1}|| = \max\{2/|1-k|, (1+|k|)/|1-k|\}$. Thus if $k > 1$, $c(A) = (1+k)^2/k-1$. Solutions to

$c(A) = 100$ are $k = 1.0417$ and $k = 96.9583$. $c(A) > 100$ if k is in the interval

(1, 1.0417) or $k > 96.9583$. If $-1 \le k < 1$, $c(A) = 4/(1-k)$, which equals 100 if $k = .96$. $c(A) > 100$ if $.96 < k < 1$. If $k < -1$, $c(A) = 1 - k$, which is greater than 100 if $k < -99$.

8. (b) $1 = \|AA^{-1}\| \le \|A\|\,\|A^{-1}\| = c(A)$.

10. If A is a diagonal matrix with diagonal elements a_{ii}, then A^{-1} is a diagonal matrix with diagonal elements $1/a_{ii}$. $\|A\| = \max\{a_{ii}\}$ and $\|A^{-1}\| = \max\{1/a_{ii}\} = 1/\min\{a_{ii}\}$, so $c(A) = (\max\{a_{ii}\})(1/\min\{a_{ii}\})$.

11. $\begin{bmatrix} 0 & -1 & 2 \\ 1 & 2 & -1 \\ 1 & 2 & 2 \end{bmatrix}\begin{bmatrix} x_1 \\ x_2 \\ x_3 \end{bmatrix} = \begin{bmatrix} 4 \\ 1 \\ 4 \end{bmatrix}$. $\begin{bmatrix} 0 & -1 & 2 & 4 \\ 1 & 2 & -1 & 1 \\ 1 & 2 & 2 & 4 \end{bmatrix} \underset{R1 \leftrightarrow R2}{\approx} \begin{bmatrix} 1 & 2 & -1 & 1 \\ 0 & -1 & 2 & 4 \\ 1 & 2 & 2 & 4 \end{bmatrix}$

$\underset{R3+(-1)R1}{\approx} \begin{bmatrix} 1 & 2 & -1 & 1 \\ 0 & -1 & 2 & 4 \\ 0 & 0 & 3 & 3 \end{bmatrix} \underset{\substack{(-1)R2 \\ (1/3)R3}}{\approx} \begin{bmatrix} 1 & 2 & -1 & 1 \\ 0 & 1 & -2 & -4 \\ 0 & 0 & 1 & 1 \end{bmatrix} \underset{R1+(-2)R2}{\approx} \begin{bmatrix} 1 & 0 & 3 & 9 \\ 0 & 1 & -2 & -4 \\ 0 & 0 & 1 & 1 \end{bmatrix}$

$\underset{\substack{R1+(-3)R3 \\ R2+2R3}}{\approx} \begin{bmatrix} 1 & 0 & 0 & 6 \\ 0 & 1 & 0 & -2 \\ 0 & 0 & 1 & 1 \end{bmatrix}$, so the exact solution is $\begin{bmatrix} x_1 \\ x_2 \\ x_3 \end{bmatrix} = \begin{bmatrix} 6 \\ -2 \\ 1 \end{bmatrix}$.

12. $\begin{bmatrix} -1 & 1 & 2 \\ 2 & 4 & -1 \\ 1 & 2 & 2 \end{bmatrix}\begin{bmatrix} x_1 \\ x_2 \\ x_3 \end{bmatrix} = \begin{bmatrix} 8 \\ 10 \\ 2 \end{bmatrix}$. $\begin{bmatrix} -1 & 1 & 2 & 8 \\ 2 & 4 & -1 & 10 \\ 1 & 2 & 2 & 2 \end{bmatrix} \underset{R1 \leftrightarrow R2}{\approx} \begin{bmatrix} 2 & 4 & -1 & 10 \\ -1 & 1 & 2 & 8 \\ 1 & 2 & 2 & 2 \end{bmatrix}$

$\underset{(1/2)R1}{\approx} \begin{bmatrix} 1 & 2 & -1/2 & 5 \\ -1 & 1 & 2 & 8 \\ 1 & 2 & 2 & 2 \end{bmatrix} \underset{\substack{R2+R1 \\ R3+(-1)R1}}{\approx} \begin{bmatrix} 1 & 2 & -1/2 & 5 \\ 0 & 3 & 3/2 & 13 \\ 0 & 0 & 5/2 & -3 \end{bmatrix}$

$\underset{\substack{(1/3)R2 \\ (2/5)R3}}{\approx} \begin{bmatrix} 1 & 2 & -1/2 & 5 \\ 0 & 1 & 1/2 & 4.33 \\ 0 & 0 & 1 & -6/5 \end{bmatrix} \underset{R1+(-2)R2}{\approx} \begin{bmatrix} 1 & 0 & -3/2 & -3.66 \\ 0 & 1 & 1/2 & 4.33 \\ 0 & 0 & 1 & -1.2 \end{bmatrix}$

(Here is the first round-off error.) (The round-off error gets multiplied here.)

$$\underset{\substack{R1+(1.5)R3\\R2+(-.5)R3}}{\approx}\begin{bmatrix}1&0&0&-5.46\\0&1&0&4.93\\0&0&1&-1.2\end{bmatrix}, \text{ so } \begin{bmatrix}x_1\\x_2\\x_3\end{bmatrix}=\begin{bmatrix}-5.46\\4.93\\-1.2\end{bmatrix}.$$ Substituting into the original

equations $5.46 + 4.93 - 2.4 = 7.99$, $-10.92 + 19.72 + 1.2 = 10$, and $-5.46 + 9.86 - 2.4 = 2$. The exact solution is $x_1 = -82/15$, $x_2 = 74/15$, $x_3 = -6/5$.

15. First we multiply equation 2 by 1/.01, then let $.1x_3 = y_3$, $x_2 = y_2$, $x_1 = y_1$. The matrix

equation becomes $\begin{bmatrix}0&1&-1\\2&1&0\\1&-4&1\end{bmatrix}\begin{bmatrix}y_1\\y_2\\y_3\end{bmatrix}=\begin{bmatrix}2\\1\\2\end{bmatrix}$.

$$\begin{bmatrix}0&1&-1&2\\2&1&0&1\\1&-4&1&2\end{bmatrix}\underset{R1\leftrightarrow R2}{\approx}\begin{bmatrix}2&1&0&1\\0&1&-1&2\\1&-4&1&2\end{bmatrix}\underset{(1/2)R1}{\approx}\begin{bmatrix}1&1/2&0&1/2\\0&1&-1&2\\1&-4&1&2\end{bmatrix}$$

$$\underset{R3+(-1)R1}{\approx}\begin{bmatrix}1&1/2&0&1/2\\0&1&-1&2\\0&-9/2&1&3/2\end{bmatrix}\underset{R2\leftrightarrow R3}{\approx}\begin{bmatrix}1&1/2&0&1/2\\0&-9/2&1&3/2\\0&1&-1&2\end{bmatrix}$$

$$\underset{(-2/9)R2}{\approx}\begin{bmatrix}1&1/2&0&1/2\\0&1&-.2222&-.3333\\0&1&-1&2\end{bmatrix}\underset{\substack{R1+(-1/2)R2\\R3+(-1)R2}}{\approx}\begin{bmatrix}1&0&.1111&.6667\\0&1&-.2222&-.3333\\0&0&-.7778&2.3333\end{bmatrix}$$

(Four decimal places for **y** will result in three decimal places for **x**.)

$$\underset{(-1/.7778)R3}{\approx}\begin{bmatrix}1&0&.1111&.6667\\0&1&-.2222&-.3333\\0&0&1&-2.9999\end{bmatrix}\underset{\substack{R1+(-.1111)R3\\R2+(.2222)R3}}{\approx}\begin{bmatrix}1&0&0&1.0000\\0&1&0&-.9999\\0&0&1&-2.9999\end{bmatrix}, \text{ so}$$

$\begin{bmatrix}y_1\\y_2\\y_3\end{bmatrix}=\begin{bmatrix}1.0000\\-.9999\\-2.9999\end{bmatrix}$. Therefore $x_1 = 1.000$, $x_2 = -1.000$, and $x_3 = -29.999$.

The exact solution is $x_1 = 1$, $x_2 = -1$, and $x_3 = -30$.

17. $\begin{bmatrix} 1 & -2 & -1 \\ 1 & -.001 & -1 \\ 1 & 3 & -.002 \end{bmatrix} \begin{bmatrix} x_1 \\ x_2 \\ x_3 \end{bmatrix} = \begin{bmatrix} 1 \\ 2 \\ -1 \end{bmatrix}$.

$$\begin{bmatrix} 1 & -2 & -1 & 1 \\ 1 & -.001 & -1 & 2 \\ 1 & 3 & -.002 & -1 \end{bmatrix} \underset{\substack{R2+(-1)R1 \\ R3+(-1)R1}}{\approx} \begin{bmatrix} 1 & -2 & -1 & 1 \\ 0 & 1.999 & 0 & 1 \\ 0 & 5 & .998 & -2 \end{bmatrix}$$

$$\underset{R2 \leftrightarrow R3}{\approx} \begin{bmatrix} 1 & -2 & -1 & 1 \\ 0 & 5 & .998 & -2 \\ 0 & 1.999 & 0 & 1 \end{bmatrix} \underset{(1/5)R2}{\approx} \begin{bmatrix} 1 & -2 & -1 & 1 \\ 0 & 1 & .200 & -.4 \\ 0 & 1.999 & 0 & 1 \end{bmatrix}$$

(The first round-off error is here.)

$$\underset{\substack{R1+2R2 \\ R3+(-1.999)R2}}{\approx} \begin{bmatrix} 1 & 0 & -.600 & .2 \\ 0 & 1 & .200 & -.4 \\ 0 & 0 & -.400 & 1.800 \end{bmatrix} \underset{(-1/.400)R3}{\approx} \begin{bmatrix} 1 & 0 & -.600 & .2 \\ 0 & 1 & .200 & -.4 \\ 0 & 0 & 1 & -4.500 \end{bmatrix}$$

$\underset{\substack{R1+.600R3 \\ R2+(-.200)R3}}{\approx} \begin{bmatrix} 1 & 0 & 0 & -2.500 \\ 0 & 1 & 0 & .500 \\ 0 & 0 & 1 & -4.500 \end{bmatrix}$, so $\begin{bmatrix} x_1 \\ x_2 \\ x_3 \end{bmatrix} = \begin{bmatrix} -2.500 \\ .500 \\ -4.500 \end{bmatrix}$. Substituting in the original

equations $-2.500 - 1.000 + 4.500 = 1$, $-2.500 - .001 + 4.500 = 2 - .001 = 1.999$,

and $-2.500 + 1.500 + .009 = -1 + .009 = -.991$.

Exercise Set 9.3, page 412

For exercises 1–6 we give iterations found using a computer program. We stop when for all variables two successive iterations have given the same value to two decimal places.

1.

x	y	z
1	2	3
2.25	0	1.9375
2.484375	0.425	1.98515625
2.390039063	0.4059375	2.003974609
2.399509277	0.3984101562	1.99972522
2.400328766	0.4001099121	1.999945287

Thus to two decimal places x = 2.40, y = 0.40, and z = 2.00. Substitution into the given equations will show that in fact this is the exact solution.

3.

x	y	z
0	0	0
4	5.5	2.875
4.525	5.2375	2.678125
4.511875	5.2440625	2.683046875
4.512203125	5.243898438	2.682923828

Thus to two decimal places x = 4.51, y = 5.24, and z = 2.68. These values are the two-decimal-place approximations to the exact solution x = 185/41, y = 215/41, z = 110/41.

5.

x	y	z
20	30	−40
30	−22.5	11.3
−1.02	5.835	−0.571
9.3954	−2.34045	3.14717
6.273042	0.1502715	2.0245541
7.22023266	−0.603977805	2.364842093
6.933267602	−0.3754232776	2.261738176
7.020220074	−0.4446754931	2.292979113
6.993873256	−0.4236918496	2.283513021
7.001856422	−0.4300499555	2.286381275
6.999437499	−0.4281234305	2.285512186

Thus to two decimal places x = 7.00, y = −0.43, and z = 2.29. These values are the two-decimal-place approximations to the exact solution x = 7, y = −3/7, z = 16/7.

6.

x	y	z	w
0	0	0	0
3.333333333	3.333333333	6.666666667	2.333333333
2.388888888	1.784722222	5.989583333	2.319097222
2.246006944	1.971853299	5.939326534	2.332596571
2.283321699	1.979753154	5.939729234	2.32525443
2.285794915	1.979343327	5.942657121	2.325638153
2.285174675	1.978688885	5.942534914	2.325725539

Thus to two decimal places x = 2.29, y = 1.98, z = 5.94, and w = 2.33. These values are the two-decimal-place approximations to the exact solution x = 7310/3199, y = 6330/3199, z = 19010/3199, w = 7440/3199.

Exercise Set 9.4, page 420

In exercises 1 and 3 $\mathbf{x} = \begin{bmatrix} 1 \\ 2 \\ 1 \end{bmatrix}$ and in exercises 5 and 7 $\mathbf{x} = \begin{bmatrix} 1 \\ 2 \\ 1 \\ 2 \end{bmatrix}$. Results are recorded to five decimal places. The process is stopped when successive results agree to three decimal places.

1.

$\frac{A\mathbf{x} \cdot \mathbf{x}}{\mathbf{x} \cdot \mathbf{x}}$	Scaled $A\mathbf{x}$
2	$\begin{bmatrix} 1 \\ 0.5 \\ -0.5 \end{bmatrix}$
8	$\begin{bmatrix} 1 \\ 0.125 \\ -0.875 \end{bmatrix}$
8.31579	$\begin{bmatrix} 1 \\ 0.03125 \\ -0.96875 \end{bmatrix}$
8.09063	$\begin{bmatrix} 1 \\ 0.00781 \\ -0.99219 \end{bmatrix}$
8.02325	$\begin{bmatrix} 1 \\ 0.00195 \\ -0.99805 \end{bmatrix}$
8.00585	$\begin{bmatrix} 1 \\ 0.00049 \\ -0.99951 \end{bmatrix}$
8.00146	$\begin{bmatrix} 1 \\ 0.00012 \\ -0.99988 \end{bmatrix}$
8.00037	$\begin{bmatrix} 1 \\ 0.00003 \\ -0.99997 \end{bmatrix}$

Hence $\lambda = 8$ and the dominant

eigenvector is $r\begin{bmatrix} 1 \\ 0 \\ -1 \end{bmatrix}$.

3.

$\frac{A\mathbf{x} \cdot \mathbf{x}}{\mathbf{x} \cdot \mathbf{x}}$	Scaled $A\mathbf{x}$
6	$\begin{bmatrix} 1 \\ -0.1 \\ 1 \end{bmatrix}$
5.26866	$\begin{bmatrix} 1 \\ 0.01887 \\ 1 \end{bmatrix}$
6.13081	$\begin{bmatrix} 1 \\ -0.00308 \\ 1 \end{bmatrix}$
5.97843	$\begin{bmatrix} 1 \\ 0.00051 \\ 1 \end{bmatrix}$
6.00360	$\begin{bmatrix} 1 \\ -0.00008 \\ 1 \end{bmatrix}$
5.99940	$\begin{bmatrix} 1 \\ 0.00001 \\ 1 \end{bmatrix}$

Hence $\lambda = 6$ and the dominant

eigenvector is $r\begin{bmatrix} 1 \\ 0 \\ 1 \end{bmatrix}$.

5.

$\frac{A\mathbf{x}\cdot\mathbf{x}}{\mathbf{x}\cdot\mathbf{x}}$	Scaled $A\mathbf{x}$
4.8	$\begin{bmatrix} 0.28571 \\ 0.14286 \\ 0.85714 \\ 1 \end{bmatrix}$
5.46667	$\begin{bmatrix} 0.10526 \\ 0.05263 \\ 0.94737 \\ 1 \end{bmatrix}$
5.86087	$\begin{bmatrix} 0.03636 \\ 0.01818 \\ 0.98182 \\ 1 \end{bmatrix}$
5.95964	$\begin{bmatrix} 0.01227 \\ 0.00613 \\ 0.99387 \\ 1 \end{bmatrix}$
5.98728	$\begin{bmatrix} 0.00411 \\ 0.00205 \\ 0.99795 \\ 1 \end{bmatrix}$
5.99584	$\begin{bmatrix} 0.00137 \\ 0.00069 \\ 0.99931 \\ 1 \end{bmatrix}$
5.99862	$\begin{bmatrix} 0.00046 \\ 0.00023 \\ 0.99977 \\ 1 \end{bmatrix}$
5.99954	$\begin{bmatrix} 0.00015 \\ 0.00008 \\ 0.99992 \\ 1 \end{bmatrix}$

Hence $\lambda = 6$ and the dominant

eigenvector is $r\begin{bmatrix} 0 \\ 0 \\ 1 \\ 1 \end{bmatrix}$.

7.

$\frac{A\mathbf{x}\cdot\mathbf{x}}{\mathbf{x}\cdot\mathbf{x}}$	Scaled $A\mathbf{x}$
14.2	$\begin{bmatrix} 1 \\ 0 \\ 0.03030 \\ 0.20202 \end{bmatrix}$
82.22351	$\begin{bmatrix} 1 \\ 0.00976 \\ -0.00940 \\ 0.04785 \end{bmatrix}$
84.30649	$\begin{bmatrix} 1 \\ 0.01118 \\ -0.01117 \\ 0.04011 \end{bmatrix}$
84.31129	$\begin{bmatrix} 1 \\ 0.01125 \\ -0.01125 \\ 0.03976 \end{bmatrix}$
84.31129	$\begin{bmatrix} 1 \\ 0.01126 \\ -0.01126 \\ 0.03975 \end{bmatrix}$

Hence $\lambda = 84.311$ and the dominant

eigenvector is $r\begin{bmatrix} 1 \\ 0.011 \\ -0.011 \\ 0.040 \end{bmatrix}$.

In exercises 9–11 we choose $\mathbf{x} = \begin{bmatrix} 1 \\ 2 \\ 1 \end{bmatrix}$. Results are recorded to five decimal places. The process is stopped when successive results agree to three decimal places.

9.

$\frac{A\mathbf{x}\cdot\mathbf{x}}{\mathbf{x}\cdot\mathbf{x}}$	9.16667	9.99083	9.99991	10.00000	10.00000
Scaled $A\mathbf{x}$	$\begin{bmatrix} 0.9375 \\ 1 \\ 0.5 \end{bmatrix}$	$\begin{bmatrix} 0.99359 \\ 1 \\ 0.5 \end{bmatrix}$	$\begin{bmatrix} 0.99936 \\ 1 \\ 0.5 \end{bmatrix}$	$\begin{bmatrix} 0.99994 \\ 1 \\ 0.5 \end{bmatrix}$	$\begin{bmatrix} 0.99999 \\ 1 \\ 0.5 \end{bmatrix}$

Thus the dominant eigenvalue is $\lambda = 10$ and the dominant eigenvector is $r\begin{bmatrix} 1 \\ 1 \\ 0.5 \end{bmatrix}$.

The unit eigenvector is $\mathbf{y} = \begin{bmatrix} 2/3 \\ 2/3 \\ 1/3 \end{bmatrix}$, and $B = A - \lambda \mathbf{y}\mathbf{y}^t = \frac{1}{9}\begin{bmatrix} 5 & -4 & -2 \\ -4 & 5 & -2 \\ -2 & -2 & 8 \end{bmatrix}$.

Let $C = 9B$.

$\frac{C\mathbf{x}\cdot\mathbf{x}}{\mathbf{x}\cdot\mathbf{x}}$	0.83333	9	9
$C\mathbf{x}$	$\begin{bmatrix} -1 \\ 0.8 \\ 0.4 \end{bmatrix}$	$\begin{bmatrix} -1 \\ 0.8 \\ 0.4 \end{bmatrix}$	$\begin{bmatrix} -1 \\ 0.8 \\ 0.4 \end{bmatrix}$

Thus the dominant eigenvalue of C is $\lambda = 9$ with eigenvector $\begin{bmatrix} -1 \\ 0.8 \\ 0.4 \end{bmatrix}$, so the dominant eigenvalue of B (and therefore an eigenvalue of A) is 1 with eigenvector $\begin{bmatrix} -1 \\ 0.8 \\ 0.4 \end{bmatrix}$.

The eigenspace of $\lambda = 1$ has dimension 2 with basis $\begin{bmatrix} -1 \\ 1 \\ 0 \end{bmatrix}$ and $\begin{bmatrix} 0 \\ -1 \\ 2 \end{bmatrix}$, so there are no more eigenvalues.

11.

$\frac{A\mathbf{x}\cdot\mathbf{x}}{\mathbf{x}\cdot\mathbf{x}}$	5.33333	11.47368	11.98459	11.99957	11.99999	12.00000
Scaled $A\mathbf{x}$	$\begin{bmatrix} 1 \\ 0.33333 \\ 1 \end{bmatrix}$	$\begin{bmatrix} 1 \\ 0.05556 \\ 1 \end{bmatrix}$	$\begin{bmatrix} 1 \\ 0.00926 \\ 1 \end{bmatrix}$	$\begin{bmatrix} 1 \\ 0.00154 \\ 1 \end{bmatrix}$	$\begin{bmatrix} 1 \\ 0.00026 \\ 1 \end{bmatrix}$	$\begin{bmatrix} 1 \\ 0.00004 \\ 1 \end{bmatrix}$

Thus $\lambda = 12$ with dominant eigenvector $r\begin{bmatrix} 1 \\ 0 \\ 1 \end{bmatrix}$. The unit eigenvector is $\mathbf{y} = \begin{bmatrix} 1/\sqrt{2} \\ 0 \\ 1/\sqrt{2} \end{bmatrix}$, and

$B = A - \lambda\mathbf{y}\mathbf{y}^t = \begin{bmatrix} 1 & 0 & -1 \\ 0 & 2 & 0 \\ -1 & 0 & 1 \end{bmatrix}$. Three iterations of the process confirms that $\lambda = 2$ with

dominant eigenvector $\begin{bmatrix} 0 \\ 1 \\ 0 \end{bmatrix}$. The eigenspace of $\lambda = 2$ has dimension 2 with basis

vectors $\begin{bmatrix} 0 \\ 1 \\ 0 \end{bmatrix}$ and $\begin{bmatrix} 1 \\ 0 \\ -1 \end{bmatrix}$, so there are no additional eigenvalues of A.

12. We choose $\mathbf{x} = \begin{bmatrix} 1 \\ 2 \\ 1 \\ 2 \end{bmatrix}$ and record successive values of $\frac{A\mathbf{x}\cdot\mathbf{x}}{\mathbf{x}\cdot\mathbf{x}}$ to five decimal places.

$\frac{A\mathbf{x}\cdot\mathbf{x}}{\mathbf{x}\cdot\mathbf{x}}$: 3.8, 9.36066, 9.92232, 9.97852, 9.99251, 9.99731, 9.99903, 9.99965.

$\lambda = 10$ and the dominant eigenvector is $r\begin{bmatrix} 1 \\ 0 \\ 0 \\ 1 \end{bmatrix}$. The unit eigenvector is $\mathbf{y} = \begin{bmatrix} 1/\sqrt{2} \\ 0 \\ 0 \\ 1/\sqrt{2} \end{bmatrix}$

and $B = A - \lambda\mathbf{y}\mathbf{y}^t = \begin{bmatrix} -1 & 0 & 0 & 1 \\ 0 & 2 & -4 & 0 \\ 0 & -4 & 2 & 0 \\ 1 & 0 & 0 & -1 \end{bmatrix}$.

$\frac{B\mathbf{x}\cdot\mathbf{x}}{\mathbf{x}\cdot\mathbf{x}}$: -0.7, 1.78947, 5.12088, 5.89175, 5.98783, 5.99865, 5.99985, 5.99998.

$\lambda = 6$ and the dominant eigenvector is $r\begin{bmatrix} 0 \\ 1 \\ -1 \\ 0 \end{bmatrix}$. The unit eigenvector is $\mathbf{z} = \begin{bmatrix} 0 \\ 1/\sqrt{2} \\ -1/\sqrt{2} \\ 0 \end{bmatrix}$

and $C = B - \lambda\mathbf{z}\mathbf{z}^t = \begin{bmatrix} -1 & 0 & 0 & 1 \\ 0 & -1 & -1 & 0 \\ 0 & -1 & -1 & 0 \\ 1 & 0 & 0 & -1 \end{bmatrix}$. $\frac{C\mathbf{x}\cdot\mathbf{x}}{\mathbf{x}\cdot\mathbf{x}}$: -1, -2, -2.

$\lambda = -2$ with dominant eigenvector $r\begin{bmatrix} -1/3 \\ 1 \\ 1 \\ 1/3 \end{bmatrix}$. The eigenspace of $\lambda = -2$ is two-

dimensional with basis vectors $\begin{bmatrix} -1 \\ 0 \\ 0 \\ 1 \end{bmatrix}$ and $\begin{bmatrix} 0 \\ 1 \\ 1 \\ 0 \end{bmatrix}$, so there are no more eigenvalues.

14. (a) $B = A + I = \begin{bmatrix} 1 & 1 & 0 & 0 \\ 1 & 1 & 1 & 0 \\ 0 & 1 & 1 & 1 \\ 0 & 0 & 1 & 1 \end{bmatrix}$ and $\mathbf{x} = \begin{bmatrix} 1 \\ 2 \\ 2 \\ 1 \end{bmatrix}$.

$\frac{B\mathbf{x}\cdot\mathbf{x}}{\mathbf{x}\cdot\mathbf{x}}$	2.6	2.6176	2.6180
Scaled B**x**	$\begin{bmatrix} 0.6 \\ 1 \\ 1 \\ 0.6 \end{bmatrix}$	$\begin{bmatrix} 0.615 \\ 1 \\ 1 \\ 0.615 \end{bmatrix}$	$\begin{bmatrix} 0.618 \\ 1 \\ 1 \\ 0.618 \end{bmatrix}$

The sum of the entries is 3.236, so the Gould accessibility indices are .618/3.236 = .191, 1/3.236 = .309, 1/3.236 = .309, and .618/3.236 = .191. The given initial vector yields three decimal place accuracy after only three iterations.

(b) $B = A + I = \begin{bmatrix} 1 & 0 & 1 & 0 & 0 & 0 \\ 0 & 1 & 1 & 0 & 0 & 0 \\ 1 & 1 & 1 & 1 & 0 & 0 \\ 0 & 0 & 1 & 1 & 1 & 1 \\ 0 & 0 & 0 & 1 & 1 & 0 \\ 0 & 0 & 0 & 1 & 0 & 1 \end{bmatrix}$ and $\mathbf{x} = \begin{bmatrix} 1 \\ 1 \\ 3 \\ 3 \\ 1 \\ 1 \end{bmatrix}$.

$\dfrac{B\mathbf{x}\cdot\mathbf{x}}{\mathbf{x}\cdot\mathbf{x}}$	2.90909	3	3
Scaled $B\mathbf{x}$	$\begin{bmatrix} 0.5 \\ 0.5 \\ 1 \\ 1 \\ 0.5 \\ 0.5 \end{bmatrix}$	$\begin{bmatrix} 0.5 \\ 0.5 \\ 1 \\ 1 \\ 0.5 \\ 0.5 \end{bmatrix}$	$\begin{bmatrix} 0.5 \\ 0.5 \\ 1 \\ 1 \\ 0.5 \\ 0.5 \end{bmatrix}$

The sum of the entries is 4, so the Gould accessibility indices are 1/8,1/8,1/4,1/4,1/8, and 1/8. The given initial vector yields an eigenvector for the largest positive eigenvalue immediately.

Chapter 9 Review Exercises, page 421

1. From the first equation $x_1 = 2$, and from the second equation $6 - x_2 = 7$, so $x_2 = -1$.

   From the third equation $2 - 3 + 2x_3 = 1$, so $x_3 = 1$.

2. The "rule" is Ri – aRj causes the (i,j)th position in L to be a. $L = \begin{bmatrix} 1 & 0 & 0 \\ 3/2 & 1 & 0 \\ -2/7 & 4 & 1 \end{bmatrix}$.

3. The matrix equation $A\mathbf{x} = \mathbf{b}$ is $\begin{bmatrix} 6 & 1 & -1 \\ -6 & 1 & 1 \\ 12 & 12 & 1 \end{bmatrix} \begin{bmatrix} x_1 \\ x_2 \\ x_3 \end{bmatrix} = \begin{bmatrix} 5 \\ 1 \\ 52 \end{bmatrix}$.

$$\begin{bmatrix} 6 & 1 & -1 \\ -6 & 1 & 1 \\ 12 & 12 & 1 \end{bmatrix} \underset{\substack{R2+R1 \\ R3+(-2)R1}}{\approx} \begin{bmatrix} 6 & 1 & -1 \\ 0 & 2 & 0 \\ 0 & 10 & 3 \end{bmatrix} \underset{R3+(-5)R2}{\approx} \begin{bmatrix} 6 & 1 & -1 \\ 0 & 2 & 0 \\ 0 & 0 & 3 \end{bmatrix} = U, \text{ and}$$

$$L = \begin{bmatrix} 1 & 0 & 0 \\ -1 & 1 & 0 \\ 2 & 5 & 1 \end{bmatrix}. \ L\mathbf{y} = \begin{bmatrix} 5 \\ 1 \\ 52 \end{bmatrix} \text{ gives } \mathbf{y} = \begin{bmatrix} 5 \\ 6 \\ 12 \end{bmatrix}, \text{ and } U\mathbf{x} = \mathbf{y} \text{ gives } \mathbf{x} = \begin{bmatrix} 1 \\ 3 \\ 4 \end{bmatrix}.$$

4. The arrays below show the numbers of multiplications and additions needed to compute each entry in the matrix LU = A.

$$\begin{matrix} 1 & 1 & 1 & \cdots & 1 & 1 \\ 1 & 2 & 2 & \cdots & 2 & 2 \\ 1 & 2 & 3 & \cdots & 3 & 3 \\ \vdots & \vdots & \vdots & & \vdots & \vdots \\ 1 & 2 & 3 & \cdots & n-1 & n-1 \\ 1 & 2 & 3 & \cdots & n-1 & n \end{matrix} \qquad\qquad \begin{matrix} 0 & 0 & 0 & \cdots & 0 & 0 \\ 0 & 1 & 1 & \cdots & 1 & 1 \\ 0 & 1 & 2 & \cdots & 2 & 2 \\ \vdots & \vdots & \vdots & & \vdots & \vdots \\ 0 & 1 & 2 & \cdots & n-2 & n-2 \\ 0 & 1 & 2 & \cdots & n-2 & n-1 \end{matrix}$$

multiplications   additions

In the multiplication array, for any i the sum of the numbers in the ith column down to a_{ii} and the numbers in the ith row to and including a_{ii} is i^2:

$$a_{i1} + a_{1i} + a_{i2} + a_{2i} + \ldots + a_{i\,i-1} + a_{i-1\,i} + a_{ii} = 2x1 + 2x2 + \ldots + 2(i-1) + i$$

$= 2(1+2+\ldots+(i-1)) + i = 2\dfrac{(i-1)i}{2} + i = i^2$. Thus the number of multiplications needed to calculate LU is the sum of the squares of the numbers from 1 to n:

$1^2 + 2^2 + \ldots + (n-1)^2 + n^2 = \dfrac{n(n+1)(2n+1)}{6} = \dfrac{2n^3+3n^2+n}{6}$. Likewise, the number of additions is the sum of the squares of the numbers from 1 to n−1:

$1^2 + 2^2 + \ldots + (n-2)^2 + (n-1)^2 = \dfrac{(n-1)n(2n-1)}{6} = \dfrac{2n^3-3n^2+n}{6}$. Thus the total number of operations is $\dfrac{2n^3+3n^2+n}{6} + \dfrac{2n^3-3n^2+n}{6} = \dfrac{2n^3+n}{3}$.

If one assumes the elements on the diagonal of L are all 1, the number of multiplications for each term on or above the diagonal is decreased by 1. There are

$n + (n-1) + \ldots + 2 + 1 = \frac{n(n+1)}{2}$ such terms, so the number of multiplications in this

case is $\frac{2n^3+3n^2+n}{6} - \frac{n(n+1)}{2} = \frac{n^3-n}{3}$ and the total number of operations is

$$\frac{2n^3-3n^2+n}{6} + \frac{n^3-n}{3} = \frac{4n^3-3n^2-n}{6}.$$

5. $A = \begin{bmatrix} 250 & 401 \\ 125 & 201 \end{bmatrix}$ and $A^{-1} = \begin{bmatrix} 201/125 & -401/125 \\ -1 & 2 \end{bmatrix}$, so $||A|| = \max\{375,602\} = 602$ and

$||A^{-1}|| = \max\{326/125, 651/125\} = 651/125$. Thus $c(A) = 3135.216 = .3135216\text{x}10^4$.

The solution of a system of equations AX = B can have four fewer significant digits of accuracy than the elements of A or B.

6. $A = \begin{bmatrix} 1 & 1 & -1 \\ 1 & 0 & 2 \\ 0 & 3 & 2 \end{bmatrix}$ and $A^{-1} = \begin{bmatrix} 6/11 & 5/11 & -2/11 \\ 2/11 & -2/11 & 3/11 \\ -3/11 & 3/11 & 1/11 \end{bmatrix}$, so $||A|| = \max\{2,4,5\} = 5$ and

$||A^{-1}|| = \max\{1, 10/11, 6/11\} = 1$. Thus $c(A) = 5 = .5\text{x}10^1$. The solution of a system of equations AX = B can have one fewer significant digit of accuracy than the elements of A.

$A = \begin{bmatrix} 1 & -1 & -1 & 0 & 0 \\ 0 & 0 & 1 & -1 & -1 \\ 1 & 1 & 0 & 0 & 0 \\ 1 & 0 & 0 & 2 & 0 \\ 0 & 0 & 0 & 2 & 2 \end{bmatrix}$ and $A^{-1} = \begin{bmatrix} 1/2 & 1/2 & 1/2 & 0 & 1/4 \\ -1/2 & -1/2 & 1/2 & 0 & -1/4 \\ 0 & 1 & 0 & 0 & 1/2 \\ -1/4 & -1/4 & -1/4 & 1/2 & -1/8 \\ 1/4 & 1/4 & 1/4 & -1/2 & 5/8 \end{bmatrix}$,

so $||A|| = \max\{3,2,2,5,3\} = 5$ and $||A^{-1}|| = \max\{3/2, 5/2, 3/2, 1, 7/4\} = 5/2$.
Thus $c(A) = 25/2 = 12.5 = .125\text{x}10^2$. The solution of a system of equations AX = B can have two fewer significant digits of accuracy than the elements of A.

7. $c(A) = ||A||\,||A^{-1}|| = ||A^{-1}||\,||A|| = ||A^{-1}||\,||(A^{-1})^{-1}|| = c(A^{-1})$.

8. $c(A)$ is defined only for nonsingular matrices, so it is not a mapping of M_{nn}.

9. (a) Multiply the first equation by 10, then let $100x_1 = y_1$, $x_2 = y_2$, $.01x_3 = y_3$. The matrix equation becomes $\begin{bmatrix} 0 & -4 & 2 \\ 1 & 1 & -1 \\ 2 & 2 & -3 \end{bmatrix}\begin{bmatrix} y_1 \\ y_2 \\ y_3 \end{bmatrix} = \begin{bmatrix} 2 \\ 1 \\ 1 \end{bmatrix}$.

$$\begin{bmatrix} 0 & -4 & 2 & 2 \\ 1 & 1 & -1 & 1 \\ 2 & 2 & -3 & 1 \end{bmatrix} \underset{R1 \leftrightarrow R3}{\approx} \begin{bmatrix} 2 & 2 & -3 & 1 \\ 1 & 1 & -1 & 1 \\ 0 & -4 & 2 & 2 \end{bmatrix} \underset{(1/2)R1}{\approx} \begin{bmatrix} 1 & 1 & -3/2 & 1/2 \\ 1 & 1 & -1 & 1 \\ 0 & -4 & 2 & 2 \end{bmatrix}$$

$$\underset{R2+(-1)R1}{\approx} \begin{bmatrix} 1 & 1 & -3/2 & 1/2 \\ 0 & 0 & 1/2 & 1/2 \\ 0 & -4 & 2 & 2 \end{bmatrix} \underset{R2 \leftrightarrow R3}{\approx} \begin{bmatrix} 1 & 1 & -3/2 & 1/2 \\ 0 & -4 & 2 & 2 \\ 0 & 0 & 1/2 & 1/2 \end{bmatrix}$$

$$\underset{(-1/4)R2}{\approx} \begin{bmatrix} 1 & 1 & -3/2 & 1/2 \\ 0 & 1 & -1/2 & -1/2 \\ 0 & 0 & 1/2 & 1/2 \end{bmatrix} \underset{R1+(-1)R2}{\approx} \begin{bmatrix} 1 & 0 & -1 & 1 \\ 0 & 1 & -1/2 & -1/2 \\ 0 & 0 & 1/2 & 1/2 \end{bmatrix}$$

$$\underset{(2)R3}{\approx} \begin{bmatrix} 1 & 0 & -1 & 1 \\ 0 & 1 & -1/2 & -1/2 \\ 0 & 0 & 1 & 1 \end{bmatrix} \underset{\substack{R1+R3 \\ R2+(1/2)R3}}{\approx} \begin{bmatrix} 1 & 0 & 0 & 2 \\ 0 & 1 & 0 & 0 \\ 0 & 0 & 1 & 1 \end{bmatrix}, \text{ so } \begin{bmatrix} y_1 \\ y_2 \\ y_3 \end{bmatrix} = \begin{bmatrix} 2 \\ 0 \\ 1 \end{bmatrix}.$$

Thus $x_1 = 1/50$, $x_2 = 0$, and $x_3 = 100$.

(b) Multiply the second equation by 10 and the third equation by 100, then let $.001x_3 = y_3$, $x_1 = y_1$, $x_2 = y_2$. The matrix equation becomes

$$\begin{bmatrix} 0 & -1 & 1 \\ 1 & 0 & 2 \\ 1 & 1 & 3 \end{bmatrix}\begin{bmatrix} y_1 \\ y_2 \\ y_3 \end{bmatrix} = \begin{bmatrix} 6 \\ -2 \\ 2 \end{bmatrix}. \quad \begin{bmatrix} 0 & -1 & 1 & 6 \\ 1 & 0 & 2 & -2 \\ 1 & 1 & 3 & 2 \end{bmatrix} \underset{R1 \leftrightarrow R2}{\approx} \begin{bmatrix} 1 & 0 & 2 & -2 \\ 0 & -1 & 1 & 6 \\ 1 & 1 & 3 & 2 \end{bmatrix}$$

$$\underset{R3+(-1)R1}{\approx} \begin{bmatrix} 1 & 0 & 2 & -2 \\ 0 & -1 & 1 & 6 \\ 0 & 1 & 1 & 4 \end{bmatrix} \underset{(-1)R2}{\approx} \begin{bmatrix} 1 & 0 & 2 & -2 \\ 0 & 1 & -1 & -6 \\ 0 & 1 & 1 & 4 \end{bmatrix} \underset{R3+(-1)R2}{\approx} \begin{bmatrix} 1 & 0 & 2 & -2 \\ 0 & 1 & -1 & -6 \\ 0 & 0 & 2 & 10 \end{bmatrix}$$

$$\underset{(1/2)R3}{\approx} \begin{bmatrix} 1 & 0 & 2 & -2 \\ 0 & 1 & -1 & -6 \\ 0 & 0 & 1 & 5 \end{bmatrix} \underset{\substack{R1+(-2)R3 \\ R2+R3}}{\approx} \begin{bmatrix} 1 & 0 & 0 & -12 \\ 0 & 1 & 0 & -1 \\ 0 & 0 & 1 & 5 \end{bmatrix}, \text{ so } \begin{bmatrix} y_1 \\ y_2 \\ y_3 \end{bmatrix} = \begin{bmatrix} -12 \\ -1 \\ 5 \end{bmatrix}.$$

Thus $x_1 = -12$, $x_2 = -1$, and $x_3 = 5000$.

10. These iterations were found using a computer program. We stop when for all variables two successive iterations have given the same value to two decimal places.

x	y	z
1	2	-3
3.666666667	-1.047619048	3.845238095
5.31547619	1.414965986	2.817389456
4.733737245	1.140670554	3.03139805
4.815121249	1.20913595	2.9939357
4.797466625	1.198124836	3.001102135
4.800496217	1.200330567	2.999793304

Thus to two decimal places x = 4.80, y = 1.20, and z = 3.00. Substitution of these values into the given equations will show that in fact this is the exact solution.

11. We choose $\mathbf{x} = \begin{bmatrix} 1 \\ 2 \\ 1 \end{bmatrix}$. Results are recorded to 5 decimal places. The process is stopped when successive results agree to three significant digits.

$\frac{A\mathbf{x}\cdot\mathbf{x}}{\mathbf{x}\cdot\mathbf{x}}$	Scaled $A\mathbf{x}$
9.83333	$\begin{bmatrix} 0.61111 \\ 0.83333 \\ 1 \end{bmatrix}$
11.71045	$\begin{bmatrix} 0.64 \\ 1 \\ 0.955 \end{bmatrix}$
11.65458	$\begin{bmatrix} 0.64954 \\ 0.98520 \\ 1 \end{bmatrix}$
11.72116	$\begin{bmatrix} 0.65327 \\ 1 \\ 0.99933 \end{bmatrix}$
11.71197	$\begin{bmatrix} 0.65228 \\ 0.99646 \\ 1 \end{bmatrix}$
11.71629	$\begin{bmatrix} 0.65256 \\ 0.99752 \\ 1 \end{bmatrix}$
11.71567	$\begin{bmatrix} 0.65246 \\ 0.99723 \\ 1 \end{bmatrix}$
11.71594	$\begin{bmatrix} 0.65248 \\ 0.99731 \\ 1 \end{bmatrix}$

Hence $\lambda = 11.716$ and the dominant eigenvector is $r\begin{bmatrix} 0.652 \\ 0.997 \\ 1 \end{bmatrix}$.

12. We choose $\mathbf{x} = \begin{bmatrix} 1 \\ 2 \\ 1 \end{bmatrix}$. Values of $\dfrac{\mathbf{Ax \cdot x}}{\mathbf{x \cdot x}}$ are 4, 5.5, 5.74194, 5.87645, 5.94314, 5.97432, 5.98851, 5.99487, 5.99772, 5.99899. Thus $\lambda = 6$. The dominant eigenvector is $r\begin{bmatrix} 1 \\ 2 \\ -1 \end{bmatrix}$ and the unit eigenvector is $\mathbf{y} = \begin{bmatrix} 1/\sqrt{6} \\ 2/\sqrt{6} \\ -1/\sqrt{6} \end{bmatrix}$.

$B = A - \lambda \mathbf{yy}^t = \begin{bmatrix} 2 & 0 & 2 \\ 0 & 0 & 0 \\ 2 & 0 & 2 \end{bmatrix}$. Three iterations of the process confirm that $\lambda = 4$ with dominant eigenvector $s\begin{bmatrix} 1 \\ 0 \\ 1 \end{bmatrix}$. The unit eigenvector is $\mathbf{z} = \begin{bmatrix} 1/\sqrt{2} \\ 0 \\ 1/\sqrt{2} \end{bmatrix}$, and $C = A - \lambda \mathbf{zz}^t$ is the zero matrix. Thus the remaining eigenvalue is zero. The corresponding eigenvector is $t\begin{bmatrix} -1 \\ 1 \\ 1 \end{bmatrix}$.

Chapter 10

Exercise Set 10.1, page 431

For exercises 1 - 14 in this section we refer to the drawing at the right. The points A, B, C, E, and F will be identified, and the objective function will be evaluated at the three nonzero vertices A, B, and C of the feasible region.

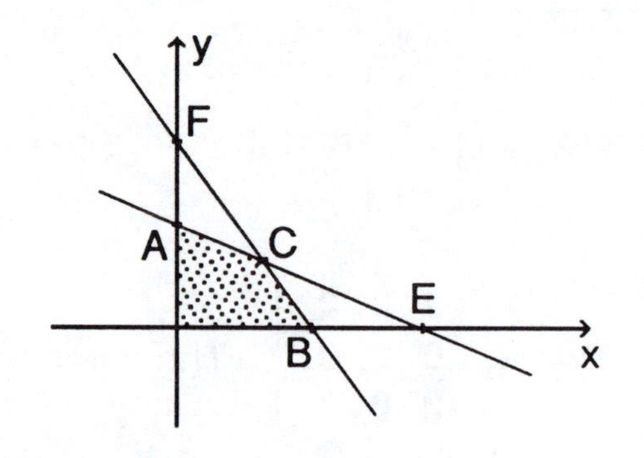

1. $4x + y = 36$ F = (0,36) B = (9,0)

   $4x + 3y = 60$ A = (0,20) E = (15,0) C = (6,12)

   At A, $f = 2(0) + 20 = 20$; at B, $f = 2(9) + 0 = 18$; and at C, $f = 2(6) + 12 = 24$, so the maximum value of f is 24 and it occurs at the point C.

3. $x + 3y = 15$ A = (0,5) E = (15,0)

   $2x + y = 10$ F = (0,10) B = (5,0) C = (3,4)

   At A, $f = 4(0) + 2(5) = 10$; at B, $f = 4(5) + 2(0) = 20$; and at C, $f = 4(3) + 2(4) = 20$, so the maximum value of f is 20 and it occurs at every point on the line segment joining B and C.

5. $2x + y = 4$ A = (0,4) E = (2,0)

   $6x + y = 8$ F = (0,8) B = (4/3,0) C = (1,2)

   At A, $f = 4(0) + 4 = 4$; at B, $f = 4(4/3) + 0 = 16/3$; and at C, $f = 4(1) + 2 = 6$, so the maximum value of f is 6 and it occurs at the point C.

6. $x + y = 150$ A = (0,150) E = (150,0)

$4x + y = 450$ F = (0,450) B = (112.5,0) C = (100,50)

At A, $-f = 3(0) + 150 = 20$; at B, $-f = 3(112.5) + 0 = 337.5$; and at C, $-f = 3(100) + 50 = 350$, so the maximum value of $-f$ is 350 and it occurs at the point C. Thus the minimum value of f is -350 and it occurs at the point C.

8. $x + 2y = 4$ F = (0,2) B = (4,0)

$x + 4y = 6$ A = (0,3/2) E = (6,0) C = (2,1)

At A, $-f = 0 - 2(3/2) = -3$; at B, $-f = 4 - 2(0) = 4$; and at C, $-f = 2 - 2(1) = 0$, so the maximum value of $-f$ is 4 and it occurs at the point B. Thus the minimum value of f is -4 and it occurs at the point B.

9.

	hours	cost($)	profit($)	number to manufacture
C1	1	30	10	x
C2	4	20	8	y
totals	≤1600	≤18000		

We are to maximize profit, $f = 10x + 8y$, subject to the constraints

$x + 4y \leq 1600$ (hours),

$30x + 20y \leq 18000$ (divide by 10: $3x + 2y \leq 1800$) (cost),

$x \geq 0$, and $y \geq 0$.

$x + 4y = 1600$ A = (0,400) E = (1600,0)

$3x + 2y = 1800$ F = (0,900) B = (600,0) C = (400,300)

At A, $f = 10(0) + 8(400) = 3200$; at B, $f = 10(600) + 8(0) = 6000$; and at C, $f = 10(400) + 8(300) = 6400$, so the maximum value of f is 6400 and it occurs at the point C. To ensure the maximum profit of $6400, the company should manufacture 400 of model C1 and 300 of model C2.

11.

	hours	cost($)	profit($)	number to ship
X	20	60	40	x
Y	10	10	20	y
totals	≤1200	≤2400		

We are to maximize profit, $f = 40x + 20y$, subject to the constraints

$20x + 10y \le 1200$ (divide by 10: $2x + y \le 120$) (hours),

$60x + 10y \le 2400$ (divide by 10: $6x + y \le 240$) (cost),

$x \ge 0$, and $y \ge 0$.

$2x + y = 120$ $A = (0,120)$ $E = (60,0)$

$6x + y = 240$ $F = (0,240)$ $B = (40,0)$ $C = (30,60)$

At A, $f = 40(0) + 20(120) = 2400$; at B, $f = 40(40) + 20(0) = 1600$; and at C, $f = 40(30) + 20(60) = 2400$, so the maximum value of f is 2400 and it occurs at all points on the line segment joining points A and C. To ensure the maximum profit of $2400, the company should ship x refrigerators from the plant at town X and $120 - 2x$ refrigerators from the plant at town Y, where $0 \le x \le 30$.

13.

	cotton(sq. yd.)	wool(sq. yd.)	number to make	income($)
suit	2	1	x	90
dress	1	3	y	90
totals	≤80	≤120		

We are to maximize income, $f = 90x + 90y$, subject to the constraints

$2x + y \le 80$ (sq. yd. cotton),

$x + 3y \le 120$ (sq. yd. wool),

$x \ge 0$, and $y \ge 0$.

$2x + y = 80$ $F = (0,80)$ $B = (40,0)$

$x + 3y = 120$ $A = (0,40)$ $E = (120,0)$ $C = (24,32)$

At A, $f = 90(0) + 90(40) = 3600$; at B, $f = 90(40) + 90(0) = 3600$; and at C, $f = 90(24) + 90(32) = 5040$, so the maximum value of f is 5040 and it occurs at the point C. To ensure the maximum income of $5040, the tailor should make 24 suits and 32 dresses.

15.

	from A	from B	time
cars to C	x	120 − x	2x + 6(120 − x)
cars to D	y	180 − y	4y + 3(180 − y)
totals	≤100	≤200	≤1030

We are to maximize $f = 120 - x$, subject to the constraints

$x + y \le 100,$

$120 - x + 180 - y \le 200$ (simplified: $x + y \ge 100$)

(the first two constraints imply $x + y = 0$),

$2x + 6(120 - x) + 4y + 3(180 - y) \le 1030$ (simplified: $4x - y \ge 230$),

$x \ge 0$, and $y \ge 0$.

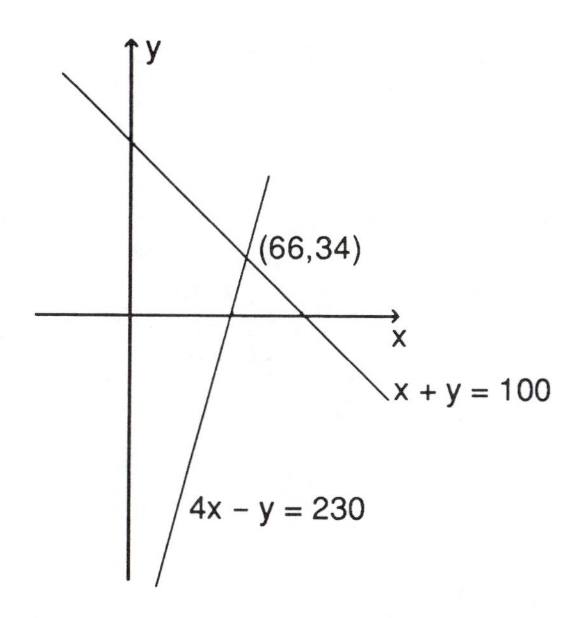

The only point in the feasible region is the point (66,34). Thus from port A 66 cars should be moved to city C and 34 cars should be moved to city D. From port B 54 cars should be moved to city C and 146 cars should be moved to city D.

17.

	A (tons)	B (tons)	tons of fertilizer
X	.8	.2	x
Y	.6	.4	y
totals	≤100	≤50	

We are to maximize the amount of fertilizer, $f = x + y$, subject to the constraints

$.8x + .6y \le 100$ (multiply by 5: $4x + 3y \le 500$) (amount of A),

$.2x + .4y \le 50$ (multiply by 5: $x + 2y \le 250$) (amount of B),

$x \ge 30$, and $y \ge 50$.

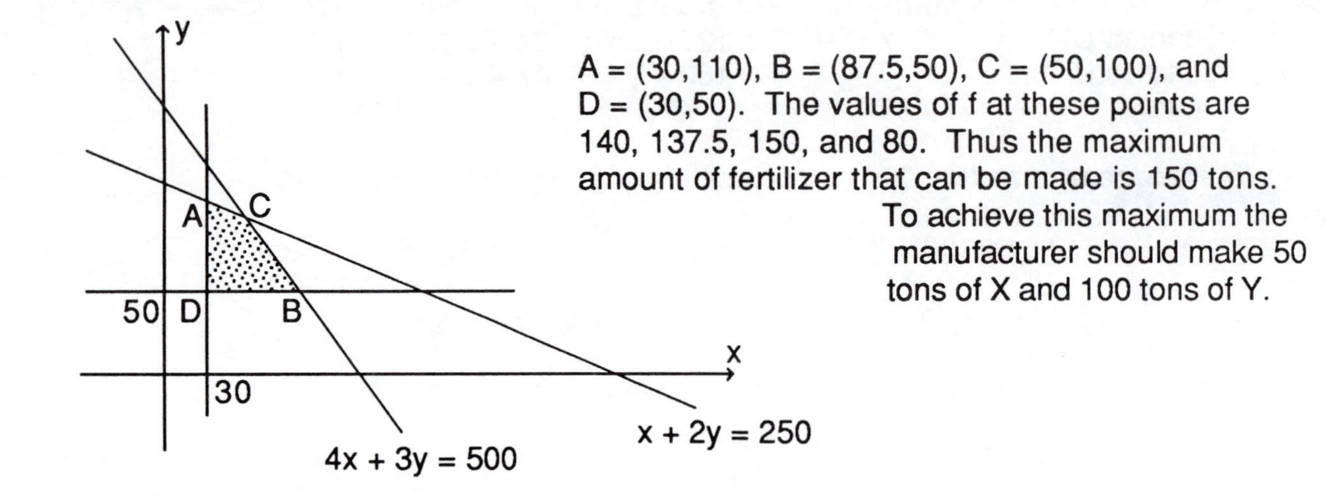

A = (30,110), B = (87.5,50), C = (50,100), and D = (30,50). The values of f at these points are 140, 137.5, 150, and 80. Thus the maximum amount of fertilizer that can be made is 150 tons. To achieve this maximum the manufacturer should make 50 tons of X and 100 tons of Y.

18.

	units of A per oz.	units of B per oz.	cost(cents per oz.)	oz.
M	1	2	8	x
N	1	1	12	y
totals	≥ 7	≥ 10		

We are to minimize the cost, $f = 8x + 12y$, subject to the constraints

$x + y \geq 7$ (units of A),

$2x + y \geq 10$ (units of B),

$x \geq 0$, and $y \geq 0$.

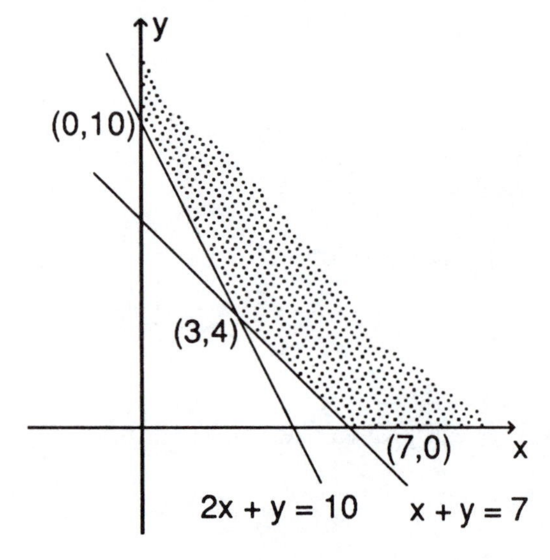

The values of f are $8(0) + 12(10) = 120$, $8(3) + 12(4) = 72$, and $8(7) + 12(0) = 56$. Thus the hospital can achieve the minimum cost of 56 cents by serving 7 oz. of item M and no item N.

20. (a)

	pounds	cu. ft.	profit($)	number of packages
Pringle	5	5	.30	x
Williams	6	3	.40	y
totals	≤12000	≤9000		

We are to maximize profit, $f = .3x + .4y$, subject to the constraints

$5x + 6y \le 12000$ (weight in pounds),

$5x + 3y \le 9000$ (volume in cu. ft.),

$x \ge 0$, and $y \ge 0$.

At A, B, and C, f = 800, 540, and 760. Thus the profit is maximized if the shipper carries no packages for Pringle and 2000 packages for Williams.

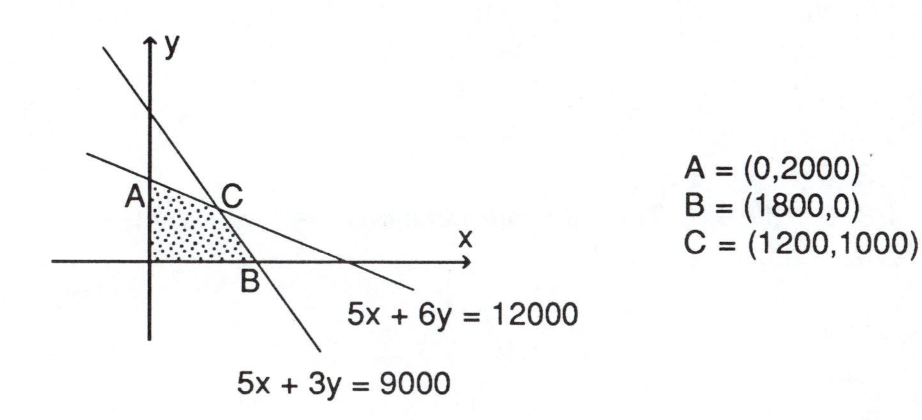

Exercise Set 10.2, page 438

1. We indicate the row and column of the initial pivot element with arrows.

$$\begin{array}{c} \\ \rightarrow \\ \\ \\ \end{array}\begin{array}{c} \begin{array}{cccccc} x & y & u & v & f & \end{array} \\ \begin{bmatrix} 4 & 1 & 1 & 0 & 0 & 36 \\ 4 & 3 & 0 & 1 & 0 & 60 \\ -2 & -1 & 0 & 0 & 1 & 0 \end{bmatrix} \\ \uparrow \end{array} \approx \begin{bmatrix} 1 & 1/4 & 1/4 & 0 & 0 & 9 \\ 4 & 3 & 0 & 1 & 0 & 60 \\ -2 & -1 & 0 & 0 & 1 & 0 \end{bmatrix} \approx \begin{bmatrix} 1 & 1/4 & 1/4 & 0 & 0 & 9 \\ 0 & 2 & -1 & 1 & 0 & 24 \\ 0 & -1/2 & 1/2 & 0 & 1 & 18 \end{bmatrix}$$

$$\approx \begin{bmatrix} 1 & 1/4 & 1/4 & 0 & 0 & 9 \\ 0 & 1 & -1/2 & 1/2 & 0 & 12 \\ 0 & -1/2 & 1/2 & 0 & 1 & 18 \end{bmatrix} \approx \begin{bmatrix} 1 & 0 & 3/8 & -1/8 & 0 & 6 \\ 0 & 1 & -1/2 & 1/2 & 0 & 12 \\ 0 & 0 & 1/4 & 1/4 & 1 & 24 \end{bmatrix}.$$ The elements in the last row are all positive, so this is the final tableau. The maximum value of f is 24 when u and v are both zero. With u = v = 0, the first two rows give x = 6 and y = 12.

2.

$$\begin{array}{c} \\ \rightarrow \\ \\ \\ \end{array}\begin{array}{cccccc} x & y & u & v & f & \\ 1 & 2 & 1 & 0 & 0 & 4 \\ 1 & 6 & 0 & 1 & 0 & 8 \\ -1 & 4 & 0 & 0 & 1 & 0 \\ \uparrow & & & & & \end{array} \approx \begin{bmatrix} 1 & 2 & 1 & 0 & 0 & 4 \\ 0 & 4 & -1 & 1 & 0 & 4 \\ 0 & 6 & 1 & 0 & 1 & 4 \end{bmatrix}$$

. The maximum value of f is 4 when $y = u = 0$. The first row gives $x = 4$.

4.

$$\begin{array}{cccccc} x & y & u & v & f & \\ 1 & 1 & 1 & 0 & 0 & 180 \\ 3 & 2 & 0 & 1 & 0 & 480 \\ -10 & -5 & 0 & 0 & 1 & 0 \\ \uparrow & & & & & \end{array} \begin{array}{c} \\ \\ \leftarrow \\ \\ \end{array} \approx \begin{bmatrix} 1 & 1 & 1 & 0 & 0 & 180 \\ 1 & 2/3 & 0 & 1/3 & 0 & 160 \\ -10 & -5 & 0 & 0 & 1 & 0 \end{bmatrix}$$

$$\approx \begin{bmatrix} 0 & 1/3 & 1 & -1/3 & 0 & 20 \\ 1 & 2/3 & 0 & 1/3 & 0 & 160 \\ 0 & 5/3 & 0 & 10/3 & 1 & 1600 \end{bmatrix}$$

. The maximum value of f is 1600 when $y = v = 0$. From the second row $x = 160$.

6.

$$\begin{array}{cccccccc} x & y & z & u & v & w & f & \\ 5 & 5 & 10 & 1 & 0 & 0 & 0 & 1000 \\ 10 & 8 & 5 & 0 & 1 & 0 & 0 & 2000 \\ 10 & 5 & 0 & 0 & 0 & 1 & 0 & 500 \\ -100 & -200 & -50 & 0 & 0 & 0 & 1 & 0 \\ & \uparrow & & & & & & \end{array} \begin{array}{c} \\ \\ \\ \leftarrow \\ \\ \\ \end{array} \approx \begin{bmatrix} 5 & 5 & 10 & 1 & 0 & 0 & 0 & 1000 \\ 10 & 8 & 5 & 0 & 1 & 0 & 0 & 2000 \\ 2 & 1 & 0 & 0 & 0 & 1/5 & 0 & 100 \\ -100 & -200 & -50 & 0 & 0 & 0 & 1 & 0 \end{bmatrix}$$

$$\approx \begin{bmatrix} -5 & 0 & 10 & 1 & 0 & -1 & 0 & 500 \\ -6 & 0 & 5 & 0 & 1 & -8/5 & 0 & 1200 \\ 2 & 1 & 0 & 0 & 0 & 1/5 & 0 & 100 \\ 300 & 0 & -50 & 0 & 0 & 40 & 1 & 20000 \end{bmatrix}$$

$$\approx \begin{bmatrix} -1/2 & 0 & 1 & 1/10 & 0 & -1/10 & 0 & 50 \\ -6 & 0 & 5 & 0 & 1 & -8/5 & 0 & 1200 \\ 2 & 1 & 0 & 0 & 0 & 1/5 & 0 & 100 \\ 300 & 0 & -50 & 0 & 0 & 40 & 1 & 20000 \end{bmatrix}$$

$$\approx \begin{bmatrix} -1/2 & 0 & 1 & 1/10 & 0 & -1/10 & 0 & 50 \\ -7/2 & 0 & 0 & -1/2 & 1 & -11/10 & 0 & 950 \\ 2 & 1 & 0 & 0 & 0 & 1/5 & 0 & 100 \\ 275 & 0 & 0 & 5 & 0 & 35 & 1 & 22500 \end{bmatrix}.$$ The maximum value of f is 22500 when

$x = u = w = 0$. Row 3 gives $y = 100$ and row 1 gives $z = 50$.

7.

$$\begin{array}{c} \\ \rightarrow \\ \\ \\ \end{array} \begin{array}{c} \begin{matrix} x & y & z & u & v & w & f & \end{matrix} \\ \begin{bmatrix} -1 & 2 & 3 & 1 & 0 & 0 & 0 & 6 \\ -1 & 4 & 5 & 0 & 1 & 0 & 0 & 5 \\ -1 & 5 & 7 & 0 & 0 & 1 & 0 & 7 \\ -2 & -4 & -1 & 0 & 0 & 0 & 1 & 0 \end{bmatrix} \\ \uparrow \text{ (column } y) \end{array} \approx \begin{bmatrix} -1 & 2 & 3 & 1 & 0 & 0 & 0 & 6 \\ -1/4 & 1 & 5/4 & 0 & 1/4 & 0 & 0 & 5/4 \\ -1 & 5 & 7 & 0 & 0 & 1 & 0 & 7 \\ -2 & -4 & -1 & 0 & 0 & 0 & 1 & 0 \end{bmatrix}$$

$$\approx \begin{bmatrix} -1/2 & 0 & 1/2 & 1 & -1/2 & 0 & 0 & 7/2 \\ -1/4 & 1 & 5/4 & 0 & 1/4 & 0 & 0 & 5/4 \\ 1/4 & 0 & 3/4 & 0 & -5/4 & 1 & 0 & 3/4 \\ -3 & 0 & 4 & 0 & 1 & 0 & 1 & 5 \end{bmatrix} \approx \begin{bmatrix} -1/2 & 0 & 1/2 & 1 & -1/2 & 0 & 0 & 7/2 \\ -1/4 & 1 & 5/4 & 0 & 1/4 & 0 & 0 & 5/4 \\ 1 & 0 & 3 & 0 & -5 & 4 & 0 & 3 \\ -3 & 0 & 4 & 0 & 1 & 0 & 1 & 5 \end{bmatrix}$$

$$\approx \begin{bmatrix} 0 & 0 & 2 & 1 & -3 & 2 & 0 & 5 \\ 0 & 1 & 2 & 0 & -1 & 1 & 0 & 2 \\ 1 & 0 & 3 & 0 & -5 & 4 & 0 & 3 \\ 0 & 0 & 13 & 0 & -14 & 12 & 1 & 14 \end{bmatrix}.$$ We can do no more.

There are no positive terms in the pivot column. This means that the feasible region is unbounded and the objective function is unbounded. That is, there is no maximum.

10.

	I	II	III	profit($)	number to make
X	2	4		10	x
Y	3		6	8	y
Z	1	2	3	12	z
totals	≤360	≤360	≤360		

We are to maximize profit, $f = 10x + 8y + 12z$, subject to the constraints $2x + 3y + z \le 360$, $4x + 2z \le 360$, $6y + 3z \le 360$, $x \ge 0$, $y \ge 0$, $z \ge 0$.

$$\begin{array}{c} \\ \\ \rightarrow \\ \\ \end{array} \begin{array}{c} \begin{matrix} x & y & z & u & v & w & f & \end{matrix} \\ \begin{bmatrix} 2 & 3 & 1 & 1 & 0 & 0 & 0 & 360 \\ 4 & 0 & 2 & 0 & 1 & 0 & 0 & 360 \\ 0 & 6 & 3 & 0 & 0 & 1 & 0 & 360 \\ -10 & -8 & -12 & 0 & 0 & 0 & 1 & 0 \end{bmatrix} \\ \uparrow \text{ (column } z) \end{array} \approx \begin{bmatrix} 2 & 3 & 1 & 1 & 0 & 0 & 0 & 360 \\ 4 & 0 & 2 & 0 & 1 & 0 & 0 & 360 \\ 0 & 2 & 1 & 0 & 0 & 1/3 & 0 & 120 \\ -10 & -8 & -12 & 0 & 0 & 0 & 1 & 0 \end{bmatrix}$$

$$\approx \begin{bmatrix} 2 & 1 & 0 & 1 & 0 & -1/3 & 0 & 240 \\ 4 & -4 & 0 & 0 & 1 & -2/3 & 0 & 120 \\ 0 & 2 & 1 & 0 & 0 & 1/3 & 0 & 120 \\ -10 & 16 & 0 & 0 & 0 & 4 & 1 & 1440 \end{bmatrix} \approx \begin{bmatrix} 2 & 1 & 0 & 1 & 0 & -1/3 & 0 & 240 \\ 1 & -1 & 0 & 0 & 1/4 & -1/6 & 0 & 30 \\ 0 & 2 & 1 & 0 & 0 & 1/3 & 0 & 120 \\ -10 & 16 & 0 & 0 & 0 & 4 & 1 & 1440 \end{bmatrix}$$

$$\approx \begin{bmatrix} 0 & 3 & 0 & 1 & -1/2 & 0 & 0 & 180 \\ 1 & -1 & 0 & 0 & 1/4 & -1/6 & 0 & 30 \\ 0 & 2 & 1 & 0 & 0 & 1/3 & 0 & 120 \\ 0 & 6 & 0 & 0 & 5/2 & 7/3 & 1 & 1740 \end{bmatrix}$$

. The maximum value of f is 1740 when

$y = v = w = 0$. Row 2 gives $x = 30$ and row 3 gives $z = 120$. So the company should produce 30 of item X, no item Y, and 120 of item Z to make maximum profit of $1740 per day.

12.

	cost($)	time(hrs)	profit($)	number to transport
A	10	6	12	x
B	20	4	20	y
C	40	2	16	z
totals	≤6000	≤4000		

We are to maximize profit, $f = 12x + 20y + 16z$, subject to the constraints $10x + 20y + 40z \le 6000$, $6x + 4y + 2z \le 4000$, $x \ge 0$, $y \ge 0$, $z \ge 0$.

$$\begin{matrix} & x & y & z & u & v & f & \\ \to & \end{matrix}\begin{bmatrix} 10 & 20 & 40 & 1 & 0 & 0 & 6000 \\ 6 & 4 & 2 & 0 & 1 & 0 & 4000 \\ -12 & -20 & -16 & 0 & 0 & 1 & 0 \end{bmatrix} \approx \begin{bmatrix} 1/2 & 1 & 2 & 1/20 & 0 & 0 & 300 \\ 6 & 4 & 2 & 0 & 1 & 0 & 4000 \\ -12 & -20 & -16 & 0 & 0 & 1 & 0 \end{bmatrix}$$

$$\uparrow$$

$$\approx \begin{bmatrix} 1/2 & 1 & 2 & 1/20 & 0 & 0 & 300 \\ 4 & 0 & -6 & -1/5 & 1 & 0 & 2800 \\ -2 & 0 & 24 & 1 & 0 & 1 & 6000 \end{bmatrix} \approx \begin{bmatrix} 1 & 2 & 4 & 1/10 & 0 & 0 & 600 \\ 4 & 0 & -6 & -1/5 & 1 & 0 & 2800 \\ -2 & 0 & 24 & 1 & 0 & 1 & 6000 \end{bmatrix}$$

$$\approx \begin{bmatrix} 1 & 2 & 4 & 1/10 & 0 & 0 & 600 \\ 0 & -8 & -22 & -3/5 & 1 & 0 & 400 \\ 0 & 4 & 32 & 6/5 & 0 & 1 & 7200 \end{bmatrix}$$

. The maximum value of f is 7200 when

$y = z = u = 0$. Row 1 then gives $x = 600$. The maximum profit of $7200 is attained if 600 washing machines are transported from A to P and none is transported from B or C to P.

13.

	material (sq.yd.)	cost($)	profit	number to make
Aspen	60	32	12	x
Alpine	30	20	8	y
Cub	15	12	4	z
totals	≤7800	≤8320		

We are to maximize profit, $f = 12x + 8y + 4z$, subject to the constraints $60x + 30y + 15z \le 7800$, $32x + 20y + 12z \le 8320$, $x \ge 0, y \ge 0, z \ge 0$.

$$\begin{matrix} \\ \rightarrow \\ \\ \\ \end{matrix}\begin{array}{c} x \quad y \quad z \quad u \quad v \quad f \\ \left[\begin{array}{cccccc|c} 60 & 30 & 15 & 1 & 0 & 0 & 7800 \\ 32 & 20 & 12 & 0 & 1 & 0 & 8320 \\ -12 & -8 & -4 & 0 & 0 & 1 & 0 \end{array}\right] \\ \uparrow \qquad\qquad\qquad\qquad \end{array} \approx \left[\begin{array}{ccccccc} 1 & 1/2 & 1/4 & 1/60 & 0 & 0 & 130 \\ 32 & 20 & 12 & 0 & 1 & 0 & 8320 \\ -12 & -8 & -4 & 0 & 0 & 1 & 0 \end{array}\right]$$

$$\approx \begin{matrix} \rightarrow \\ \\ \\ \end{matrix}\left[\begin{array}{ccccccc} 1 & 1/2 & 1/4 & 1/60 & 0 & 0 & 130 \\ 0 & 4 & 4 & -8/15 & 1 & 0 & 4160 \\ 0 & -2 & -1 & 1/5 & 0 & 1 & 1560 \end{array}\right] \approx \left[\begin{array}{ccccccc} 2 & 1 & 1/2 & 1/30 & 0 & 0 & 260 \\ 0 & 4 & 4 & -8/15 & 1 & 0 & 4160 \\ 0 & -2 & -1 & 1/5 & 0 & 1 & 1560 \end{array}\right]$$

$\qquad\quad\uparrow$

$$\approx \left[\begin{array}{ccccccc} 2 & 1 & 1/2 & 1/30 & 0 & 0 & 260 \\ -8 & 0 & 2 & -2/3 & 1 & 0 & 3120 \\ 4 & 0 & 0 & 4/15 & 0 & 1 & 2080 \end{array}\right].$$ The maximum value of f is 2080 when

$x = u = 0$. From row 1, $y + z/2 = 260$. Thus the maximum profit of \$2080 will be achieved if no Aspens are manufactured and the number of Alpines manufactured plus one-half the number of Cubs manufactured is 260.

15. To minimize $f = -2x + y$, maximize $-f = 2x - y$.

$$\begin{array}{c} x \quad y \quad u \quad v \quad f \\ \left[\begin{array}{cccccc} 2 & 2 & 1 & 0 & 0 & 8 \\ 1 & -1 & 0 & 1 & 0 & 2 \\ -2 & 1 & 0 & 0 & 1 & 0 \end{array}\right] \end{array} \approx \left[\begin{array}{cccccc} 0 & 4 & 1 & -2 & 0 & 4 \\ 1 & -1 & 0 & 1 & 0 & 2 \\ 0 & -1 & 0 & 2 & 1 & 4 \end{array}\right] \approx \left[\begin{array}{cccccc} 0 & 1 & 1/4 & -1/2 & 0 & 1 \\ 1 & -1 & 0 & 1 & 0 & 2 \\ 0 & -1 & 0 & 2 & 1 & 4 \end{array}\right]$$

$$\approx \left[\begin{array}{cccccc} 0 & 1 & 1/4 & -1/2 & 0 & 1 \\ 1 & 0 & 1/4 & 1/2 & 0 & 3 \\ 0 & 0 & 1/4 & 3/2 & 1 & 5 \end{array}\right].$$ The maximum value of $-f$ is 5 when $u = v = 0$.

Row 2 gives $x = 3$ and row 1 gives $y = 1$. Thus the minimum value of f is -5 at $x = 3$, $y = 1$.

16. To minimize f = 2x + y − z, maximize −f = −2x − y + z.

$$\begin{array}{c} \begin{matrix} x & y & z & u & v & w & f & \end{matrix} \\ \begin{bmatrix} 1 & 2 & -2 & 1 & 0 & 0 & 0 & 20 \\ 2 & 1 & 0 & 0 & 1 & 0 & 0 & 10 \\ 1 & 3 & 4 & 0 & 0 & 1 & 0 & 15 \\ 2 & 1 & -1 & 0 & 0 & 0 & 1 & 0 \end{bmatrix} \end{array} \approx \begin{bmatrix} 1 & 2 & -2 & 1 & 0 & 0 & 0 & 20 \\ 2 & 1 & 0 & 0 & 1 & 0 & 0 & 10 \\ 1/4 & 3/4 & 1 & 0 & 0 & 1/4 & 0 & 15/4 \\ 2 & 1 & -1 & 0 & 0 & 0 & 1 & 0 \end{bmatrix}$$

$$\approx \begin{bmatrix} 3/2 & 7/2 & 0 & 1 & 0 & 1/2 & 0 & 55/2 \\ 2 & 1 & 0 & 0 & 1 & 0 & 0 & 10 \\ 1/4 & 3/4 & 1 & 0 & 0 & 1/4 & 0 & 15/4 \\ 9/4 & 7/4 & 0 & 0 & 0 & 1/4 & 1 & 15/4 \end{bmatrix}.$$ The maximum value of −f is 15/4 when

x = y = w = 0. Row 3 gives z = 15/4. Thus the minimum value of f is −15/4 at x = 0, y = 0, z = 15/4.

<u>Exercise Set 10.3</u>, page 446

1.

$$\begin{array}{c} \begin{matrix} x & y & u & v & f & \end{matrix} \\ \rightarrow \begin{bmatrix} 4 & 1 & 1 & 0 & 0 & 36 \\ 4 & 3 & 0 & 1 & 0 & 60 \\ -2 & -1 & 0 & 0 & 1 & 0 \end{bmatrix} \\ \uparrow \end{array} \approx \begin{bmatrix} 1 & 1/4 & 1/4 & 0 & 0 & 9 \\ 4 & 3 & 0 & 1 & 0 & 60 \\ -2 & -1 & 0 & 0 & 1 & 0 \end{bmatrix} \approx \begin{bmatrix} 1 & 1/4 & 1/4 & 0 & 0 & 9 \\ 0 & 2 & -1 & 1 & 0 & 24 \\ 0 & -1/2 & 1/2 & 0 & 1 & 18 \end{bmatrix}$$

basic:	u,v	x,v
nonbasic:	x,y	y,u
entering:	x	y
departing:	u	v

$$\approx \begin{bmatrix} 1 & 1/4 & 1/4 & 0 & 0 & 9 \\ 0 & 1 & -1/2 & 1/2 & 0 & 12 \\ 0 & -1/2 & 1/2 & 0 & 1 & 18 \end{bmatrix} \approx \begin{bmatrix} 1 & 0 & 3/8 & -1/8 & 0 & 6 \\ 0 & 1 & -1/2 & 1/2 & 0 & 12 \\ 0 & 0 & 1/4 & 1/4 & 1 & 24 \end{bmatrix}.$$

basic:	x,y
nonbasic:	u,v

The maximum value of f is 24 when u and v are both zero. With u = v = 0, the first two rows give x = 6 and y = 12. Thus the optimal solution is f = 24 at x = 6, y = 12.

3.

$$
\begin{array}{c}
\\
\rightarrow \\
\\
\\
\end{array}
\begin{array}{ccccccc}
x & y & z & u & v & f & \\
\end{array}
$$

$$
\rightarrow\left[\begin{array}{rrrrrrr}
3 & 1 & 1 & 1 & 0 & 0 & 3 \\
1 & -10 & -4 & 0 & 1 & 0 & 20 \\
-1 & -2 & -1 & 0 & 0 & 1 & 0
\end{array}\right]
\approx
\left[\begin{array}{rrrrrrr}
3 & 1 & 1 & 1 & 0 & 0 & 3 \\
31 & 0 & 6 & 10 & 1 & 0 & 50 \\
5 & 0 & 1 & 2 & 0 & 1 & 6
\end{array}\right].
$$

$\uparrow$ (under column y)

basic:	u,v	y,v
nonbasic:	x,y,z	x,z,u
entering:	y	
departing:	u	

The maximum value of f is 6 when x = z = u = 0. The first two rows give y = 3 and v = 50. Thus the optimal solution is f = 6 at x = 0, y = 3, z = 0. (x and z are nonbasic variables.)

6.

$$
\begin{array}{cccccccc}
x & y & z & w & u & v & f & \\
\end{array}
$$

$$
\rightarrow\left[\begin{array}{rrrrrrrr}
5 & 0 & 4 & 6 & 1 & 0 & 0 & 20 \\
4 & 2 & 2 & 8 & 0 & 1 & 0 & 40 \\
-1 & -2 & -4 & 1 & 0 & 0 & 1 & 0
\end{array}\right]
\approx
\left[\begin{array}{rrrrrrrr}
5/4 & 0 & 1 & 3/2 & 1/4 & 0 & 0 & 5 \\
4 & 2 & 2 & 8 & 0 & 1 & 0 & 40 \\
-1 & -2 & -4 & 1 & 0 & 0 & 1 & 0
\end{array}\right]
$$

$\uparrow$ (under column z)

basic:	u,v
nonbasic:	x,y,z,w
entering:	z
departing:	u

$$
\approx\left[\begin{array}{rrrrrrrr}
5/4 & 0 & 1 & 3/2 & 1/4 & 0 & 0 & 5 \\
3/2 & 2 & 0 & 5 & -1/2 & 1 & 0 & 30 \\
4 & -2 & 0 & 7 & 1 & 0 & 1 & 20
\end{array}\right]
\approx
\left[\begin{array}{rrrrrrrr}
5/4 & 0 & 1 & 3/2 & 1/4 & 0 & 0 & 5 \\
3/4 & 1 & 0 & 5/2 & -1/4 & 1/2 & 0 & 15 \\
4 & -2 & 0 & 7 & 1 & 0 & 1 & 20
\end{array}\right]
$$

basic:	z,v
nonbasic:	x,y,w,v
entering:	y
departing:	v

$$
\approx\left[\begin{array}{rrrrrrrr}
5/4 & 0 & 1 & 3/2 & 1/4 & 0 & 0 & 5 \\
3/4 & 1 & 0 & 5/2 & -1/4 & 1/2 & 0 & 15 \\
11/2 & 0 & 0 & 12 & 1/2 & 1 & 1 & 50
\end{array}\right].
$$

basic:	z,y
nonbasic:	x,w,u,v

The maximum value of f is 50 when x = w = u = v = 0. The first row then yields z = 5 and the second row gives y = 15. Thus the optimal solution is f = 50 at x = 0, y = 15, z = 5, w = 0. (x and w are nonbasic variables.)

Chapter 10 Review Exercises, page 446

In exercises 1 - 4 we refer to the drawing at the right. The points A, B, C, E, and F will be identified, and the objective function will be evaluated at the three nonzero vertices A, B, and C of the feasible region.

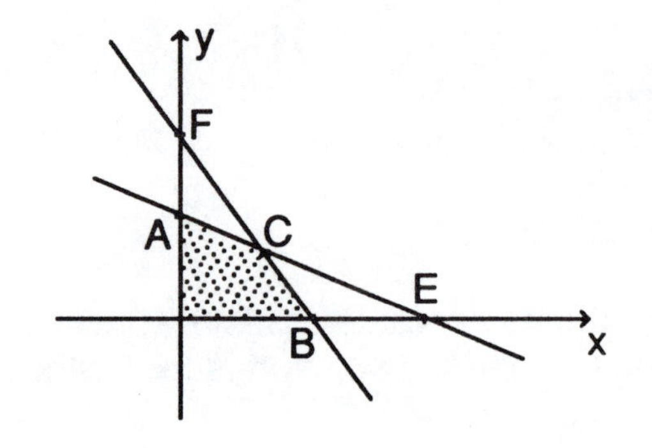

1. $2x + 4y = 16$ A = (0,4)   E = (8,0)

   $3x + 2y = 12$ F = (0,6)   B = (4,0)   C = (2,3)

   At A, f = 2(0) + 3(4) = 12; at B, f = 2(4) + 3(0) = 8; and at C, f = 2(2) + 3(3) = 13, so the maximum value of f is 13 and it occurs at the point C.

2. $x + 2y = 16$ A = (0,8)   E = (16,0)

   $3x + 2y = 24$ F = (0,12)   B = (8,0)   C = (4,6)

   At A, f = 6(0) + 4(8) = 32; at B, f = 6(8) + 4(0) = 48; and at C, f = 6(4) + 4(6) = 48, so the maximum value of f is 48 and it occurs at every point on the line segment joining B and C.

3. $3x + 2y = 21$ F = (0,21/2)   B = (7,0)

   $x + 5y = 20$ A = (0,4)   E = (20,0)   C = (5,3)

   At A, –f = –4(0) – 4 = –3; at B, –f = –4(7) – 0 = –28; and at C, –f = –4(5) – 3 = –23, so the maximum value of –f is zero and it occurs at the origin.Thus the minimum value of f is zero and it occurs at the origin.

4.

	acres	picking time(hrs)	profit($)
strawberries	x	8	700
tomatoes	y	6	600
total	≤40	≤300	

We are to maximize profit, $f = 700x + 600y$, subject to the constraints

$x + y \leq 40$, $8x + 6y \leq 300$ (divide by 2: $4x + 3y \leq 150$),

$x \geq 0$, and $y \geq 0$.

$x + y = 40$ $A = (0,40)$ $E = (40,0)$

$4x + 3y = 150$ $F = (0,50)$ $B = (75/2,0)$ $C = (30,10)$

At A, $f = 700(0) + 600(40) = 24000$; at B, $f = 700(75/2) + 600(0) = 26250$; and at C, $f = 700(30) + 600(10) = 27000$, so the maximum value of f is 27000 and it occurs at the point C. To ensure the maximum profit of $27000, the farmer should plant 30 acres of strawberries and 10 acres of tomatoes.

5.

	cu. ft.	sq. ft.	cost($)	number
X	36	6	54	x
Y	44	8	60	y
totals		≤256	≤2100	

We are to maximize volume, $f = 36x + 44y$, subject to the constraints

$6x + 8y \leq 256$ (divide by 2: $3x + 4y \leq 128$),

$54x + 60y \leq 2100$ (divide by 6: $9x + 10y \leq 350$),

$x \geq 0$, and $y \geq 20$.

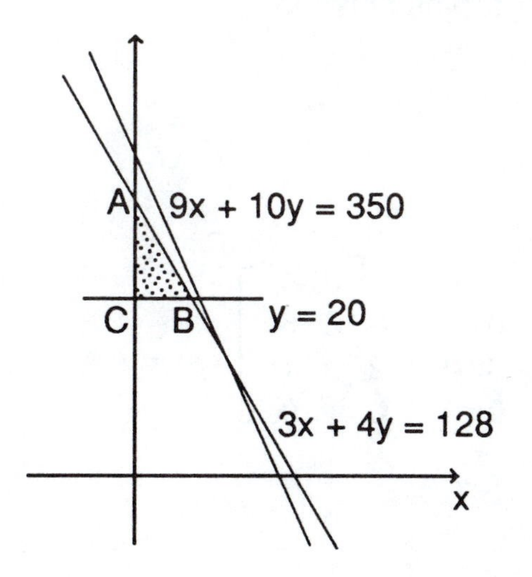

A = (0,32), B = (16,20), and C = (0,20). The values of f at these three points are 1408, 1456, and 880. Thus volume will be maximized if the company purchases 16 lockers of type X and 20 of type Y.

6.

$$\begin{array}{c} \\ \rightarrow \\ \\ \\ \\ \end{array}\begin{array}{ccccccc c} x & y & z & u & v & w & f & \\ \end{array}$$

$$\rightarrow\begin{bmatrix} 1 & 2 & 4 & 1 & 0 & 0 & 0 & 20 \\ 2 & 4 & 4 & 0 & 1 & 0 & 0 & 60 \\ 3 & 4 & 1 & 0 & 0 & 1 & 0 & 90 \\ -2 & -1 & -1 & 0 & 0 & 0 & 1 & 0 \end{bmatrix} \approx \begin{bmatrix} 1 & 2 & 4 & 1 & 0 & 0 & 0 & 20 \\ 0 & 0 & -4 & -2 & 1 & 0 & 0 & 20 \\ 0 & -2 & -11 & -3 & 0 & 1 & 0 & 30 \\ 0 & 3 & 7 & 2 & 0 & 0 & 1 & 40 \end{bmatrix}.$$

$\uparrow$ (column x)

The maximum value of f is 40 when y = z = u = 0. Row 1 gives x = 20.

7. Let x be the number of X tables and y be the number of Y tables. We are to maximize profit, f = 8x + 4y, subject to the constraints $10x + 5y \leq 300$, $8x + 4y \leq 300$, $4x + 8y \leq 300$, $x \geq 0$, and $y \geq 0$.

$$\begin{array}{cccccc} x & y & u & v & w & f \end{array}$$

$$\rightarrow\begin{bmatrix} 10 & 5 & 1 & 0 & 0 & 0 & 300 \\ 8 & 4 & 0 & 1 & 0 & 0 & 300 \\ 4 & 8 & 0 & 0 & 1 & 0 & 300 \\ -8 & -4 & 0 & 0 & 0 & 1 & 0 \end{bmatrix} \approx \begin{bmatrix} 1 & 1/2 & 1/10 & 0 & 0 & 0 & 30 \\ 8 & 4 & 0 & 1 & 0 & 0 & 300 \\ 4 & 8 & 0 & 0 & 1 & 0 & 300 \\ -8 & -4 & 0 & 0 & 0 & 1 & 0 \end{bmatrix}$$

$\uparrow$ (column x)

$$\approx \begin{bmatrix} 1 & 1/2 & 1/10 & 0 & 0 & 0 & 30 \\ 0 & 0 & -4/5 & 1 & 0 & 0 & 60 \\ 0 & 6 & -2/5 & 0 & 1 & 0 & 180 \\ 0 & 0 & 4/5 & 0 & 0 & 1 & 240 \end{bmatrix}.$$ The maximum value of f is 240 when u = 0.

Row 1 gives x + y/2 = 30. Thus the maximum daily profit of $240 is realized when the number of X tables plus one-half the number of Y tables finished is 30.

8. To minimize f = x − 2y + 4z, maximize −f = −x + 2y − 4z.

$$\begin{array}{ccccccc} x & y & z & u & v & w & f \end{array}$$

$$\begin{matrix} \\ \\ \rightarrow \\ \\ \end{matrix}\begin{bmatrix} 1 & -1 & 3 & 1 & 0 & 0 & 0 & 4 \\ 2 & 2 & -3 & 0 & 1 & 0 & 0 & 6 \\ -1 & 2 & 3 & 0 & 0 & 1 & 0 & 2 \\ 1 & -2 & 4 & 0 & 0 & 0 & 1 & 0 \end{bmatrix} \approx \begin{bmatrix} 1 & -1 & 3 & 1 & 0 & 0 & 0 & 4 \\ 2 & 2 & -3 & 0 & 1 & 0 & 0 & 6 \\ -1/2 & 1 & 3/2 & 0 & 0 & 1/2 & 0 & 1 \\ 1 & -2 & 4 & 0 & 0 & 0 & 1 & 0 \end{bmatrix}$$

$\uparrow$ (column y)

$$\approx \begin{bmatrix} 1/2 & 0 & 9/2 & 1 & 0 & 1/2 & 0 & 5 \\ 3 & 0 & -6 & 0 & 1 & -1 & 0 & 4 \\ -1/2 & 1 & 3/2 & 0 & 0 & 1/2 & 0 & 1 \\ 0 & 0 & 7 & 0 & 0 & 1 & 1 & 2 \end{bmatrix}.$$ The maximum value of $-f$ is 2 when $z = w = 0$.

Row 3 then gives $-x/2 + y = 1$, and rows 1 and 2 give $x/2 + u = 5$ and $3x + v = 4$. Thus the minimum value of f is -2 for all x,y,z where $z = 0$ and $y = 1 + x/2$, with $0 \le x \le 4/3$.